Advances in X-ray Absorption Fine Structure Analysis

Advances in X-ray Absorption Fine Structure Analysis

Zhongrui (Jerry) Li

CWP

Central West Publishing

This edition has been published by Central West Publishing, Australia

For more information about the books published by Central West Publishing, please visit https://centralwestpublishing.com

Disclaimer
Every effort has been made by the publisher and author while preparing this book, however, no warranties are made regarding the accuracy and completeness of the content. The publisher and author disclaim without any limitation all warranties as well as any implied warranties about sales, along with fitness of the content for a particular purpose. Citation of any website and other information sources does not mean any endorsement from the publisher and author. For ascertaining the suitability of the contents contained herein for a particular lab or commercial use, consultation with the subject expert is needed. In addition, while using the information and methods contained herein, the practitioners and researchers need to be mindful for their own safety, along with the safety of others, including the professional parties and premises for whom they have professional responsibility. To the fullest extent of law, the publisher and author are not liable in all circumstances (special, incidental, and consequential) for any injury and/or damage to persons and property, along with any potential loss of profit and other commercial damages due to the use of any methods, products, guidelines, procedures contained in the material herein.

NATIONAL LIBRARY OF AUSTRALIA

A catalogue record for this book is available from the National Library of Australia

ISBN (print): 978-1-925823-88-2

Contents

Preface

X-ray absorption fine structure (XAFS) refers to the oscillating feature of the X-ray absorption coefficient, as a function of the energy of incident X-ray photons when the energy exceeds the binding energy of the core-level electron in a specific element. Its most prominent features are the short-range order sensitivity and elemental specificity, which enable XAFS spectroscopy to selectively detect the surrounding environment of specific elements in matter and provide information on the electronic, structural and magnetic properties of solids, liquids and even gases. As a powerful tool for structural characterization, XAFS evolves with the development of effective scattering theory and the availability of synchrotron radiation light sources. Although the X-ray absorption phenomenon was first observed in 1920s, the underlying physics was preliminarily disclosed by Dale Sayers, Ed Stern and Farrel Lytle in the 1970s, introducing of Fourier analysis to the point-scattering theory of X-ray absorption fine structures, which allows for quantitatively determining structural parameters such as bond distance, coordination number, and thermal and disorder parameters. The XAFS field has grown considerably since 1990. After the synchrotron radiation light source with high brightness has been established, numerous measurement methods have been developed to obtain high-quality XAFS data under various conditions, such as transmission XAFS for ordinary bulk materials, fluorescent XAFS for trace elements, and surface grazing incidence XAFS studies, magnetic XAFS for spin studies, in-situ and operational XAFS for high temperature / high pressure / in situ reaction studies, time-resolved XAFS for kinetic studies, and space-resolved XAFS for micro area studies. XAFS has evolved into a mature and powerful technology for studying the local atomic / electronic structure of specific types of atoms in condensed states. The number of XAFS experiments conducted has grown exponentially since the 1970s, and today XAFS is commonly used as a local structure probe in many scientific fields, including physics, chemistry, biology, materials science, geo- and environmental science, etc. The structural information provided by XAFS helps to solve many important scientific problems.

More recently, the XAFS science and technology have significantly progressed thanks to the developments of synchrotron radiation sources and storage rings with lower beam emittance and higher brilliance. More sophisticated instrumentation with extremely high resolution has promised for better tools and new techniques. Notably, new detectors are as equally important as new experimental methods. The state-of-the-art pixel detectors play an increasingly important role for high energy-resolution fluorescence-detected synchrotron radiation XAFS spectroscopy experiments. Since the local structure in XAFS analysis is always limited to the average structure, combining two or more in situ characterization methods in one setting would overcome such a limitation, which would allow for great strides in XAFS. Various fields of synchrotron radiation applications have developed well and fast, with the emergency of new experimental methods and numerous research results. Many techniques have marked great research developments in XAFS applications over these years, such as time and space resolved technique, especially dynamical research on the relationship between structure and function during operation, namely operando techniques; many novel imaging (like photoemission electron microscopy) techniques, and new X-ray diffraction technique for dynamical structural studies on increasingly complex systems. With the advent of many new-generation synchrotron radiation sources and more advanced XAFS beamlines, there is no doubt that the scope of XAFS applications will continue to expand.

The purpose of this book is to provide a comprehensive, theoretical, practical, and thorough guide on current XAFS spectroscopy and modern applications. This book contains the experimental and theoretical aspects of XAFS in 14 chapters based on recent developments in the XAFS approach to science and technology, encompassing the basic principles and theories, calculation methods, computer programs, advanced methodologies and techniques, experiments, data analysis and interpretation, and applications. Especially, recent experiments increasingly rely on the polarity and time structure of synchrotron radiation, so the basic synchrotron physics is included. The book also illustrates the enormous potential of spectroscopic techniques with many important applications. More specifically, the fundamentals and the advanced research results are provided. Among the broad XAFS subjects of synchrotron radiation applications, this book covers soft X-rays, hard X-rays, and X-ray free electron laser, and various synchrotron-based techniques and methods, such as EXAFS (extended X-ray absorption fine structure), XANES (X-ray absorption near edge structure) and NEXAFS (near edge X-ray absorption fine structure), AXAFS (atomic X-ray absorption fine structure),

DAFS (diffraction anomalous fine structure), XMCD (X-ray magnetic circular dichroism), MEXAFS (magnetic extended x-ray absorption fine structure), XMLD (X-ray magnetic linear dichroism), XNCD (X-ray natural circular dichroism), and XNLD (X-ray natural linear dichroism).

This book may serve as a reference book for researchers and technicians taking up synchrotron radiation application research and postgraduates majoring in this field. The author hopes this book would possibly provide the beginners with an exhaustive guide in the field of XAFS. Also experienced users active in particular subfields of XAFS spectroscopies could rely upon this book about the enormous potential of XAFS for other applications, which would inspire to design new experiments and thus to enrich science.

In the preparation of this book, the author is fortunate to have the encouragement and help from many colleagues and friends. Special thanks are given to Professors Shiqiang Wei, Kunquan Lu, Hiroyuki Oyanagi, Wensheng Yan, Zhihu Sun, Klaus Baberschke, Panagiotis Poulopoulos, Andrei Rogalev, Fabrice Wilhelm, Robert Scott, John Rehr, Hubert Ebert, Yves Joly, Joachim Stohr and Graham George for their guidance, inspiration, and encouragement. The author is also grateful to all Central West Publishing staff, especially the Editor and Managing Director Dr. Vikas Mittal, for continuous encouragement, useful suggestions and careful production throughout the XAFS book project. Hopefully, this book could contribute toward present and future fundamental and practical research of the related domains and would meet the needs of the broad researchers and students working in the XAFS field or related topics.

Zhongrui (Jerry) Li
EMAL, University of Michigan
July 30, 2020, Ann Arbor

Chapter 1

Historical Perspective on X-ray Absorption Fine Structure

1.1 X-ray History and X-ray Generation

X-ray is sometimes viewed as a kind of electromagnetic waves with very short wavelengths (10~0.01nm, corresponding to frequencies in the range 30 petahertz to 30 exahertz), and sometimes treated as quantum particles (photons) with an energy in the range 124 eV to 124 keV. The X-ray photon energy relates to its frequency (ν) through the Planck postulate $h\nu$, where h is Planck's constant (about 6.6×10^{-34} Joule·second). The wave nature of X-ray can be found in interference and diffraction phenomena while its particle property is often seen in X-ray absorption and X-ray emission.[1] The dual nature of X-ray is comprehensively described and harmonically unified by the theory of quantum electrodynamics.

X-rays can be produced by accelerating (or decelerating) charged particles. For example, by colliding a metal target (anode) with high-speed electrons, the X-ray tubes produce a continuous radiation spectrum with several characteristic spike lines, which are respectively related to braking-induced (Bremsstrahlung) radiation and characteristic X-ray emission. Bremsstrahlung radiation is a continuous spectrum of electromagnetic wave generated by decelerating electrons scattered by a strong electric field near a high Z (proton number) nucleus. The lose kinetic energy of the moving electrons is converted into radiation (that is, photons) by the law of conservation of energy. Thus, the highest frequency of the induced radiation is determined by the energy of the incident electrons. When incident electrons with enough kinetic energy, they can knock the orbiting electrons out of the core shells of target atoms, creating vacancies, which are filled by electrons from higher energy levels, the released energy emits in form of X-ray photons. This transition produces X-ray emission spectra at some discrete frequencies, sometimes called spectral/characteristic lines. Neither Bremsstrahlung radiation nor characteristic X-ray emission is energy efficient, usually less than one percent of the electrical energy used by the X-ray tube can be converted into X-ray photons. The vast majority of the electrical energy is released as a form of heat, detrimental to the tube. Therefore, the

X-ray tube needs to be designed to efficiently dissipate excess heat in order to maintain the proper operation. Restricted by the lower intensity of the X-ray tube along with other drawbacks, applications of the X-ray tubes are very limited.

High-intensity X-ray beam can be generated by synchrotron facilities or free electron lasers. When accelerated electrons with high energy are forced to move in a curved path, synchrotron radiation is produced accordingly.[2] Essentially, the relativistic velocity of electrons will change the observed frequency due to the Doppler effect by the Lorentz factor γ. Relativistic length contraction then increases the observed frequency by another γ, thereby doubling the GHz frequency of the resonant cavity, which elevates the radiation into the X-ray range. The radiant power is given by the relativistic Lamor formula, and the force on the emitted electrons is given by Abraham-Lorentz-Dirac force. The radiation pattern can be distorted from an isotropic dipole to a polar forwarded cone. When viewed on the orbital plane, the acceleration geometry appears to polarize the radiation, and when viewed at a small angle to this plane, it is circularly polarized. However, the amplitude and frequency are concentrated on the extremely ecliptic. Synchrotron radiation produced by electron storage ring features high flux, high brilliance, high stability, excellent collimation and controllable polarization (both linear and circular), broad continuous wavelength, extremely small beam cross-section area, pulsed time structure.[3] Particularly, the third-generation synchrotron sources are more brilliant than the early lab-based sources by a factor of approximately 10^{12}. Synchrotron radiation light source has become the best X-ray light source for scientific research.

Since its discovery (generally believed) by Wilhelm Conrad Roentgen in 1895, X-rays immediately triggered enormous new discoveries and applications. Around 1906, physicist Charles Barkla discovered a characteristic X-ray spectrum of elements in gas when scattering X-rays. For this discovery, he won the Nobel Prize in Physics in 1917. In 1912, the X-ray diffraction of crystals was first observed by Max von Lloyd, Paul Kempp, and Walter Friedrich. Along with Paul Peter Ewald, William Henry Prague, and William Lawrence Prague, they established the field of X-ray crystallography. In 1913, based on the crystallographic experiments with X-rays emitted from various metals, Henry Moseley formulated the Mosley's Law, which relates the frequency of emitted X-rays to the atomic number of metals. In 1914, in order to quickly perform X-ray examinations of injured soldiers in World War I, Marie Curie developed a radiation vehicle which enabled battlefield surgeons to perform surgery quickly and accurately. X-ray microscopy gets significantly developed since 1950s, phase contrast X-ray imaging techniques can map soft tissue using phase information from a coherent X-ray beam. X-ray microscopy has become an important tool for visualizing cellular and histological structures in various biological and medical research. In the X-ray phase contrast imaging, the phase change of X-rays emitted from an object is converted into a change in intensity by using various principles. These include propagation-based phase contrast, talbot interferometry, refraction enhanced imaging, and X-ray interferometry. Compared to conventional absorption contrast X-ray imaging, these methods enable to see much smaller details with higher contrast, thanks to the advance in X-ray sources, X-ray optics, and X-ray detectors. Many X-ray related groundbreaking discoveries were awarded Nobel prizes, some are listed in Table 1.1.

Now X-ray has been widely used for crystal structure analysis, and chemical composition analysis and atomic structure research. It significantly impacts human's understanding of this world and instantly revolutionized the fields of physics, chemistry, materials, biology, geology, medicine and so on.

1.2 X-ray Interaction with Matter

Based on its wavelength, X-ray is often classified into two categories: hard X-ray with wavelength below 1 Ångstrom (0.1 nm, above 12.3984 keV), and soft X-ray with longer wavelength (and lower energy). With its strong penetration capability, hard X-rays are widely used to image the interior of objects, such as in medical radiography and airport security. They can also be utilized to determine crystal structure by X-ray crystallography as the wavelengths of hard X-rays are comparable to the size of atoms. In contrast, soft X-rays are easily absorbed by the air, for instance, 600 eV (~ 2 nm) X-rays in water have an attenuation length of less than 1 micron.[4] Thus soft X-ray experiments are commonly performed in ultrahigh vacuum chambers.

Table 1.1 Nobel prizes for research in X-ray

Year/field	Laureate	Contribution
1901 Physics	Wilhelm Conrad Röntgen	"in recognition of the extraordinary services he has rendered by the discovery of the remarkable rays subsequently named after him"
1902 Physics	Hendrik Lorentz Pieter Zeeman	"in recognition of the extraordinary service they rendered by their researches into the influence of magnetism upon radiation phenomena"
1914 Physics	Max von Laue	"For his discovery of the diffraction of X-rays by crystals"
1915 Physics	William Henry Bragg William Lawrence Bragg	"For their services in the analysis of crystal structure by means of X-rays"
1917 Physics	Charles Glover Barkla	"For his discovery of characteristic roentgen radiation of elements"
1924 Physics	Manne Siegbahn	"for his discoveries and research in the field of X-ray spectroscopy"
1927 Physics	Arthur holly Compton	"for his discovery of the effect named after him"
1936 Chemistry	Peter J.W DEBYE	"For his contributions to our knowledge of molecular structure through his investigations on dipole moments & on the Diffraction of x rays and electrons in gases"
1979 Physiology/ Medicine	Allan McLeod Cormack Godfrey Newbold Hounsfield	"For the development of computed axial Tomography"
1981 Physics	Kai Manne Börje Siegbahn	"for his contribution to the development of high-resolution electron spectroscopy"
1985 Chemistry	Herbert A. Hauptman Jerome Karle	"in recognition of the extraordinary service they rendered by their researches into the influence of magnetism upon radiation phenomena"
2002 Physics	Raymond Davis Jr. Riccardo Giacconi Masatoshi Koshiba	"For pioneering contributions to astrophysics, in particular for detection of cosmic neutrinos and discovery of cosmic X-ray sources"

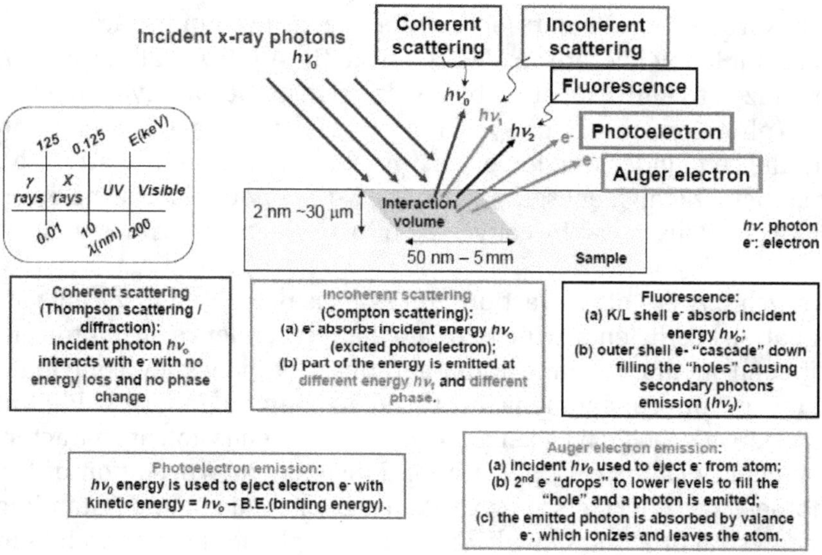

Figure 1.1 Some typical interactions between X-ray and matter.

A photon in an X-ray beam that passes through a medium is likely to undergo one of three fates: complete absorption, scatter, or transmission. The first two processes contribute to X-ray beam attenuation. When electromagnetic waves interact with matter, the incident wave can simply be partially or completely absorbed by the sample, and can be scattered elastically or inelastically, coherently (Thomson scattering, diffraction) or incoherently (like Compton scattering), and Rayleigh scattering (Figure 1.1). Five main interactions can cause attenuation of photons are (1) coherent scattering, (2) photoelectric effect, (3) Compton scattering, (4) pair production, and (5) photodisintegration. How much of the beam gets attenuated depends on which interaction process dominates. The strength of these interactions is determined by the X-ray photon energy and the elemental composition of the material, but has little to do with the chemical properties, because the photon energy is much higher than the chemical binding energy. Broadly speaking, coherent scattering is most important at low photon energies and photon disintegration at high energies, whereas the other interactions are likely to occur at intermediate photon energies. The dominant interaction mechanism in the soft X-ray and the lower hard X-ray regime is photoabsorption or photoelectric absorption, while Compton scattering dominates at higher energies. When the X-ray photon energy is much too high to be absorbed in electron transitions between states of the atom, they will completely liberate an electron from the atom, *i.e.*, all X-rays can be utilized to ionize an atom by donating all of the energy to an electron (photoionization), or by giving part of the energy to the electron and the remainder to a lower energy photon (Compton scattering). At sufficiently high energies, when the X-ray photon is passing by an atomic nucleus, it can be converted into an electron-positron pair.

The rich physics of interaction processes is presented in various attenuation coefficients, which explain their dependence on photon energy and medium composition, and describe how they contribute many exciting phenomena observed in material physics. Photon emission at specific wavelengths and electron emission at characteristic energy can occur. Based upon Einstein's interpretation of the photoelectric effect, great progress has been made in understanding the physical processes involved in all these phenomena, as well as in using these methods (usually spectroscopy). They allow the analysis of various materials, solids, liquids or gases, ordered or disordered substances used in many fields, such as biology, geology, chemistry or condensed matter physics.

1.3 Basics of X-ray Absorption

From the discovery of X-ray in 1896, the most important uses of X-rays are probably due to their high penetration depth of substances and substances with different absorption rates depending on their composition. Therefore, it is an excellent tool for studying the structure and composition of all types of materials. For homogeneous isotropic materials, the transmitted beam intensity simply decreases exponentially in terms of sample thickness (t) by following the Beer–Lambert law: $I = I_0 e^{-\mu t}$, where I_0 and μ are the incident intensity and the total linear absorption coefficient μ, respectively. This absorption coefficient, for a crystalline solid, is related to the absorption cross section (σ_i), of the *n* different chemical elements of the unit cell: $\mu = \frac{1}{V}\sum_{i=1}^{n} \sigma_i$, where *V* is the volume of the unit cell. The absorption cross section is usually expressed in megabarns (1 Mbarn = 10^{-18} cm^2). However, XAFS users are often only interested in spectral shapes. It is then normalized to the one after the edge, so it is given in arbitrary units.

When absorbing X-ray energy close to the binding energy of electrons, a feature called an absorption edge is observed. In the region of the X-ray spectrum, incident photons interact with matter through predominant photoelectric absorption. The process in which a photon is absorbed and a core electron is excited to an unoccupied state. Core energy E_0 is different for every element, thus X-ray absorption spectroscopy is element-specific. The basic physical quantity measured in XAFS (X-ray absorption fine structure) is the X-ray absorption coefficient $\mu(E)$, which describes how the X-ray absorptivity of a sample changes with the change in incident X-ray photon energy E. At specific energies, X-ray photons have sufficient energy to liberate electrons from the low-energy bound states in the absorbing atoms and cause a sudden increase in the absorption coefficient $\mu(E)$. This energy is called the element-specific X-ray absorption edge. The excited electrons from 1s, 2s, $2p_{1/2}$ and $2p_{3/2}$ correspond to the edges of K, L_1, L_2 and L_3, respectively (Figure 1.2). At the entire absorption edge, the X-ray absorption coefficient $\mu(E)$ shows a jump, and then an oscillating structure appears at higher energy.[5] XAFS then describes the details of X-ray absorption at energy near and above the absorption edge.

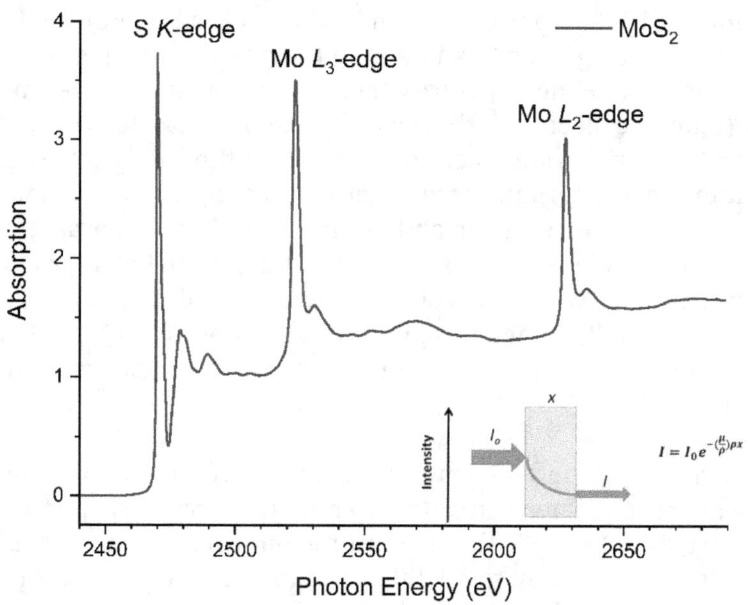

Figure 1.2 Sulfur K (at about 2470 eV) and the Mo L_3, L_2 edges (2521eV and 2625eV, respectively) for MoS_2.

When measuring the energy dependence of the photoelectric absorption cross section of X-ray interaction with matter, it can be seen that at some energies, the cross section increases sharply. Edge energy is characteristic of absorbing atoms and their chemical bonds. Careful observation of the spectrum shows that for the same element but in different chemical environments, a small shift in threshold energy (up to a few eV) can be observed. When the oxidation state is increased, the binding energy of the core electron also increases, the edges tend to move toward higher energy since more energy is needed to liberate the electron. This fact is often used to check the valence of metal atoms in oxides.

Since the first X-ray absorption spectrum was observed by De Broglie in 1913,[6,7,8] X-ray absorption spectroscopy (XAS) has been evolving more than one hundred years and has come a long way. Table 1.2 summarizes the history and progress of XAFS, showing principal achievements and landmarks for the theories, calculation methods, computer programs, methodologies and techniques. In its early age, there is no obvious practical application possibility, because no solid and comprehensive theory could describe and predict experimental observations on gas, liquid and solid (crystalline and amorphous) systems. In 1971, Sayers, Stern, and Lytle made a breakthrough progress by applying Fourier analysis to the point-scattering theory of X-ray absorption fine structures, thereby transforming experimental data (mainly collected in the vector space of photoelectron waves) into radial distribution function. For the first time, they were able to quantitatively determine structural parameters such as bond distance, coordination number, and thermal and disorder parameters.[9] In the four decades since this key publication, the field has grown exponentially, and XAFS has evolved into a powerful structural characterization tool. It has many significant advantages, such as element specificity, sensitivity to short range-order (usually several Å) and chemical state.[10]

For data analysis and interpretation, it is conventional to divide the X-ray absorption spectrum into two regions: X-ray absorption near-edge structure (XANES, the spectral region from (or slightly below) the absorption edge up to 30–50 eV) and extended X-ray absorption fine structure (EXAFS, typically cover approximately 30–50 eV above the edge up to 1000 eV or higher). The interval of these regions may vary for various compounds. The boundary between XANES and EX-

Table 1.2 The evolution history of X-ray absorption fine structure

Year	Theory/simulation, technique	Authors	Publication
1896	Discovery of X-ray	Rontgen	*Nature*, 53, 274-277(1896)
1913	Absorption edge measurement	Broglie	*Comptes Rendus* 157, 924-926.
1920	K-edge absorption	Fricke	*Phys. Rev*, 16, 202-215(1920)
1920	First Theory of XANES	Kossel	*Z. Phys.* 1, 119-134(1920)
1931	First Theory of EX-AFS	Kronig	*Z. Phys*, 70, 1585-1589(1954)
1970	XAFS point scatter-ing theory	Sayers *et al.*	*Adv. X-ray Anal.* 13, 248-271(1970)
1971	EXAFS Fourier analysis	Sayers *et al.*	*Phys. Rev. Lett.*, 27, 1204-1208(1971)
1972	Total electron NEXAFS	Gudat *et al.*	*Phys. Rev. Lett.* 29, 169-172(1972)
1974	EXAFS theory	Stern	*Phys. Rev. B* 10, 3027-3039(1974)
1977	Auger electron yield NEXAFS	Bianconi *et al.*	*Solid State Commun.*, 24, 539-542(1977)
1978	Surface EXAFS	Stohr *et al.*	Phys. Rev. *B*, 18, 4132-4135(1978)
1979	White line inten-sity	Lytle *et al.*	*J.Chem. Phys.* 70, 4849-4855(1979)
1980	First-principle XANES	Natoli *et al.*	*Phys. Rev. B*, 22, 1104-1108(1980)
1980	DXAFS	Matsu-shita *et al.*	*J.Appl. Cryst.*, 13, 465-471(1980)
1981	Reflectivity EXAFS	Martens *et al.*	*J.Phys. C*, 14, 1523-1534(1981)
1982	Multiple scattering XAS calculation	Durham *et al.*	*Comput.Phys. Commun.*, 25, 193-205(1982)
1986	EXCURVE	Gurman *et al.*	*J.Phys. C*, 19, 1845-1861(1986)
1986	CONTINUUM	Natoli *et al.*	*Phys. Rev.* A34, 4682-4694(1986)
1986	Dispersive and to-tal reflection XAS	Dartyge *et al.*	*Phys. Lett.* 113A, 384-388(1986)
1986	X-ray magnetic lin-ear dichroism	van der Laan *et al.*	*Phys. Rev. B* 34, 6529 (1986)
1987	X-ray magnetic cir-cular dichroism	Schutz *et al.*	*Phys. Rev. Lett.* 58, 737-740(1987)

1988	X-ray photoemission electron microscopy	Tonner *et al.*	*Rev. Sci. Instrum.* 59, 853 (1988)
1989	Magnetic EXAFS	Schutz *et al.*	*Phys. Rev. Lett. 62 , p. 2620(1989)*
1990	FEFF	Rehr *et al.*	*Phys. Rev. B*41, 8139-8149(1990)
1992	Polarized total reflection florescence EXAFS	Asakura *et al.*	*Catal .Lett.*, 15, 247-254(1992)
1992	Diffraction anomalous fine structure	Stragier *et al.*	*Phys. Rev, Lett.*, 21, 3064-3067(1992)
1995	GNXAS	Cicco	*Phys. B*, 208-209, 125-128(1995)
1997	FDNES	Joly	*J. Phys. IV*7, C2-111-115(1997)
1998	X-ray natural circular dichroism	Alagna *et al.*	*Phys. Rev. Lett.* 80, 4799(1998)
2000	SPRKKR	Ebert	*Electr.Struct.&Phys.Prop. Solids,* Springer, 191-246(2000)
2002	180s time-resolved QXAFS and DAXFS	Dent	*Top. Catal.*, 18, 27-35(2002)
2004	5μm(KB mirror) μ-XAFS	Paktune *et al.*	*Geochim. Cosmo. Acta,* 68, 969-983(2004)
2004	0.8μm(Fresnel zone plate) μ-xAFS	Foriel *et al.*	*Geochim. Cosmo. Acta,* 68, 1561-1569(2004)
2009	Free electron laser NEXAFS	Bernstein *et al.*	*App. Phys. Lett.*, 95, 134102(2009)
2011	Computed Tomographic XANES	Meirer *et al.*	*J. Synchrotron Rad.*, 18, 773-781(2011)
2014	Back illumination fluorescence XAFS	Asakua *et al.*	*Phys. Chem. Chem. Phys.*, 16, 13748-13754(2014)

AFS is somewhat arbitrary, no fundamental difference in physics gives rise to the fine structures. Therefore, the term XAFS now often refers to the entire oscillating component beyond the absorption edge. However, the complexity and analysis methods are quite different for these two regions. At characteristic wavelengths, X-ray photon can pump core electron into the vacancy in the high orbital/valence band, this dramatic change in the X-ray absorption of the element is called absorption edge. Thus, the absorption spectrum near the absorption edge is significantly affected by the electronic and geometric environment of the absorbing atoms. In the XANES region, the photoelectron energy is close to the absorption edge, and the potential of adjacent atoms has an important effect on photoelectron scattering. Therefore, the multiple scattering process contributes to the fine structure. The distant EXAFS region is controlled by a single scattering process. Its upper limit is determined by the signal-to-noise ratio and/or another absorption edge. In the EXAFS region, excited photoelectrons have higher kinetic energy and are less affected by the potential of neighboring atoms. Therefore, in most cases, the single scattering of photoelectrons by the neighbors dominates.

XAFS can provide quantitative structural information on the local environment around the absorbing atoms at the atomic scale. More specifically, EXAFS is sensitive to the distance, coordination number, and species of atoms immediately adjacent to the absorbing atom. The XANES region is determined by the local density of the vacant atomic states and multiple scattering effects (*i.e.*, the scattering of excited photoelectrons on several atoms). XANES can distinguish between electronic state and three-dimensional coordination geometry (including symmetry, atom type, bond angle and distance). The sensitivity of XAFS to short-range order and element specificity give XAFS unique ability to determine the local structure of a variety of systems, whether the sample being studied is solid (crystalline or amorphous), liquid or even gaseous.

1.4 Applications of XAFS

Since its development in the early 1970s, synchrotron-based X-ray absorption fine structure spectroscopy (XAFS) has been widely used to study the local environment of specific elements in various environments, including crystalline solids, glass, and high-temperature liquids. Aqueous adsorption systems involving metal complexes that bind (or adsorb) to the solid/water interface. XAFS can provide information that basically complements the results of other experimental methods, such as X-ray and neutron diffraction (scattering), photoelectron and emission X-ray spectra. The basic advantages of XAFS are: (1) the selectivity of the type of chemical element, which enables people to obtain information about the local environment of each element of the material under study; (2) the partial density of the empty state near the Fermi level; (3) When using synchrotron radiation, it has high density sensitivity (10-100 particles per mole), and the detection time of the experimental spectrum is relatively short (from milliseconds to tens of minutes); (4) the required sample size is small (usually, less than 30 mg/cm^2 is sufficient). Because of these advantages, the use of XAFS is particularly attractive for studying crystalline and disordered (amorphous, glassy, liquid and gas) multicomponent materials, as well as for in situ studies of kinetic processes (phase transitions and chemical reactions). In many cases, including short-range ordering of atoms in crystalline solids, atomic environments in poorly ordered materials such as gels, glass, melts, and radiation-damaged materials, and adsorption at solid/water interfaces that are important to the environment structure and composition of matter, XAFS provides unique structural data that cannot be replicated by other methods.

XAFS spectroscopy is very important for understanding the local structure of many materials, including biological materials, new materials with special electronic properties (such as superconductivity), rare materials in fluids, and complex heterogeneous materials. It can provide information on bioremediation processes, study tiny minerals returned from space missions, and can be used to understand chemical reactions such as heterogeneous catalysis and hydrothermal synthesis of industrial materials. XAFS and related technologies have been established as leading scientific characterization tools. Without the contribution of XAFS and related technologies, it is impossible to imagine today's achievements in materials science, solid state physics and chemistry, catalysis, chemistry, biology, medicine, earth sciences, environmental sciences, cultural heritage, nanoscience, and so on.[11]

Although XAFS has powerful functions as a local structure detection technology, it also has limitations. First, it can only provide information about the overall average structure around the absorbing atoms, but it is difficult to distinguish target elements that are mixed in different phases or located in different special positions. Second, its sensitivity to local structures is limited to the radial range within the first few coordination shells.

1.5 Synchrotron-based New XAFS Techniques

Since the design of the first synchrotron radiation source in Russia's VEPP-3 storage ring in 1971, synchrotron radiation sources have been developed for three generations. Currently, the third-generation synchrotron radiation sources ESRF (France), APS (USA) and SPRING8 (Japan) with radiances up to ~ 10^{19} photons are running. There are now more than 50 sources running and more than 10 are being designed. Synchrotron radiation has high brightness and unique linear or circular polarization, providing unique research possibilities. The development of XAFS scientific and technological methods is inseparable from the development of synchrotron radiation sources and storage rings with higher brightness, and the generation of soft X-rays (especially hard X-rays). The high intensity of synchrotron X-rays makes it possible to detect very low-concentration elements. The tunability of synchrotron radiation allows XAFS spectra of almost all elements to be collected in a short period of time. Combining high brightness with the usable small spot size allows inspection of samples with very high spatial resolution. In particular, new XAFS methods have been developed to study the atomic and electronic structures in bulk and surface layers of different thickness values. These methods include: (1) energy-resolved fluorescence XAFS; (2) surface EXAFS (SEXAFS) spectroscopy, by measuring Auger electron, total or partial yield of photoelectron current, yield of photoionized ion desorption and total internal reflection;

After decades of development, this element-specific, short-range, core-level X-ray absorption spectroscopy techniques go beyond the conventional X-ray Absorption Near-Edge Structure (XANES), Extended X-ray Absorption Fine Structure (EXAFS). The emergence of new techniques such as X-ray excited optical emission (XEOL), Resonant Inelastic X-ray Scattering (RIXS), X-ray Emission Spectroscopy (XES), and Bragg peak diffraction anomalous fine structure (DAFS), has overcome some limits of conventional XAFS. So various high-quality studies can be performed in physics, chemistry, biological science, materials science and engineering, energy science, environmental science, earth science, metallurgy, mineralogy and other scientific fields.

Thanks to the polarization tunability of synchrotron radiation, dichroism in X-ray absorption with linearly polarized light leads to X-ray natural linear dichroism (XNLD) and X-ray magnetic linear dichroism (XMLD), or with circularly polarized light, one can obtain X-ray natural circular dichroism (XNCD) and X-ray magnetic circular dichroism (XMCD).[12] Over three decades, these new experimental techniques provide unique capabilities for studying magnetic phenomena, owing to the advantage of being directly coupled to spin-resolved electronic status. X-ray magnetic circular or linear dichroism (XMCD or XMLD) spectroscopy is a unique tool for measuring element-specific 3d, 4f and 5d magnetic moments which are usually divided into spin and orbital components. With the advent of new instruments, powerful inelastic X-ray scattering in resonant and nonresonant modes emerging spectral probes provide extensive information about electronic states in highly relevant materials or in materials under high electric and strong magnetic fields. Such an experiment requires high brightness of third-generation synchrotron light sources, such as APS, ESRF or SSRL, etc.

With the assistance of focusing optics such as Kirkpatrick-Baez (K-B) mirror or Fresnel zone plate, synchrotron radiation sources of the third generation allows to perform magnetic imaging using XAFS techniques on a sub-micrometer level. This Photoemission Electron Microscopy (PEEM) technique combines X-ray microscopy or X-ray photoelectron microscopy with spectroscopy and permits imaging of magnetic domain structures with a lateral spatial resolution of a few 10 nm. A new development is lens-less imaging by Fourier transform X-ray holography, where the diffraction pattern of a coherently illuminated sample is recorded in Fourier space. These methods provide a key technique for research on small structures important in microelectronics, which are often heterogeneous and composed of several elements. Together with the temporal structure of the synchrotron radiation, they permit element-sensitive time-domain studies, which are of prime importance for magnetic recording. Examples are the dynamics of domain-wall displacements and transformation or dynamics of the magnetization of mesoscopic magnetic structures.[13] The underlying processes occur at times in the nano- to femto-second range. Understanding is limited due to the lack of a microscopic theory.

In this book, no attempt is made to cover all of the information necessary for an inexperienced scientist to understand all aspects of the design and execution of an XAFS experiment, data analysis and interpretation. Instead, this book specializes in the main basic phenomena and aspects of X-ray absorption fine structure, describes modern theoretical concepts of electromagnetic wave and matter interactions, and focuses on core-level X-ray spectroscopy, and synchrotron sources and equipment. Comprehensive understanding about theory, instrumentation and applications of XAFS is discussed at great lengths in the book which could serve as unique diagnostics or a go-to reference to study almost any condensed matter. Certainly, graduates and researchers across multi-disciplines working on X-rays and related synchrotron sources could rely upon it as a primer.

---

[1] J. A. Van Bokhoven, C. Lamberti, (2015) *X-ray absorption and X-ray emission spectroscopy: theory and applications*, edited by Jeroen A. van Bokhoven and Carlo Lamberti. ISBN 978-1-118-84423-6.

[2] S. Mobilio, Federico Boscherini, Carlo Meneghini, (2014) *Synchrotron Radiation: Basics, Methods and Applications*. Springer, ISBN 978-3-642-55315-8.

[3] H. Winick, (1998) Synchrotron radiation sources—present capabilities and future directions. *J. Synchrotron Rad.* 5, 168–175.

[4] D. Attwood, (1999) *Soft X-rays and extreme ultraviolet radiation*. Cambridge University. p. 2. ISBN 978-0-521-65214-8.

[5] D.C. Koningsberger, R. Prins, (1988) *X-Ray Absorption: Principles, Applications, Techniques of EXAFS, SE XAFS and XANES*. New York: Wiley.

[6] M. de Broglie, (1913) Sur une nouveau proced´e permettant d'obtenir la photographie des spectres de raies des rayons Röntgen, *C. R. Acad. Sci.*, 157, 924–926.

[7] L. de Broglie, (1920) Sur le calcul de frequences limites d'absorption K et L des ´el´ements lourds, *C. R. Acad. Sci.*, 170, 585.

[8] L. de Broglie, (1920) Sur l'absorption des rayons R¨ontgen par la mati`ere, *C. R. Acad. Sci.*, 171, 1137.

[9] D.E. Sayers, E.A. Stern, & F.W. Lytle, (1971) New Technique for Investigating Noncrystalline Structures: Fourier Analysis of the Extended X-Ray Absorption Fine Structure, *Phys. Rev. Lett.*, 27, 1204–1207.

[10] Scott Calvin, (2013) *XAFS for Everyone*, By, CRC Press, ISBN 9781439878637.

[11] Y. Iwasawa, K. Asakura, M. Tada (Eds.) (2016) XAFS Techniques for Catalysts, Nanomaterials, and Surfaces, Editors: Springer, ISBN 978-3-319-43866-5

[12] J. Stoehr H.C. Siegmann, (2006) *Magnetism From Fundamentals to Nanoscale Dynamics*, Springer Berlin Heidelberg New York, ISBN-10 3-540-30282-4.

[13] E. Beaurepaire, H. Bulou, F. Scheurer, J.-P. Kappler, (2010) *Magnetism and Synchrotron Radiation: New Trends*, Springer Heidelberg Dordrecht London New York, ISBN 978-3-642-04497-7.

Chapter 2

Multiple-Scattering Theory of EXAFS

2.1 Introduction

X-ray absorption fine structure (XAFS) refers to the oscillations in the energy-dependent photo-absorption coefficient of an element at energies near and above the core-level binding energies of that element in condensed matters due to the chemical and physical state of the element. As the fine structure at all energies, the acronym XAFS encompasses both extended X-ray absorption fine structure (EXAFS) and X-ray absorption near edge spectra (XANES), where the terms refer, respectively, to the structure in the X-ray absorption spectrum $\mu(\omega)$ at high and low energies relative to the absorption edge with the crossover typically at about 30~50 eV above the edge. Shortly after the discovery of the XAFS in 1920, physicists started to explore the mysterious energy-dependent structure that appeared at the energy above the absorption edge in the X-ray spectrum. Finally the revolutionary breakthrough was made by Dale Sayers, Ed Stern, and Farrel Lytle in 1971,[1] the underlying physical processes of these XAFS features can be satisfactorily described. The application of Fourier transforms provides an easy-to-understand connection between the oscillation structure of the absorption spectrum and the local geometry of the absorbing element in the material under study. This discovery immediately made EXAFS a powerful tool to probe local structure of specific elements in various materials. To realize the extraordinary capabilities, these spectroscopies heavily rely on quantitative simulations of the spectra. Fortunately, theories of XAFS spectroscopy have become increasingly accurate and sophisticated, generally overcoming this limitation. [2,3] The element specific and short-range nature makes this core-level X-ray absorption spectroscopy a routine tool to elucidate the local structural, electronic, vibrational, magnetic, and other physical properties of condensed matters.

Underlying the main process for XAFS is photoelectric absorption (photoelectric effect) and scattering (elastic, inelastic).[4] A photon with energy above the core-shell ionization energy can be completely absorbed and liberates a core electron from the absorbing atom leaving behind a core hole. Although this excited atom with a core hole may produce further excitations, this probability is often small and ignored in the simple picture. XAFS is ***element specific*** because the X-ray absorption edges of different elements have different energies. This photoelectron will be ejected with energy equal to energy of the incoming photon minus its binding energy (E_0), when in the core. This photoelectron will interact with the surrounding atoms (Figure 2.1). For low photoelectron energies, the correspondingly long de Broglie wavelengths commensurate with the interatomic distance (R) in a crystal lattice, which corresponds to the XANES region, the multiple-scattering contribution usually far exceeds the single-scattering one. As photoelectron energy increases, the de Broglie wavelengths are getting very shorter and, therefore, single scattering dominates. Considering the wave nature of the ejected photoelectron and regarding the atoms as point scatterers, in experiment, as photon energy is scanned above E_0, the kinetic energy (E_{kin}) of photoelectron is varied, and consequently the momentum of photoelectron (k) and the de Broglie wavelength λ. A simple picture can be seen in which the backscattered waves interfere with the forward wave either constructively or destructively, depending on the ratio between λ and neighboring distance R. The interference between outgoing and incoming waves gives rise to the sinusoidal variation of absorption coefficient $\mu(\omega)$. That is to say, the self-interference of photoelectron excitations result in these XAFS oscillations. The oscillation frequency in XAFS spectrum is associated with the distance from neighbors, the amplitude is related to the number and type of neighbors. The spatially dependent elastic and inelastic electron scattering coefficients of the material directly modulate an atom's X-ray absorption probability, therefore the XAFS spectrum is a precise function of the complex electronic potential within a critical space near the X-ray absorbing atom.

In a quantum mechanics view, this is an interference effect on the final state. Since the transition probability is given by a matrix element between the final and initial states and the absorption coefficient is related to the transition probability, this interference affects the absorption coefficient which is measured in an XAFS experiment. The matrix element is only non-zero where the initial core state is non-zero, thus the interference need only be calculated at the center of the absorbing atom. As the energy of photoelectron determines its wavelength, the phase of the backscattered wave at the central atom varies with the energy of the incoming X-ray photon. This leads to the oscillatory nature of the interference effect. Since backscattering amplitude and phase are dependent on the type of atom doing the backscattering and the distance it is from the central atom, information regarding the coordination environment of the absorbing atom can be obtained by analyzing the EXAFS.

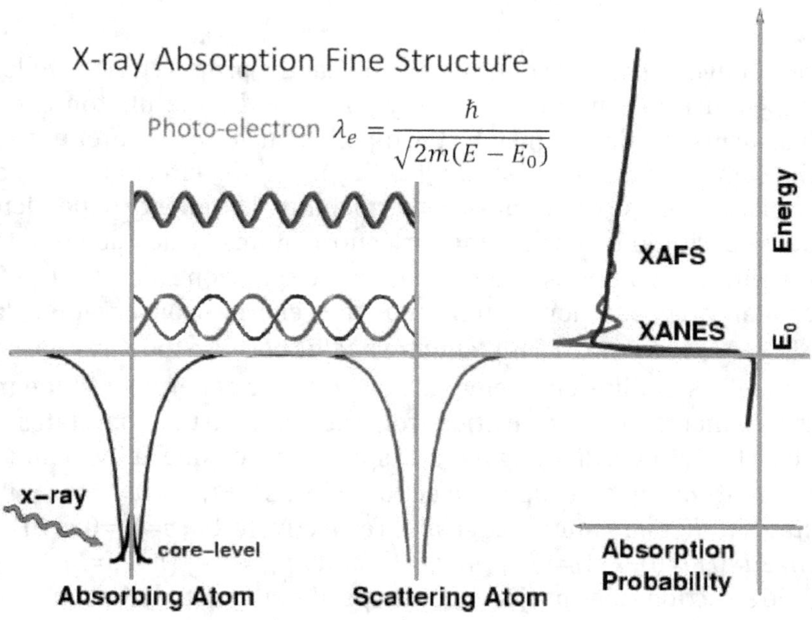

Figure 2.1 XAFS occurs because the photoelectron can scatter from a neighboring atom. The scattered photoelectron can return to the absorbing atom, modulating the amplitude of the photo-electron wave-function at the absorbing atom. This in turn modulates the absorption coefficient μ(E), causing the EXAFS.

Since X-ray absorption is a transition between two quantum states, *i.e.*, from an initial state with an X-ray plus a core electron to a final state with a core hole and a photoelectron, the formular of X-ray absorption spectroscopy is based on the Fermi's golden rule

$$\mu(\omega) \sim \Sigma_f \left|\langle f|\hat{O}|i\rangle\right|^2 \delta(E_f - E_i - \hbar\omega), \tag{2.1}$$

which requires a summation over exact many-body final states $|f\rangle$ with energies E_f, i.e., superposition of the ejected wave and backscattered waves. $\hat{O}$ is the transition operator, which can be electric dipole, electric quadruple and magnetic dipole transitions. In terms of the polarization vector ϵ =**E**/|E| and its normalized wavevector k, within independent-particle approximation, the absorption cross section for the dipole and quadrupole transitions has the form

$$\sigma(\omega) = 4\pi\alpha^2\hbar\omega \sum_f \langle f|\epsilon \cdot r|i\rangle \delta(E_f - E_i - \hbar\omega) \text{ and}$$
$$\sigma(\omega) = \pi\alpha^2\hbar\omega \sum_f \langle f|(\epsilon \cdot r)(k \cdot r)|i\rangle \delta(E_f - E_i - \hbar\omega),$$

respectively. These formulas indicate that the process has the following features: excitation selectivity governed by the photon energy $\hbar\omega$; dichroism determined by the dipole-transition operator **e·r** with the components **e·z** or $(\Sigma x - i\Sigma y)(\Sigma x + i\Sigma y)$; and the effect of final-state relaxation on the total transition probability. The energy dependence of the excitation cross section for neutral atoms, which do not interact with each other, is determined by the selection rule $\Delta l = \pm 1$ in the orbital angular momentum and the energy conservation law $\delta(E_f - E_i - \hbar\omega)$. XAFS is ***element specific*** because the X-ray absorption edges of different elements have different energies. As the matrix elements for the transition from the initial to the final states can be factor into a radial, angular and spin parts, looking at the non-zero elements one get the dipole selection rules: $\Delta l = \pm 1$, $\Delta s = 0$; As the initial states of K, L_1, L_2 and L_3-edge are, respectively 1s (n=1,l=0,m=0), 2s (n=2,l=0,m=0), 2p(j=1/2) (n=2,l=1), and 2p(j=3/2) (n=2,l=1), the dipole selection rules project out the specific symmetry components of final state wavefunction – K, L_1 edges probe p part of final states, while $L_{2,3}$ edges probe d and s part of final states. Note that these rules are for atomic potentials, in crystal field, can have violations/mixing near the absorption edge.

And the sum is over unoccupied many-body final quasi-particle states |f⟩ which are often calculated with a self-consistent potential including a core-hole (*i.e.*, ΔSCF approximation). The final state effects are also needed to be included, such as the core-hole potential and lifetime, as well as inelastic losses (both extrinsic and intrinsic), and vibrational effects. However, practical calculations depend on the reduction of this relation to an effective, independent particle theory that takes into account several important many-body effects. Another difficulty is that deep core levels are involved, thus requiring an all-electron treatment in the presence of a core-hole, with relativistic corrections that include spin-orbit interactions. Thus, approaches to core-level spectra differ qualitatively and quantitatively from ground state theories. As a result, band structure and quantum chemistry codes with ground state potentials and the neglect of vibrational damping and inelastic losses are usually inappropriate for XAFS calculations. With an improved understanding of these effects and accurate new approximations, in particular for inelastic losses and vibrational damping,[5] the XAFS theory and calculation advance significantly in recent years. The new physics for XAFS goes beyond the conventional independent-particle approaches of quantum chemistry and condensed matter such as Hartree–Fock or density functional theory (DFT). The new development adopts an energy-dependent self-energy to replace the exchange-correlation functional in DFT.[6] Corrections to the quasi-particle theory can also be included approximately in terms of a spectral function which further broadens the spectra.[7] In addition, different limits and time-scales are involved, XAFS calculation needs to be efficiently simplified by a variety of approximations and computational methods like Green's function and Lanczos techniques. The recent advances can correctly treat multiple-electron excitations and the effects of screening of the X-ray field. Finally, the theory can be employed to interpret the XAFS spectra in terms of the geometrical, electronic and vibrational properties of a material. In addition to X-ray absorption spectroscopy, these advances can be applied to many core-spectroscopies involving electronic excitations, such as non-resonant inelastic X-ray scattering (NRIXS, also known as X-ray Raman scattering or XRS),[8] and electron energy-loss spectroscopy (EELS).[9]

2.2 The EXAFS Equation

One of the key developments in the history of EXAFS was the famous equation of Sayers, Stern, and Lytle, which represents the normalized fine structure $\chi(\omega) = [\mu(\omega) - \mu_0]/\mu_0$ in XAFS in terms of oscillatory contributions from near-neighbor atoms. Here μ_0 is the jump in the XAFS at the edge. Although their model was heuristic, it included two key many-body effects, namely the mean free path of the photoelectron and vibrational damping. This key observation demonstrates both the failure of any independent particle interpretation and the need to include both electronic and vibrational effects in quantitative theories. It also shows that XAFS is a short-range order phenomenon and cannot be described using conventional band-structure methods. Remarkably, an exact treatment based on the multiple-scattering path expansion can be cast in a similar form,[10] namely

$$\chi(k) = S_0^2 \sum_R \frac{|f_{eff}(k)|}{kR^2} \sin(2kR + \Phi_k)\, e^{-2R/\lambda_k} e^{-2\sigma_R^2 k^2}\,, \tag{2.2}$$

except that all quantities are redefined to include curved-wave and many-body effects implicitly. In particular f_{eff} is the effective curved wave scattering amplitude for a given photoelectron wave number k defined relative to the absorption threshold (or Fermi level E_F) for X-ray absorption, *i.e.*, with $k^2 = 2(\hbar\omega - E_F)$, λ_k the inelastic mean free path, and σ_R^2 the mean square relative vibrational amplitude of path R. The sum is over all multiple-scattering paths, R being the half-path length, S_0^2 the many-body reduction factor which is typically about 0.9 as discussed below, and Φ_k a smoothly varying phase shift. Remarkably, the expansion generally converges rapidly away from the edge, with only of order 10^2 multiple-scattering paths in the EXAFS. This simplification is a consequence of the smallness of λ_k which is generally of order 5–20 Å. This representation also explains why the dominant peaks in the EXAFS Fourier transform $\chi(R) = FT[\chi(k)]$ are close to the geometrical near-neighbor distances in a material, but shifted by system dependent phase shifts, $\delta R \approx \frac{1}{2} d\Phi_k/dk$. Moreover, this multiple-scattering path representation permits an analysis of experimental XAFS signals in terms of geometrical and vibrational properties of a material, provided one has a reasonable approximation of the phase shifts and λ_k. The theoretical dependence of the backward scattering amplitude and phase shift for various elements demonstrates that $f_{eff}(k)$ rapidly decreases with the electron momentum for small-Z atoms and has a large value for large-Z atoms. This difference in the scattering amplitudes can be employed to separate various contributions to the EXAFS spectra. They are scattering properties of the neighboring atom, which make EXAFS sensitive to the atomic species of the neighboring atom.

In the above sections, polarization dependence of EXAFS was not taken into account. Within the plane-wave approximation, Gaussian distribution model, and single scattering approximations,[11], the K or L_1-edge EXAFS function is simply given as

$$\chi(k) = S_0^2 \frac{N^* F(k,r)}{kr^2} e^{-2r/\lambda(k)} e^{-2k^2\sigma^2} \sin[2kr + \Phi(k,r)],$$

where N_j^*, r_j, φ^2_j and E_{0j} represent the effective coordination number, the bond distance, the Debye–Waller factor, and the difference of the threshold energy between reference and sample, respectively. $F(k,r)$ and $\Phi(k,r)$ represent amplitude and phase shift functions, respectively (Figure 2.2). N^* depends on the X-ray incident angle θ, and is given as $N^*(\theta) = \sum_{j=1}^{N} 3\cos^2\delta_j$, where N is the real coordination number and δ_j is the angle between the electric field vector of the X-ray and the bond direction of the absorber and j-th scattering pair. Polarization dependent surface EXAFS of adsorbate-substrate systems allows one to determine the adsorption geometry of adsorbate atoms with respect to the substrate single crystal. For the L_2 or L_3–edge EXAFS, the description of the effective coordination number is more complicated and less drastically dependent on $N^*(\theta) = \frac{N}{2} + \frac{3}{2}\sum_{j=1}^{N} 3\cos^2\delta_j$. In this expression the contribution of the p→s transition is neglected. However, it is not appropriate to ignore the p→s contribution in the polarization dependent EXAFS spectra since the first-order cross term does not vanish in contrast to the polarization averaged spectra. This contribution is estimated as larger as 20%. So care should thus be taken when analyzing polarization dependent L_2 and L_3-edge EXAFS data.

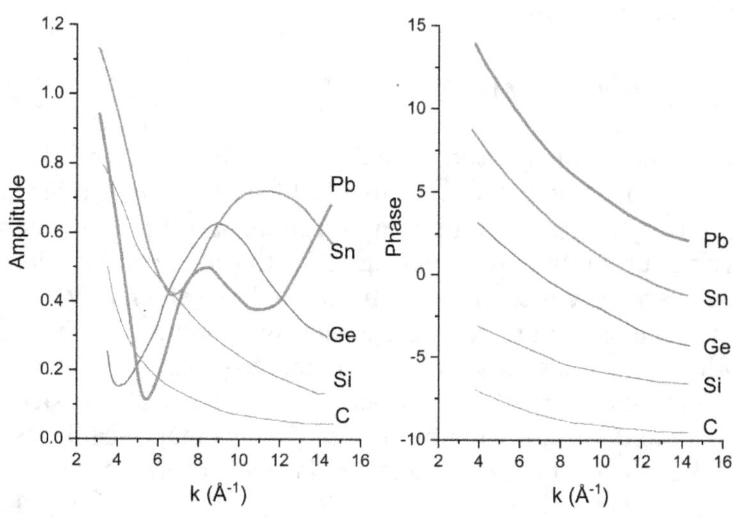

Figure 2.2 Amplitude and phase shift functions of IVA elements: C, Si, Ge, Sn and Pb.

2.3 Many-Body XAFS Formula

As mentioned above, XAFS process encompasses two steps, (1) the production of the photoelectron by photoexcitation from a particular core-state, and (2) the inelastic losses and secondary excitations. The many-body effects can be easily incorporated by following the Campbell's approach. [12] The first step is represented by an effective one-body (*i.e.*, quasi-particle) absorption $\mu'(\omega)$, while the effect of inelastic losses and secondary excitations (such as plasmons and electron–hole pairs) in the second step, can be given by an energy-dependent "spectral function" A(ω,ω'). This yields an exact representation of the many-body XAFS in terms of a convolution

$$\mu(\omega) = \int A(\omega,\omega')\mu'(\omega-\omega')d\omega' \tag{2.3}$$

The spectral function A(ω,ω') subsequently broadens and shifts the spectrum. Its shape typically consists of a sharp "quasi-particle" peak of width related to the inverse mean free path, together with a broad satellite. Consequently the EXAFS problem can be factored into two parts: an effective quasi-particle XAFS $\mu'(\omega)$ and a spectral function part that accounts for inelastic losses. The quasi-particle XAFS can be calculated using an independent-particle Fermi's golden rule, with the ΔSCF approximation, i.e., with the final single-particle states |f⟩ calculated with the final state Hamiltonian in the presence of the core-hole

$$\mu'(\omega) \sim \sum_f |\langle f|\hat{O}\hat{P}|i\rangle|^2 \delta(\omega + E_i - E_f), \tag{2.4}$$

where for the dipole operator $\hat{O} = \vec{\epsilon}\cdot\vec{r}$ and $\hat{P}$ is the projection operator onto unoccupied states. Although this is a considerable simplification to Eq. (2.1), a serious computational bottleneck is the calculation and summation over final states with the Fermi's golden rule at high energies, which can only be carried out efficiently for small systems such as atoms, small molecules, or systems with high symmetry like periodic solids. In practice, most systems of interest lack significant symmetry. Even for crystals, reciprocal k-space (i.e., standard band structure) approaches [13,14] are inapplicable, as they ignore the effects of the core-hole and lattice vibrations, which spoil the crystal translation symmetry, although the core-hole can be treated approximately with super-cell approximation. Thus, instead of explicitly calculating the final states and energy eigenvalues in a wave-function based approach, it is computationally preferable to re-express the XAFS in terms of the photoelectron Green's function or propagator G in real space. This procedure implicitly sums over all final states, as can be seen by using the spectral representation of the Green's function G for which $-\frac{1}{\pi}ImG = \sum_f |f\rangle\delta(E - E_f)\langle f|$, so that the dipolar Fermi's golden rule can be re-expressed as a single matrix element of a non-local operator for each core-state |i⟩,

$$\mu(\omega) \approx -\frac{1}{\pi}Im\langle i \mid \vec{\epsilon}\cdot\vec{r}'G(r,r',E)\vec{\epsilon}\cdot\vec{r} \mid i\rangle\,, \qquad (2.5)$$

where $E = \hbar\omega + E_i$. This formulation turns out to be advantageous for XAFS calculations even in periodic and aperiodic materials alike, since inelastic losses (*i.e.*, the mean free path) limit the range probed by XAFS experiment to the inelastic mean-free path, which is typically between 5 and 20 Å. This range corresponds to clusters of roughly a few hundred atoms surrounding a given absorption site. Consequently, XAFS is described by a short-range order theory. More precisely the mean-free path depends on both final state and core-hole lifetimes.

2.4 Real-Space Multiple-Scattering Theory

The XAFS and related spectra can be simulated using the real-space multiple-scattering approach, which is also known as the real-space Green's function method (the real-space analog of the Koringa-Kohn-Rostoker band-structure theory).[15,16] The real-space multiple-scattering method exploits the close connection between XAFS and electronic structure theory. More importantly the approach simplifies the treatment of inelastic losses and enables one to efficiently calculate XAFS over an extended energy range through *ab initio* approach, for both periodic and amorphous systems. Moreover, in the EXAFS region (e.g.,$E \gg \omega_{ph}$), scattering becomes relatively weak or negligible, and perturbation theory in the scattering potential converges quickly.

$$v'_{coul} + \Sigma(E) = \sum_R v_R(r - R).$$

The starting point of the real-space multiple-scattering theory used for practical calculations of EXAFS is the separation of the potential into contributions from the local "scattering potentials" v_R at each atomic site R,

$$v_R(r) = v'_{coul} + \Sigma(E) = \sum_R v_R(r - R). \tag{2.6}$$

Here v'_{coul} is the Hartree potential calculated for the final state in the presence of a core hole, and $\Sigma(E)$ is the energy-dependent self-energy. In the so-called "muffin-tin" approximation, one can use a spherical approximation $v_R(r - R) = v_R(|r - R|)$, with the range r_{MT} confined within a given cell. As a consequence, the Green's function propagator is separable in terms of radial scattering states $R_L(rR) = R_{LR}(r)$ with the angular momentum variables $L = (l, m)$ at each site R. With the separation and the approximation of local spherical symmetry at each site, the propagator Green's function is also separable

$$G(r, r', E) = \sum_{L,L'} R_L(r) G_{LR,L'R'} R_{L'}(r'), \tag{2.7}$$

With this representation the expression for the XAFS can be reduced to a calculation of atomic-like dipole-matrix elements at a given absorption site, $M_L = \langle i|\varepsilon \cdot r|LR\rangle\rangle$ and a propagator matrix elements of $G(r, r', E)$.

$$\mu(E) = -4\pi\alpha e^2\omega \sum_{LL'} M_L(E) G_{LL'} M_{L'}(E), \tag{2.8}$$

Here for simplicity the site indices are suppressed, and the origin is taken as $R = 0$, $M_L = \langle i|\epsilon \cdot r|L\rangle$ and matrix $G_{L0,L'0}(E)$,

$$\mu(E) = -4\pi\alpha e^2\omega \sum_{LL'} M_L(E) G_{L0,L'0} M_{L'}(E)$$

This formulation corresponds to a separation of the XAFS process into two steps, a production part which is essentially atomic where a photoelectron is produced due to dipole excitation, and a scattering part in which the photoelectron propagates throughout the remainder of the system. Here and elsewhere we interchangeably use either the photoelectron energy E or the photon frequency ω to characterize the photoelectron energy. Here and elsewhere either the photoelectron energy E or the photon frequency ω is interchangeably utilized to characterize the energy variation of the XAFS; also unless otherwise specified, atomic units are adopted $e = \hbar = m = 1$, so fine-structure constant $\alpha \approx 1/137$.

It is worth noting that the relativistic generalization, which is needed for heavy atoms like Pt [17], is formally identical, except that the relativistic angular momentum variables (k, m) are implied. Relativistic corrections in core-level XAFS are essential for the treatment of spin-orbit effects, which can be very large in the atomic cores and hence crucially important for the transition matrix elements in heavy atoms, *e.g.*, in materials with *d*- and *f*-state materials. On the other hand, relativity has only weak effects on scattering of the non-relativistic propagating photoelectrons in EXAFS. A relativistic single-configuration Dirac-Fock approach was employed in the FEFF codes to treat the relativistic effects and spin-orbit corrections of atomic levels at high accuracy, which is automated for calculations throughout the periodic table.[18] In addition the calculation of the scattering potentials $v_R(r)$ at each site simplifies for electrons even of moderate energy (*i.e.*, above about 10 eV of kinetic energy), where chemical effects become small, the scattering depends strongly on the density in the core of an atom, and hence spherical symmetry is an excellent approximation and a spherical muffin-tin approximation is valid. In physics, scattering at such energies depends largely on the electron density in the atomic core where spherical symmetry is valid so $\rho(r) \approx \rho(|r - R|)$. Thus, in the EXAFS regime, non-spherical effects are weak. On the other hand, self-consistency turns out to be important in calculations of scattering in EXAFS when fixing the location of the Fermi level which enforces charge neutrality. It is needed, however, to obtain an accurate estimate of the threshold (or Fermi energy E_F) and fractional occupations of the various valence states. In the FEFF codes, SCF muffin-tin potentials provide a good approximation for both EXAFS calculations and the Fermi energy E_F.[19] The adequacy of this algorithm have been verified by comparing with accurate ground state electronic structure codes. Particularly, it can give efficient calculations of the electronic densities of states (DOS). Nevertheless, muffin-tin corrections may be important in the XANES region, especially in anisotropic systems. Due to deviations from spherical symmetry, some peaks will be split and peak intensities altered. Thus, the full-potential effects are often treated by other methods (e.g., LAPW, LMTO, etc.) and the development of efficient full-potential real-space multiple-scattering approaches remains a challenge. The more details about this topic will be also discussed in the next chapter.

In real-space multiple-scattering theory, with local scattering potentials, the propagator Green's function $G_{LL'}(E)$ can naturally separate local intra-atomic contributions from the central atom G_c and multiple-scattering contributions from the environment G^{sc}, so that $G = G_c + G^{sc}$. This separation yields results identical to those with the so-called path operator approach [8,42], but is conceptually somewhat simpler. In particular, the XAFS μ then factorizes in terms of an atomic-like background absorption $\mu 0$ and a scattering part χ which contains the normalized fine-structure,

$$\mu(\omega) = \mu_0(\omega)[1 + \chi(\omega)], \tag{2.9}$$

where χ is the X-ray absorption fine structure (XAFS). Thus, the net structure in the XAFS μ depends both on the atomic background $\mu_0(\omega)$ and on the fine structure χ due to multiple-scattering: $\chi = ImTr_m G^{sc}_{LL'}$. Since $\mu_0(\omega)$ varies weakly with energy, these results are consistent with the experimental definition of the normalized XAFS $\chi = (\mu - \mu_0)/\Delta\mu_0$, where $\Delta\mu_0$ is the jump in the atomic background absorption at threshold. The scattering contribution to G is expressed in terms of the free propagators with the matrix elements $G^0_{LR,L'R'} = G^0(r)(1 - \delta_{R,R'})\delta_{L,L'}$ and $\mathrm{G_{sc}} = \frac{e^{i\delta}G^0 e^{i\delta'}}{1 - G^0 T}$; where δ_ℓ are partial-wave phase shifts at the absorption site and, for simplicity, matrix indices are implicit. Note that by definition the multiple-scattering equations explicitly forbid multiple-scattering at a given site. All intra-site scattering is included implicitly in terms of the scattering T matrices, $T = t_\ell \delta_{L,L'}\delta_{R,R'}$, where $t = v + vG^0 t$, so $t_\ell = \exp(i\delta_\ell)\sin\delta_\ell$ using Rehr-Albers dimensionless units.[20] The specific values of the partial wave phase shifts δ_ℓ are obtained by matching the radial wave-functions at the muffin-tin radius to asymptotic (Bessel function) solutions. These equations are consistent with the experimental definition of XAFS $\chi = (\mu - \mu_0)/\Delta\mu_0$, where by convention $\Delta\mu_0$ is the jump in the atomic background absorption. By expanding the matrix inverse, the full propagator matrix $G^{sc}_{LL'}(E)$ can be expressed as a sum over all multiple-scattering paths $\Gamma = [R_1, R_2, \cdots R_N]$, that a photoelectron can take away from the absorbing atom and back. This sum gives rise to the multiple-scattering *path expansion* which is expressed formally by

$$G^{sc} = e^{i\delta'}[G^0 T G^0 + G^0 T G^0 T G^0 + \cdots]e^{i\delta}, \tag{2.10}$$

where the successive terms represent single, double, ... scattering processes and δ and δ' are partial-wave phase shifts. As noted above, this multiple-scattering expansion converges rapidly for EXAFS with paths of order 10^2 paths.[21] This simplicity is again indicative of the short-range order interpretation of EXAFS. For XANES where the mean-free path is long, however, the path expansion converges poorly, typically requiring many high-order paths to converge. Surprisingly, the sum sometimes converges adequately in XANES, particularly in cases with a short core-hole lifetime as in deep core levels in heavy elements. Thus, it is usually preferable in XANES to sum the multiple-scattering series to all orders, e.g., by matrix inversion [22] or approximately, using iterative, Lanczos algorithms. [23] Thus, for XAFS, one often needs a combination of the path expansion and full multiple-scattering to cover both EXAFS and XANES regions.

However, the path expansion is not always reliable near the edge. Additionally, at high energies, the large dimensions of the propagator matrix $G_{LR,L'R'}$ make direct calculations of the multiple-scattering equations computationally formidable. Due to the large dimension of $G_{LR,L'R'}(E)$ in an angular momentum-site (L,R) scattering state basis, accurate calculations of the contributions of the path expansion can only be carried out efficiently for very few low-order multiple-scattering paths.[24] To overcome this computational bottleneck, an efficient approximation was devised by Rehr and Albers (RA) based on an exact separable scattering matrix formalism,[25] namely $G_{L''R'',L'R'} = (\frac{e^{ipR}}{pR})\Sigma\gamma Y_{L'\gamma}\,(pR)\,\tilde{Y}_{L''\gamma}\,(pR)$, where p is the photoelectron momentum. This separation then allows the on-site contributions from a given scatterer to be summed to all orders of angular momentum indices $L = (\ell,m)$, where typically up to $\ell \approx 20$ for electrons of energies up to ≈ 1200 eV is important. Consequently. the RA approach yields an exact reformulation of the curved-wave multiple-scattering problem in terms of an effective scattering amplitude $f$eff (k) in terms of a separable representation of the free propagator $G_0(E)$. Here the electron wave number k is measured from the threshold as in the EXAFS experiment, so the local momentum k is given by $p^2 = k^2 + k_F^2 + \Sigma(p,\mu + \frac{1}{2}k^2) - \Sigma(k_F,\mu)$. This reformulation leads to a huge simplification of the theory and is the crucial step that makes modern EXAFS calculations practicable. With this representation the multiple-scattering expansion can be re-expressed as a sum over multiple-scattering paths R in a form essentially the same as the original EXAFS equation of Sayers, Stern, and Lytle, but with renormalized ingredients in all terms.

With this RA representation, the multiple-scattering expansion can then be given by a sum over multiple-scattering paths R in a form analogous to the original EXAFS equation of Sayers, Stern, and Lytle

$$\chi(k) = S_0^2 \sum_R \frac{|f_{eff}(k)|}{kR^2} \sin(2kR + \Phi_k)\, e^{-\frac{2R}{\lambda_k}} e^{-2\sigma_R^2 k^2}$$

However, all quantities in this EXAFS equation are redefined to include curved wave and many-body effects implicitly, and k is the photoelectron wave number measured from threshold (E_0), $k = \sqrt{2(E - E_0)}$, where E_0 is fixed by the Fermi level. In practice, a subset of order 10^2 of the most important paths is enough to represent the observed XAFS with high accuracy. Of the other terms, $\lambda_k \approx k / (|\mathrm{Im}\,\Sigma| + \Gamma / 2)$ is the XAFS mean-free path which is calculated in terms of the self-energy and core-hole lifetime. Finally, σ, which characterizes the thermal and structural disorder, is the root mean square fluctuation in the effective path length $R = R_{path}/2$. In addition, the quantity $R_{path}/2$ also corresponds to the peaks in the EXAFS Fourier transform.

2.4.1 Mean free path

The excited states in core-level spectra differs from the ground state electronic structure in two crucial aspects: (1) the demand for an energy dependent self-energy Σ(E), rather than an exchange-correlation (V_{xc}) functional like density functional theory, and (2) the need for a screened core-hole. Since the electrons repel each other, the moving electrons polarize or displace nearby electrons, and then change the potential of the moving electron field. Self-energy represents the electric potential felt by electrons due to the interaction between the surrounding medium and electrons. Essentially the self-energy is a dynamically screened Hartree–Fock exchange interaction and is the analog of the exchange-correlation potential V_{xc} in density functional theory. The real part of the self-energy varies by about 10 eV over XAFS energies, slowly turning off in the classical limit, while the imaginary part is negative and varies by about 5 eV.

A self-energy with a non-zero imaginary part describes a quasi-particle with a finite lifetime. In the spectral function, an imaginary self-energy broadens a peak, and the peak width is associated with the imaginary part of Σ(k,ω). The imaginary part accounts for (extrinsic) inelastic losses or final-state broadening, and can be expressed in terms of a mean free path and a core-hole lifetime Γ,

$$\lambda = k/[|Im\Sigma| + \Gamma]. \tag{2.11}$$

The real part of Σ(E) accounts for systematic shifts in peak positions compared to those calculated with density functional theory, leading to a stretching of the energy scale at low energies of about 10 %. The self-energy can be well described by the GW approximation ('GW' stands for the one-body Green's function G, and W is for the dynamically screened Coulomb interaction). One of the key developments in practical XAFS codes is an efficient algorithm for calculations of $\Sigma(E)$ based on the GW approximation of Hedin.[26] For example, FEFF and several other XAFS codes by default typically use a local density approximation for $\Sigma(E)$ based on the plasmon-pole dielectric constant. This approximation works well for EXAFS and reduces to the ground state exchange-correlation potential near threshold. However, the plasmon-pole self-energy tends to overestimate losses in the XANES. This deficiency can be overcome by using a many-pole model, as discussed by Kas et al. [27]

2.4.2 Many-body amplitude reduction factor S_0^2

The effect of the convolution in the exact expression for the XAFS is to average the one-electron XAFS over the spectral function, so

$$\mu = \langle\mu_0\rangle[1 + \Sigma_R\langle\chi_R\rangle]. \tag{2.12}$$

Since the atomic absorption μ_0 is smoothly varying, the average $\langle\mu_0\rangle \approx \mu_0$ has little effect. Here $\chi_R \sim Im[exp(2ikr)]$ is the XAFS contribution for a given path. Then, averaging χ_R over the quasi-particle peak gives rise to a damping factor $exp(-2R/\lambda_k)$ from the mean free path, while the broad structure in the spectral function gives an additional factor denoted by $S_0{}^2$, which turns out to be only weakly dependent on *R*. In physics, the mean free path term is dominated by the extrinsic losses, while $S_0{}^2$ is dominated by intrinsic losses. A more detailed analysis shows that this factorization is most appropriate at high energies in the EXAFS regime. Near the edge, one expects interference terms to suppress these effects and hence the energy dependence of these contributions to be important.

2.4.3 Thermal vibrations and XAFS Debye–Waller factors

In contrast to electronic structure, the effects of thermal and structural disorder lead to exponential damping of the fine structure with respect to energy, and thus are of crucial importance in XAFS. This damping is dominated by an *XAFS Debye–Waller factor* $exp(-2\sigma^2 k^2)$, where σ^2 is the mean square relative displacement of the near-neighbor bonds, and typically varies inversely with the local bond strength. This damping factor is the analog of the X-ray diffraction Debye-Waller factor $\exp(-\frac{u^2k^2}{2})$. In XANES where k is small (a few Å^{-1}), the Debye-Waller factor is of order unity, and hence these terms are usually considered unimportant.

Higher moments of the pair distribution function are sometimes important, especially in temperature dependent investigations of XAFS in anharmonic systems. One of the key theoretical developments in the theory of vibrational damping in XAFS is the use of the cumulant expansion for the Debye-Waller factor, which for small to moderate disorder systems yields an efficient parameterization of such thermal and configurational disorder [28] in terms of a few moments or *cumulants* $\sigma^{(n)}$ of the pair distribution function. Formally this expansion yields a complex Debye–Waller factor $exp[\sum_n (2ik)^n \sigma^{(n)}/n!]$ in G^{sc}, which contributes both to the amplitude and phase of XAFS. Thus, they can sometimes be approximated by including a correlated Debye-Waller factor $\exp(-2\sigma^2_{R,R'}k^2)$ in each propagator $G_{LR,L'R'}$. However, there are some cases where structural disorder can lead to additional peaks in the spectra, which are normally forbidden by dipole selection rules. The need to treat such aperiodic effects underscores the utility of real-space approaches.

The thermal contributions to σ^2 can often be fit to a correlated Debye model.[29] The corrections to the dominant second cumulant depend on anharmonicity. The first cumulant $\sigma^{(1)}$ is the net thermal expansion, while the third $\sigma^{(3)}$ characterizes the asymmetry or skew of the pair distribution function. These odd order contributions can strongly affect the fine structure phase, especially at high energy, giving a contribution $2k\sigma^{(1)} - \frac{4}{3}k^3\sigma^{(3)} + \cdots$. Relations between the cumulants have been derived [30] which show, to leading order in the anharmonicity parameter, that $\sigma^{(1)} \propto \sigma^2(T)$ and $\sigma^{(3)}$ is related to $\sigma^2(T)$, based on an effective one-dimensional model. The third cumulant is important in the bond distance determinations and in interpretations of thermal expansion. If the third cumulant is neglected in the analysis, the bond distances obtained from EXAFS typically appear improbably short. Improved treatments of XAFS Debye-Waller factors which go beyond the correlated Debye approximation,[31] have recently been developed, and based on the Lanczos calculations of the phonon Green's function and the dynamical matrices obtained from the electronic structure codes such as ABINIT. [32]

2.5 Many Body Effects in XAFS Spectra

Many body effects play an important role in XAFS theory. For example, they are responsible for the intrinsic broadenings, damping, and energy shifts due to the interaction between the photoelectron and the core-hole and the finite lifetimes of the core-hole and photoelectron caused by the interaction with the valence states, and interactions with vibrational states of the system. In addition, there are damping effects from satellites in the spectral function (shake up/off), which give rise to the amplitude reduction factor S_0^2 observed in the EXAFS compared to independent electron calculations. Finally, Hubbard model corrections are important for calculations of the near edge spectra of correlated materials such as transition metal oxides. Notably, first principles methods have been developed recently for efficient calculations of the key many-body effects in XAFS leading to parameter-free theories. [33, 34] Debye-Waller Factors and Disorder have been addressed above. To consider the many-body effects of inelastic losses, a better approximation is to use a screening approximation based on the local dielectric function so that the core-hole strength is "tuned to a given system." This procedure with the random phase approximation (RPA) for the screening, is analogous to that in the Bethe-Salpeter Equation (GW/BSE) method.

Core-hole interactions

Conventionally, the Final State Rule or similar approximations are adopted to consider the interaction between the photoelectron and the core-hole. In the Final State Rule, the single particle states are fully relaxed in the presence of a core-hole potential. However, this prescription can overestimate the strength of the core-hole and does not take into account multiplet effects (core-hole mixing). One approach to this problem is the atomic multiplet method in which the problem is reduced to a single atom in the presence of a crystal field potential.[35] While this works quite well to explain the features of the near edge for localized states, *e.g.*, in L-edges of transition metal oxides, there are many free parameters, and the continuum cannot be treated. Another option is to solve the Bethe-Salpeter equation, which treats the particle-hole state as a two-particle system with a screened interaction, but this is computationally challenging. A related method is to use time-dependent density-functional theory (TDDFT), [36] which extends the standard Kohn-Sham techniques to time-dependent potentials. TDDFT is computationally much more efficient than the Bethe-Salpeter equation (BSE), but can underestimate the excitonic effects. The two methods are formally similar in that both give the response function in the form $\chi = \frac{\chi^0}{1 - K\chi^0}$ where K is the interaction kernel. The difference lies in the approximation for K which in BSE is given by $K = v + W$ while in TDDFT $= v + f_{xc}$. Both the screened Coulomb interaction W and the exchange correlation kernel f_{xc} are in principle non-local in space and time, however, in practice, the adiabatic form for both kernels is often used, and the local density approximation is taken for the exchange correlation kernel. It is well known that calculations based on a local kernel in TDDFT (i.e., TDLDA) often underestimate the effects of a deep core-hole. In order to overcome these deficiencies, Ankudinov et al. introduced a combined BSE-TDDFT method.[37] That is, the kernel K is split into local and non-local contributions, $K_L = v + f_{xc}$ and $K_N = v - f_{xc}$. Here the screened interaction W is calculated via the adiabatic random phase approximation (RPA) following Zangwill and Soven [38]. The total response is then calculated using a projection operator method,[39] with an efficient basis of local atomic orbitals used to reduce computational time.

Many-pole self-energy

The self-energy can be calculated via the "GW" approximation of Hedin,[40] where G is the Green's function and *W* is the screened coulomb interaction, *i.e.*, with the convolution $\Sigma_k(E) = \mathrm{i}\sum_q G_{k-q} W_q$.

Here screened coulomb interaction W is closely related to the dielectric function. For EXAFS, an efficient model must be used for the self-energy, since calculations are required over a broad range of energies, and can be quite time-consuming. Commonly used phenomenological models include the plasmon pole model of Hedin and Lundqvist [41] which are an improvement over more primitive models such as the $x\alpha$ [42] and the Dirac-Hara exchange model. [43] As a further improvement on these ad hoc methods a many-pole model has been used, where the dielectric function is represented as a sum of poles,

$$Im[\epsilon^{-1}(q,\omega)] = \frac{\pi}{2}\sum_j g_j \omega_j \delta\left(\omega - \omega_j(q)\right), \tag{2.13}$$

where $g_j \approx \left(2\Delta\omega_j / \pi\omega_j\right)$ gives the strength of the pole, ω_j its position, and $\Delta\omega_j$ the pole spacing. As in the plasmon pole model, the momentum dependence of each pole is taken as a quadratic polynomial in q^2, i.e., $\omega_j(q)^2 = \omega_j^2 + \frac{vFq^2}{3} + \frac{q^4}{4}$. This model for the dielectric function is similar to that of Penn.[44] An ab initio calculation of the Loss function $L(\omega) = -Im[\epsilon(\omega)]$ is then used to obtain the pole weights g_j and excitation energies ω_j. With these inputs, the self-energy can be expressed as a weighted sum of plasmon pole self-energies, $\Sigma(E) = \sum_j g_j \sum_p(E,\omega_j)$. The main advantage of using the many-pole model instead of the plasmon pole model is to broaden the sharp, unphysical behavior of the plasmon pole model near the plasma frequency and give a self-energy which is tuned to the dielectric response of a given system.

Multi-electron excitations

Although the major effect of the self-energy is in the quasi-particle lifetimes and energy shifts, the model can also be used to estimate the effects of multi-electron excitations. These excitations give rise to the amplitude reduction factor in EXAFS but they can also lead to additional features in the near edge structure.[45] These inelastic losses can be interpreted in terms of excitations of the valence electrons caused by: (i) the core-hole (intrinsic interactions); (ii) the photoelectron (extrinsic interactions); and interference between these excitations.[46] It is convenient to use a quasi-boson approximation to estimate these effects, which gives the final spectrum as a convolution with an energy-dependent spectral function,[47] *i.e.*,

$$\mu(\omega) = \int_0^{\infty} \mu_{QP}(\omega - \omega')A(\omega')d\omega' \tag{2.14}$$

For plasmons, the spectral function $A(\omega)$ consists of a sharp quasi-particle peak, with a broad satellite feature at about one plasma frequency ω_p above the quasi-particle peak. The satellite represents a sum of intrinsic, extrinsic and interference terms, which cancel at the threshold for plasmon creation. The main effect of these plasmon satellites is to reduce the EXAFS oscillations, giving rise to the amplitude reduction factor $S_0^2 \approx 0.9$ in the EXAFS equation. The theory actually predicts that each path could have its own amplitude reduction factor, though the differences are not large in the cases tested. In addition to quasi-particle properties and multi-electron excitations, the many-pole model can be used to calculate other properties, such as the electron inelastic mean free path, $\lambda = \sqrt{\frac{E}{2}} / Im\Sigma(E)$.[48] Again this gives an improvement over the plasmon pole model in comparison to experiment, especially at intermediate energies.

Hubbard model corrections

The single particle (even with GW self-energy corrections) band structure is often inadequate in correlated materials with localized, atomic-like, valence states. One approach to obtain improved properties of these materials is to use Hubbard corrections. To this end, Ahmed et al. have implemented Hubbard model corrections in the real-space multiple-scattering calculations of electronic structure and spectra in FEFF. [49] The approach includes both the lower energy high correlation effects as well as the higher energy GW quasi-particle effects. For this a rotationally invariant form of the LDA+U (local density approximation plus Hubbard corrections) Hamiltonian is used, which is similar to that implemented in the Koringa-Kohn-Rostoker (KKR) code SPRKKR,[50] which is adapted from the form used by Anisimov et al. [51] In this approach, the LDA Kohn-Sham potential is replaced by the Hubbard corrected potential $V^{LDA+U} = V^{LDA} + V^{U}_{l\alpha\sigma}$, where the states $|l\alpha\sigma\rangle$ diagonalize the occupation matrix $n_{lmm'}$, and

$$V^{U}_{l\alpha\sigma} = U\sum_{\alpha'}(n^{-\sigma}_{l\alpha'} - n^0) + (U - J)\sum_{\alpha'\neq\alpha}(n^{\sigma}_{l\alpha} - n^0). \tag{2.15}$$

The parameters U and J are the Hubbard parameters, and can be calculated using a constrained random phase approximation similar to that of Aryasetiawan.[52] Here the Hubbard U for a given angular momentum l is related to the screened coulomb interaction, where the electrons participating in the screening are restricted so that only electrons with $l' \neq l$ participate. For example, for d-states

$$U = \int \rho_d(r)U(r,r')\rho_d(r'), \tag{2.16}$$

with $U(r,r') = \frac{v}{1 - v\chi_r^0(r,r')}$, and $\chi_r^0 = -2Im \int_{-\infty}^{E_F} \frac{\sum_{L \neq d} G_{LL}(\omega)G_{LL}(\omega)}{\pi} d\omega$.

χ_r^0 is the restricted response function. In addition, the theory was extended by using the GW many-pole quasi-particle self-energy detailed in the previous section to account for higher energy plasmon-like excitations. It was found that this prescription gives good results for several transition metal oxides.[53]

2.6 Recent Advances in Theory

In the field of EXAFS, the need for accurate scattering amplitude and phase shift has always been a crucial issue. In the earliest EXAFS analysis, these scattering amplitude and phase shift functions could only be accurately extracted from the experimental EXAFS spectra of known neighbor distances and species (generally by measuring the relevant crystals with known structures). Such experimental criteria may be very accurate but are usually limited to the first neighbor shell. It was quite common to extract amplitude and phase functions from known standard in EXAFS data analysis up until the early 1990's. In the past thirty years, the calculation of the effective amplitude function F_j(k) and phase function Φ_j(k) has become more accurate and easier to obtain, and now the experimental standards used in EXAFS analysis are very few. These advances revolutionized the analysis of EXAFS. Nowadays, the EXAFS theory in a multiple-scattering framework is well established.[54,55] Today theoretical standards eliminate the arbitrariness in the experimental analysis as it was done in the early days by means of tabulated standards. [56] The theory of EXAFS is now well understood and has been recently reviewed by Rehr and Albers. [57] The EXAFS theory is implemented in various *ab initio* codes like CONTINUUM, [58] EXCURV, [59] FEFF, [60] GNXAFS, [61] WIEN2k, [62] the Munich SPRKKR package[63] and others.[64] Even the description of the magnetic counterpart of this technique, *i.e.*

MEXAFS, is now mature.[65] There has been dramatic progress in recent years in theories of core-level spectroscopies. Here we discuss the two complementary approaches that build on theories of excited-state electronic structure and response. Both include key many-body effects, which are crucial to a quantitative description. First is a real-space Green's function approach applicable to both core-level X-ray absorption (XAFS) and electron energy loss spectra (EELS). This theory is implemented in a relativistic, all-electron code FEFF9, [66] which is applicable to periodic and amorphous materials alike, throughout the periodic table. The approach can now include ab initio treatments of many-body effects including the screened core-hole interaction, extrinsic and intrinsic inelastic losses, and Debye-Waller factors, optionally with input from auxiliary codes. For example, FEFF9 uses a many-pole GW self-energy to treat final state broadening and self-energy shifts, the spectra can be interpreted in terms of geometrical and excited state electronic structure. Second, is a core-level GW/Bethe-Salpeter Equation (GW/BSE) approach,[67] which is implemented in the hybrid code OCEAN.[68] This package use wave-functions from plane-wave pseudopotential codes (ABINIT or QuantumESPRESSO), the NIST Bethe-Salpeter Equation solver, and also the many-pole GW self-energy. The L-shell spectra and multiplet effects are also well treated in recent extensions. Though OCEAN is computationally more demanding, recent extensions have greatly improved its efficiency. While OCEAN is limited to near edge spectra, FEFF9 loses accuracy close to an edge; thus, the combination permits full-spectrum calculations from soft- to hard-X-ray energies. These approaches are illustrated with a number of applications.

---

[1] D.E. Sayers, E.A. Stern, & F.W. Lytle, (1971) New technique for investigating noncrystalline structures: Fourier analysis of the extended X-ray-absorption fine structure. *Phys. Rev. Lett.* 27, 1204.

[2] G. Onida, L. Reining, & A. Rubio, (2002) Electronic excitations: density-functional versus many-body Green's-function approaches. *Rev. Mod. Phys.* 74, 601.

[3] E. Shirley, (1998) Ab initio inclusion of electron-hole attraction: application to X-ray absorption and resonant inelastic X-ray scattering. *Phys. Rev. Lett.* 80, 794.

[4] J.J. Rehr, & R.C. Albers, (2000) Theoretical approaches to X-ray absorption fine structure. *Rev. Mod. Phys.* 72, 621.

[5] J. J.Rehr, J.J. Kas, M.P. Prange, A.P. Sorini, Y. Takimoto, & F. Vila, (2009) Ab initio theory and calculations of X-ray spectra. *C. R. Phys.* 10, 548.

[6] G. Onida, L. Reining, & A. Rubio, (2002) Electronic excitations: density-functional versus many body Green'sfunction approaches. *Phys. Rev. Lett.* 74, 601–659.

[7] L. Campbell, L. Hedin, J.J. Rehr, & W. Bardyszewski, (2002) Interference between extrinsic and intrinsic losses in X-ray absorption fine structure. *Phys. Rev. B* 65, 064107.

[8] J.J. Rehr, J.A. Soininen, & E.L. Shirley, (2005) Final-state rule vs the Bethe-Salpeter equation for deep-core X-ray absorption spectra. *Phys. Scr.* T115, 207.

[9] K. Jorissen, J.J. Rehr, & J. Verbeeck, (2010) Multiple scattering calculations of relativistic electron energy loss spectra. *Phys. Rev. B* 81, 155108.

[10] A.L. Ankudinov, B. Ravel, J.J. Rehr, & S.D. Conradson, (1998) Real-space multiple-scattering calculation and interpretation of X-ray-absorption near-edge structure. *Phys. Rev. B* 58:7565.

[11] V. L. Aksenov, A. Yu Kuzmin, J. Purans, & S. I. Tyutyunnikov, (2001) EXAFS spectroscopy at synchrotron-radiation beams. *Physics of Particles and Nuclei*, 32, 675-707.

[12] L.W. Campbell, L. Hedin, J.J. Rehr, & W. Bardyszewski, (2002) Calculations of inelastic losses and interference terms in X-ray absorption. *Phys. Rev. B* 65, 064107.

[13] J.E. Muller, O. Jepsen, & J.W. Wilkins, (1982) X-ray absorption-spectra—K-edges of 3d transition-metals, L-edges of 3d and 4d metals, and M-edges of Palladium. *Solid State Commun.* 42, 365.

[14] P. Blaha, K. Schwarz, P. Sorantin, & S.B. Trikey, (1990) Full-potential, linearized augmented plane-wave programs for crystalline systems. *Comput. Phys. Commun.* 59, 399.

[15] W.L. Schaich, (1973) Comment on the theory of extended X-ray-absorption fine structure. *Phys. Rev. B* 8, 4028–4032.

[16] P.A. Lee & J.B. Pendry, (1975) Theory of the extended X-ray absorption fine structure. *Phys. Rev. B* 11, 2795–2811.

[17] H. Ebert, (1996) Magneto-optical effects in transition metal systems. *Rep. Prog. Phys.* 59, 1665.

[18] A.L. Ankudinov, S.I. Zabinsky, & J.J. Rehr, (1996) Single configuration Dirac-Fock atom code. *Comp. Phys. Commun.* 98, 359.

[19] A.L. Ankudinov, B. Ravel, J.J. Rehr, & S.D. Conradson, (1998) Real-space multiple-scattering calculation and interpretation of X-ray-absorption near-edge structure. *Phys. Rev. B* 58, 7565.

[20] J.J. Rehr & R.C. Albers, (2000) Theoretical approaches to X-ray absorption fine structure. *Rev. Mod. Phys.* 72, 621.

[21] A.L. Ankudinov, B. Ravel, J.J. Rehr, & S.D. Conradson, (1998) Real-space multiple-scattering calculation and interpretation of X-ray-absorption near-edge structure. *Phys. Rev. B* 58, 7565.

22 J.J. Rehr, & A.L. Ankudinov, (2004) Solid state effects on X-ray absorption, emission and scattering processes. *Radiat. Phys. Chem.* 70, 453.

23 K. Jorissen, & J.J. Rehr, (2010) Calculations of electron energy loss and X-ray absorption spectra in periodic systems without a supercell. *Phys. Rev. B* 81, 245124.

24 S.J. Gurman, N. Binsted, & I. Ross, (1986) A rapid, exact, curved-wave theory for EXAFS calculations: II. The multiple-scattering contributions. *J. Phys. C Solid State* 19, 1845.

25 J.J. Rehr,& R.C. Albers, (1990) Scattering-matrix formulation of curved-wave multiple-scattering theory: application to X-ray-absorption fine structure. *Phys. Rev. B* 41, 8139.

26 L. Hedin, (1965) New method for calculating 1-particle Green's function with application to electron-gas problem. *Phys Rev* 139, A796.

27 J.J. Kas JJ et al Many-pole model of inelastic losses in X-ray absorption spectra. *Phys. Rev. B* 76, 195116 (2007).

28 G. Dalba, P. Fornasini, (1997) EXAFS Debye-Waller factor and thermal vibrations of crystals. *J. Synchrotron Radiat.* 4, 243.

29 G. Beni & P.M. Platzman, (1976) Temperature and polarization dependence of extended X-ray absorption fine structure spectra. *Phys. Rev. B* 14, 1514–1518.

30 A.I. Frenkel & J.J. Rehr, (1993) Thermal expansion and X-ray-absorption fine-structure cumulants. Phys. Rev. Lett., 48, 585–588.

31 A.V. Poiarkova & J.J. Rehr, (1999) Multiple-scattering X-ray-absorption fine-structure Debye-Waller factor calculations. *Phys. Rev. B* 59, 948.

32 F.D. Vila, J.J. Rehr, H.H. Rossner, & H.J. Krappe, (2007) Theoretical X-ray absorption Debye-Waller factors. *Phys. Rev. B* 76, 014301.

33 E.L. Shirley, (1998) Ab initio inclusion of electron-hole attraction: application to X-ray absorption and resonant inelastic X-ray scattering. *Phys. Rev. Lett.* 80, 794–797.

34 S. Sagmeister & C. Ambrosch-Draxl, (2009) Time-dependent density functional theory versus be the salpeter equation: an all-electron study. *Phys. Chem. Chem. Phys.* 11, 4451–4457.

35 F.M.F. de Groot, J.C. Fuggle, B.T. Thole, & G.A. Sawatzky, (1990) 2p X-ray absorption of 3d transition-metal compounds: an atomic multiplet description including the crystal field. *Phys. Rev. B* 42, 5459–5468.

36 G. Onida, L. Reining, & A. Rubio, (2002) Electronic excitations: density-functional versus many body Green'sfunction approaches. *Rev. Mod. Phys.* 74, 601–659.

37 A.L. Ankudinov, Y. Takimoto, & J.J. Rehr, (2005) Combined bethe-saltpeter equations and time-dependent density-functional theory approach for X-ray absorption calculations. *Phys. Rev. B* 71, 165110.

38 A. Zangwill & P. Soven, (2007) Density-functional approach to local-field effects in finite systems: photoabsorption in the rare gases. Phys. Rev. A, 21, 1561–1572.

[39] A.I. Nesvizhskii, A.L. Ankudinov, & J.J. Rehr, (2001) Normalization and convergence of X-ray absorption sum rules. *Phys. Rev. B* 63, 094412.
[40] L. Hedin & S. Lundqvist, (1970) Effects of electron-electron and electron-phonon interactions on the one-electron states of solids. Solid State Phys., 23, 1–181.
[41] B.I. Lundqvist, (1967) Single-particle spectrum of the degenerate electron gas. *Phys. Kondens. Materie.* 6, 193.
[42] J.C. Slater, (1951) A simplification of the Hartree-Fock method. *Phys. Rev.* 81, 385–390.
[43] S. Hara, (1967) The scattering of slow electrons by hydrogen molecules. *J. Phys. Soc. Jpn.* 22, 710.
[44] D.R. Penn, (1975) Electron mean free paths for free-electron-like materials. *Phys. Rev. B* 13, 5248.
[45] M. Calandra, J.P. Rueff, C. Gougoussis, D. Céolin, M. Gorgoi, S. Benedetti, P. Torelli, A. Shukla, D. Chandesris, & Ch. Brouder, (2012) K-edge X-ray absorption spectra in transition-metal oxides beyond the single-particle approximation: shake-up many-body effects. *Phys. Rev. B* 86, 165102.
[46] L. Campbell, L. Hedin, J.J. Rehr, & W. Bardyszewski, (2002) Interference between extrinsic and intrinsic losses in X-ray absorption fine structure. *Phys. Rev. B* B, 65, 064107.
[47] J.J. Kas, A.P. Sorini, M. P. Prange, L. W. Cambell, J. A. Soininen, & J. J. Rehr (1980) Many-pole model of inelastic losses in X-ray absorption spectra. *Phys. Rev. B* 76, 195116.
[48] A.P. Sorini, J.J. Kas, J J. J. Rehr, M. P. Prange, & Z. H. Levine, (2006) Ab initio calculations of electron inelastic mean free paths and stopping powers. *Phys. Rev. B* 74, 165111.
[49] T. Ahmed, J.J. Kas, & J.J. Rehr, (2012) Hubbard model corrections in real-space X-ray spectroscopy theory. *Phys. Rev. B* 85, 165123.
[50] H. Ebert, A. Perlov, & S. Mankovsky, (2003) Incorporation of the rotationally invariant LDA+U scheme into the SPR-KKR formalism: application to disordered alloys. *Solid State Commun.* 127, 443–446.
[51] V.I Anisimov, F. Aryasetiawan, & A.I. Lichtenstein, (1997) First-principles calculations of the electronic structure and spectra of strongly correlated systems: the LDA+ U method. *J. Phy.: Condensed Matter.* 9, 767.
[52] F. Aryasetiawan, K. Karlsson, O. Jepsen, & U. Schonberger, (2006) Calculations of Hubbard U from first principles. *Phys. Rev. B* 74, 125106.
[53] T. Ahmed, J.J. Kas, & J.J. Rehr, (2012) Hubbard model corrections in real-space X-ray spectroscopy theory. *Phys. Rev. B* 85, 165123.
[54] J.J. Rehr, & A.L. Ankudinov, (2001) New developments in the theory of X-ray absorption and core photoemission. *J. Electron. Spectrosc. Relat. Phenom.* **114–116,** 1115–21.
[55] C. R. Natoli, M. Benfatto, S. Della Longa, & K. Hatada, (2003) X-ray absorption spectroscopy: state-of-the-art analysis. *J. Synchrotron Radiat.* **10,** 26–42.

56 B.-K. Teo, & P.A. Lee, (1979) Ab initio calculations of amplitude and phase functions for extended X-ray absorption fine structure spectroscopy. *J. Am. Chem. Soc.* **101,** 2815–32.

57 J.J. Rehr, & R.C. Albers, (2000) Theoretical approaches to X-ray absorption fine structure. *Rev. Mod. Phys.* **72**, 621-654.

58 C.R. Natoli, D.K. Misemer, S. Doniach, & F.W. Kutzler, (1980) First-principles calculation of X-ray absorption-edge structure in molecular clusters. *Phys. Rev. A* **22,** 1104–8.

59 N. Binsted, (1998) *EXCURV98: CCLRC Daresbury Laboratory ComputerProgram,* For details see http://srs.dl.ac.uk/XRS/Computing/Programs/excurv97/intro.html

60 A.L. Ankudinov, B. Ravel, J.J. Rehr, & S.D. Conradson, (1998) Real-space multiple-scattering calculation and interpretation of X-ray-absorption near-edge structure. *Phys. Rev.* B **58,** 7565–76.

61 A. Filipponi, A. Di Cicco, & C. R. Natoli, (1995) X-ray-absorption spectroscopy and n-body distribution functions in condensed matter. I. Theory. *Phys. Rev.* B **52,** 15122–34.

62 P. Blaha, K. Schwarz, G.K.H. Madsen, D. Kvasnicka, & J. Luitz, (2001) *WIEN2k, An Augmented Plane Wave +Local Orbitals Program for Calculating Crystal Properties* (Austria: Karlheinz Schwarz, Universit at Wien) ISBN 3-9501031-1-2, For details see http://www.wien2k.at/index.html .

63 H. Ebert, (2000) *Electronic Structure and Physical Properties of Solids* (*Lecture Notes in Physics* vol 535) ed H Dreyss (Berlin: Springer) pp 191–246.

64 J.E. Mueller, O. Jepsen, & J.W. Wilkins, (1982) X-ray absorption spectra: K-edges of 3d transition metals, L-edges of 3d and 4d metals, and M-edges of palladium. *Solid State Commun.* **42,** 365–8.

65 A.L. Ankudinov, & J.J. Rehr, (1995) Calculation of X-ray magnetic circular dichroism in Gd. *Phys. Rev.* B **52,** 10214–20.

66 J.J. Rehr, J.J. Kas, M.P. Prange, A.P. Sorini, Y. Takimoto, & F. Vila, (2009) Ab initio theory and calculations of X-ray spectra. *Comptes Rendus Physique* 10, 548.

67 J. Vinson, J.J. Rehr, J.J. Kas, & E.L. Shirley, (2011) Bethe-Salpeter equation calculations of core excitation spectra. *Phys. Rev. B* 83, 115106.

68 J. Vinson, J.J. Kas, F.D. Vila, J.J. Rehr, & E. L. Shirley, (2012) Theoretical optical and X-ray spectra of liquid and solid H_2O *Phys. Rev. B* 85, 045101.

Chapter 3

Theory of X-ray Absorption Near Edge Structure

3.1 Introduction of XANES

The X-ray absorption near edge structures (XANES), is the part of XAFS domain extending up to around 50 eV above the threshold. XANES at soft X-ray regime is also often called near-edge X-ray absorption fine structure (NEXAFS). It is associated with the excitation process of a core-level electron to bound states (located below the ionization threshold) and quasi-bound states (located above the threshold) that interact with the continuum. Because the outgoing photoelectron at near absorption edge has low kinetic energy, experiences a strong interaction with the neighboring atoms due to the dominant multiple-scattering. The most obvious determinants of this interaction potential are the number and position of neighboring atoms, and particularly their associated electron densities. This short-range nature of XANES makes it an excellent tool for the study of the electronic structure of the absorbing atom as well as the three-dimensional local structural information around it. It can provide information on the local structure, not only for amorphous solids and large compounds but also for aqueous samples, nanoparticles, and biological systems such as metalloproteins. The physical structure around the absorbing element can be probed by tuning the X-ray energy to be near an inner-shell ionization energy for a specific element, and in this way, the co-ordination numbers, crystal groups, and bond lengths can be measured routinely for materials which cannot be determined effectively with other crystallographic methods.

Within the framework of single-electron approximation, an excited electron from the core-level behaves as an active electron and the remaining other electrons are in a passive state affected by multi-electron interactions. Various theoretical treatments, *e.g.* of metals, are based on the final state rule in a *single-electron approach*: the final states are calculated in the presence of an approximately screened core hole, and the many-body effects and inelastic losses are described by a complex optical potential. Unfortunately, the single-electron approach ignores atomic multiplet effects, which often makes the single-electron picture invalid in many cases, and interactions between active and passive electrons must be taken into account which make the interpretation of XANES rather complicated.

A fully quantitative treatment of the near-edge structures remains quite challenging. This is mainly due to various many-body effects, such as the photoelectron self-energy, the interaction between photoelectron and core-hole, multiplet effects, and inelastic losses. Several complementary ways have been developed, including quantum chemistry, band-structure, and multiple-scattering theories. Recently, there has been progress to go beyond the single-electron approach in order to calculate X-ray absorption spectra. [1] In some of these works a *GW* approach is used. [2] The name of this approach originates from the *electron self-energy* which involves the product of the electron Green's function *G* and the dynamically screened interaction *W*.[3] Another approach is based on a *time-dependent density-functional theory*, as it has been successfully applied to optical excitations in atoms.[4] The application of this approach to X-ray absorption spectra at the $L_{2,3}$ edges for the 3d transition metal series has been demonstrated recently. [5]

The most challenging part in interpreting XANES is how to get the final state wave function. Different approaches have been developed to calculate these final states. (1) One approach for the calculation of the near-edge structures is the *atomic multiplet theory* for the absorbing atom in crystal field,[6,7] which makes use of a crystal-field parametrization of solid state effects and ignores delocalized states.[8] (2) Another method uses the local density approximation to calculate the final states, which is performed considering infinite crystals[9] through a band structure theory or clusters using the multiple-scattering theory. [10,11] In a typical multiple-scattering theory, the muffin-tin averaging of the potential is employed for the expansion of the wave functions. In this approximation, the potential in the atomic and outer sphere regions is spherically averaged, while that in the interatomic region is volume averaged. The use of a constant interstitial potential is certainly a serious approximation, especially when the photoelectron kinetic energy is close to the value of the approximation done on the potential. Moreover, it makes the results depend on the size of the interstitial region itself. The limitations imposed by the muffin-tin approximation can be overcome by the generalization of the multiple-scattering theory.[12]

Some theoretical XANES works have already tried to go beyond the muffin-tin approximation for energy band problems. In particular, the discrete variational method[13] uses a discretization of the potential to calculate matrix elements but keeps the usual expansion in plane waves in its Koringa-Kohn-Rostoker version.[14] Ebert and co-workers have also successfully applied the Koringa–Kohn–Rostoker formalism in a full potential approach to XANES.[15,16] The full linear augmented plane wave approach[17] is also extensively used to calculate band structure without the muffin-tin approximation, but all of these methods are restricted to periodic potential. The finite-difference method is another way to avoid the muffin-tin approximation by solving the Schroedinger equation with the local density approximation. The finite-difference method requires significant computing power and its progress has followed the improvements of computer-capabilities. The finite-difference method is a general way to solve differential equations by discretizing them over a grid of points in the whole volume where the calculation is made. In the atomic core, the potential can be represented as a multipole expansion around the center. In the outer sphere, the potential is constant and the solutions are the Neumann and Bessel (or Hankel) functions (the functions resulting from Coulomb potential could be utilized in this area). The normalization of the Hankel functions comes from the density of state in the vacuum. In the treatment of the XANES region, the modern XANES codes like FDMNES not only include non-spherical parts of the scattering potential, but also consider many-body effects such as multi-electron excitations, core-hole effects and local field effects (screening of the X-ray field).[18]

3.2 X-ray Matter Interaction

Understanding the interaction of polarized light with matter is related to Maxwell equations. Using the second quantization, the vector potential $\boldsymbol{A}$, the electric field $\boldsymbol{E}$ and the magnetic field $\boldsymbol{B}$ of a propagating plane can be written as

$$\begin{aligned}\boldsymbol{A}(r,t) &= A_0^2\left(a\epsilon e^{i(\boldsymbol{k}\cdot\boldsymbol{r}-\omega t)} + a^{+}\epsilon^{*}e^{-i(\boldsymbol{k}\cdot\boldsymbol{r}-\omega t)}\right),\\ \boldsymbol{E}(r,t) &= -i\omega A_0\left(a\epsilon e^{i(\boldsymbol{k}\cdot\boldsymbol{r}-\omega t)} - a^{+}\epsilon^{*}e^{-i(\boldsymbol{k}\cdot\boldsymbol{r}-\omega t)}\right),\\ \boldsymbol{B}(r,t) &= iA_0\left(a\boldsymbol{k}\times\epsilon e^{i(\boldsymbol{k}\cdot\boldsymbol{r}-\omega t)} - a^{+}\boldsymbol{k}\times\epsilon^{*}e^{-i(\boldsymbol{k}\cdot\boldsymbol{r}-\omega t)}\right),\end{aligned} \tag{3.1}$$

where a and a^{+} are the annihilation and creation operator of the photon, $\boldsymbol{\epsilon}$ and $\boldsymbol{k}$ are its polarization and wave vector, respectively.

The amplitude A_0 is determined by imposing the energy of the photon to be $\hbar\omega$. Here the most common convention on the sign in the exponent of the exponential was used.

The non-relativistic Hamiltonian can be expressed as $H = H_0 + H_I$. H_0 is the non-interacting part, and the interaction part H_I between the field above and an electron of charge is

$$H_I = \frac{e}{m}\left[p \cdot \boldsymbol{A} + \boldsymbol{S} \cdot \boldsymbol{B} + \frac{e}{2}\boldsymbol{A}^2\right], \tag{3.2}$$

From this, a general transition operator could be developed by using Green's function.[19] This formular can describe both single and multiple photon process at the same time. In two photon process, it corresponds to the photon-in and photon-out scattering situation such as resonant scattering. This chapter only presents the one photon situation responsible for the absorption. The last term gives the Thomson scattering, and is absent in the absorption or emission transition operator.

As discussed in the previous chapter, the formal theories of XANES and EXAFS are essentially the same, and are both interpreted by Fermi's golden rule, i.e., $\mu(\omega) \sim \sum_F |\langle f|\hat{O}|i\rangle|^2 \delta(\varepsilon_f - \varepsilon_i - \omega)$, where $|\, i\rangle$ and $|\, f\rangle$ are the initial and final many-body electronic states. When an effective single-particle description of the spectrum is reasonable, the absorption rate can be expressed

$$R_{fi} = \frac{2\pi}{\hbar} |\langle f|\hat{O}|i\rangle|^2 \delta(\varepsilon_f - \varepsilon_i - \omega), \tag{3.3}$$

where δ is the Kronecker symbol, the transition between two electronic levels normalized to one $\left(\langle \varphi_{i,f}|\varphi_{i,f}\rangle = 1\right)$, the transition operator $\hat{O} = \left(\epsilon \cdot p + iS \cdot (k \times \epsilon)\right)e^{ik \cdot r}$. Using the Taylor expansion of the exponential up to the third order for the potential vector and to the first order for the magnetic field, and commutator relations, one obtains $\langle \varphi_f|\hat{O}|\varphi_i\rangle = i\frac{m}{\hbar}(\varepsilon_f - \varepsilon_i)\ \langle \varphi_f|\hat{o}|\varphi_i\rangle$ with the operator expanded as $\hat{o} = \hat{o}_{E1} + \hat{o}_{E2} + \hat{o}_{E3} + \hat{o}_{M1} + \cdots$, they correspond to the electric dipole (E1), electric quadruple (E2), magnetic dipole (M1) transition channels:

$$\hat{o}_{E1} = \epsilon \cdot r,$$

$$\hat{o}_{E2} = \frac{i}{2}(\epsilon \cdot r)(k \cdot r),$$
$$\hat{o}_{E3} = -\frac{1}{6}(\epsilon \cdot r)(k \cdot r)^2,$$
$$\hat{o}_{M1} = \frac{\hbar}{2m}\left(\varepsilon_f - \varepsilon_i\right)(k \times \epsilon) \cdot (L + 2S), \tag{3.4}$$

Note that the third order also gives a term proportional to $\hat{o}_{E1}$ but with a factor $\alpha^2(\varepsilon_f - \varepsilon_i)/12$, in atomic units, which can be ignored. In practice, dipole transitions highly dominate in X-ray absorption edges. Other electric channels increase with energy. Typically, E2 must be taken into account from 4.5 keV (above the Sc *K*-edge). It can have also an important effect at the $L_{2,3}$ edges of heavier elements. E2 typically represents some percentages of E1, but interest in them comes from the fact that this transition channel follows different selection rules. E3 and M1 remain challenging to detect in the X-ray energy range. M1 is nevertheless of fundamental interest because the crossed channels involving this could be a way to detect fundamental properties of materials.

For a plane wave, with intensity I, the relation between the cross-section σ_{fi} and the absorption rate is given by $\sigma_{fi} = \hbar\omega R_{fi}/I$, the radiation intensity in W/m^2 of a plane wave is related to its amplitude E by $I = c\varepsilon_0 n|E|^2/2$, where *n* is the index taken here equal to 1; ε_0 and c are the permittivity and the speed of light, both in a vacuum. With the relation $|E|^2 = 4\omega^2 A_0^2$, and the fine structure constant $\alpha = e^2/(2hc\varepsilon_0)$, putting all this together, one obtains:

$$\sigma_{fi} = 4\pi\alpha^2\hbar\omega\left|\langle\varphi_f|\hat{o}|\varphi_i\rangle^2\right|\delta(\varepsilon_f - \varepsilon_i - \omega). \tag{3.5}$$

The basic interpretation of this equation is that the core-level electron is kicked out of the core-level by the incident photon and into an unoccupied excited state of energy $\varepsilon_f = \varepsilon_c + \omega$, as required by energy conservation. The deep core electrons are very nearly angular momentum eigenstates, *e.g.*, the deep-core state is typically a 1s, 2s, or 2p state that corresponds to the *K*, L_1 and $L_{2,3}$ edge XAFS spectrum, respectively. As a consequence of the dipole approximation, the selection rules determine what final states are allowed for a given initial state. For a free atom, the dipole selection rules are $\Delta l = \pm 1, \Delta m = 0, \pm 1$, where *l* is the orbital angular momentum quantum number and *m* is its projection on the quantization axis. This fact provides an interpretation of the spectrum in terms of the angular momentum projected density of states (LDOS) and smoothly varying transition matrix elements. For instance, transitions from a 1s core-level are associated with the p-DOS of the system, while transitions from a 2p core-level are related to the s-DOS and the d-DOS. In many cases, the matrix element for the p → d transition is larger than the p → s transition, therefore, the $L_{2,3}$ edge absorption spectrum mainly probes the unoccupied d states. Although the dipole approximation is usually very good, there are some solid systems for which the pre-edge peaks are due to quadrupole transitions. In this case the selection rules are $\Delta l = 0, \pm 2$, e.g., transitions from the 1s state are related to the s- and d-DOS.

Table 3.1 The selection rules for discrete transitions

		Electric dipole (E1) **("allowed")**	Magnetic dipole (M1) **("forbidden")**	Electric quadrupole (E2) **("forbidden")**
Rigorous rules	1	$\Delta J = 0, \pm 1$ (except 0↔ 0)	$\Delta J = 0, \pm 1$ (except 0 ↔0)	$\Delta J = 0, \pm 1, \pm 2$ (except 0 ↔0,1/2 ↔1/2, 0 ↔1)
	2	$\Delta M = 0, \pm 1$ (except 0 ↔0 when $\Delta J = 0$)	$\Delta M = 0, \pm 1$ (except 0 ↔0 when $\Delta J = 0$)	$\Delta M = 0, \pm 1, \pm 2$
	3	Parity change	No parity change	No parity change
With negligible configuration interaction	4	One electron jumping, with $\Delta l = \pm 1$, Δn arbitrary	No change in electron configuration; i.e., for all electrons, $\Delta l = 0$, $\Delta n = 0$	*No* change in electron configuration; *or* one electron jumping with $\Delta l = 0, \pm 2$, Δn arbitrary
For LS coupling only	5	$\Delta S = 0$	$\Delta S = 0$	$\Delta S = 0$
	6	$\Delta L = 0, \pm 1$ (except 0 ↔0)	$\Delta L = 0$ $\Delta J = \pm 1$	$\Delta L = 0, \pm 1, \pm 2$ (except 0 ↔0, 0 ↔1)

https://physics.nist.gov/Pubs/AtSpec/node17.html

3.3 XANES General Formulation

3.3.1 Interaction times and the multi-electronic problem

The transition process of the core electron up to some final states, comes with a reorganization of the multi-electronic and vibrating configurations. The limit of validity of the single-electronic approach can be understood by looking at the different characteristic times in the absorption process: [20]

(1) Time of photon-absorption process (t1) is usually less than 10^{-20} s, which is given by t1 = 1/R where R is the transition rate.
(2) Lifetime of the core-hole (t2) is given by t2 $= \hbar / \Delta\varepsilon_i$, where $\Delta\varepsilon_i$ is the core level width. For the 1s level for Z = 20 up to Z = 30, $\Delta\varepsilon_i \approx$ 1eV and t2 spreads from 10^{-15} down to 10^{-16} s.
(3) Relaxation of the electrons (t3) is the effect on all the electrons of the field created by the hole and the photoelectron, is intrinsically multi-electronic. The associated time also ranges from t3 $\approx 10^{-15}$ down to 10^{-16} s.
(4) For non-bounded states, the time (t4) associated with the kinetic energy of the photoelectron is the time for the electron to go out from the atom. For the photoelectron with kinetic energy from 1 up to 100 eV, t4 spreads from around 10^{-15} down to 10^{-17} s.
(5) The time scale associated with the thermal vibration (thermal displacements) extends from about t5$\approx 10^{-11}$ down to 10^{-14} s.

As the thermal time scale is far longer than the other ones, in most cases the thermal aspect can be seen as independent of the other phenomena. For this adiabatic approximation, the photon sees a rigid but slightly disordered arrangement of atoms. Nevertheless, in the smooth X-ray range, for instance, sometimes the orders of vibration can be observed at the *K*-edge of carbon in small molecules.

More interestingly, t2, t3 and t4 can be of the same order. *This condition involves the multi-electronic process. For this, the first condition is that the photoelectron has a very low kinetic energy.* The second condition is that the final state is localized. It does not mean necessarily bound, but the fact that the overlap of the absorbing atom orbital with the neighboring atom orbital is small, for example, the 3d orbital of the 3d elements or of the 4f orbital of the rare earth. Thanks to the selection rules, these orbitals are respectively probed in the $L_{2,3}$ and $M_{4,5}$ edges. It can be verified experimentally by just looking at the edge shape. Because the localization in space comes with the localization in energy, when the edge is more or less step-like, it means that the final states are not localized, for instance, the K-edges for some light elements when the electron probes the unoccupied molecular orbital, and the $L_{2,3}$ edges of the heavy elements. In this case the simulation of such spectra can be performed using single-electronic simulations. Since the state being not localized, the electronic structure must be evaluated in a sufficiently large volume around the absorbing atom.

On the contrary, the edge shape can be more peak-like, suggesting that the probed states are localized in energy as well as in space. In this case, multi-electronic calculations are necessary. When the final state is very highly localized, as the surrounding symmetry breaking is just parameterized, the atomic multi-configuration calculations are often sufficient. This is the domain of multiplet theory.

Unfortunately, there exist intermediate situations, where the edge starts with a high peak, the absorption cross-section then decreases rather strongly but does not completely vanish after the edge. In these cases, multi-electronic and multi-atomic calculations are mandatory. The theories that account for this are actually under development, such as a multi-channel approach,[21] the Bethe-Salpeter equation,[22] the time-dependent density functional theory (TDDFT),[23] and the most complete possible multi-configuration interaction description.[24]

3.3.2 Absorption cross-section equation

A general formula for XANES in a material can be derived from the expression of the cross-section of the transition between two single-electronic levels. There is not only one but a set of core and valence electrons. Valence electrons, unfortunately, interact between themselves. It is the same between the core and valence levels, for the $L_{2,3}$ edges of 3d elements, for instance. Transitions can produce excitons. Anyway, to get the atomic absorption cross-section, one has to sum, in some ways, over all the possible transitions, that is over states: $\sigma = 4\pi\alpha^2\hbar\omega \sum_f \left|\langle\varphi_f|\hat{o}|\varphi_i\rangle^2\right|\delta(\varepsilon_f - \varepsilon_i - \omega)$. φ_i is the multi-electronic ground state theoretically containing all the electrons of the absorbing atom (or even of a molecule around the atom). It is constructed with the anti-symmetrized product of all the single-electronic wave functions. In practice, it is often limited to the core electrons of the corresponding edge and the valence electrons of the absorbing atom. In the same way φ_f are all the possible final states with a hole in the core and one more electron in the valence states. The sum over f is thus over all the possible configurations, the number of which can be very high. ε_i and ε_f are the total energy of all the electrons in the initial and final states, respectively. The multi-electronic problem can be transformed into a single-electronic one. Nevertheless, the total absorption cross-section from the material can now be calculated by summing over the atoms in the unit cell or in the cluster, when there is no periodicity. Very often there is a benefit using the symmetry operations linking equivalent atoms. So, the total absorption cross-section from the material is $\sigma = \sum_j S_j\,(\sigma_a)$, where a index is the prototypical atoms and S_j is the symmetry operations. This summation makes the angular dependence of XANES based on the crystal point group of the material. On the contrary, the XANES shape strongly depends on the atomic point group.

3.3.3 Differences between XANES and EXAFS

As mentioned before, multiple-scattering dominates XANS while single-scattering plays a major role in EXAFS region. This leads to their different sensitivities. In some way, XANES probes the complete density of state around the absorbing atom. It is also, in many cases, sensitive to multi-electronic phenomena induced by the excited state. This sensitivity potentially enables XANES to provide more information from the tested material. On the contrary, the lower energy part (typically the first 50 eV) of the edge in EXAFS is not utilized in the analysis, and the remaining spectrum shape still reflects the electronic structure of the material. This simpler dependence is important, because the higher photoelectron energy renders its interaction with the core-hole weaker, additionally the density of state can be approximated by a simple superposition of a limited number of multiple-scattering waves. This weaker sensitivity of EXAFS can then be taken as an advantage. Indeed, only a few but important parameters can be obtained from EXAFS analysis. They are essentially the radial distances between the absorbing atom and the scattering atoms in the first few shells, the number of atoms in these shells and the Debye-Waller factors. When using a wide energy range, these parameters can be determined far more precisely than using the XANES range.

In the aspects of the simulation, several of the approximations appropriate for the EXAFS regime (beyond about ~30 eV above the edge) are not valid any more in the near edge regime, with some of these related to the reduction of the many-body formulation to an effective single-particle description. For instance, in highly correlated systems such as transition metal oxides and f-electron systems, many-body effects can change the qualitative behavior of the near edge spectrum,[25] while the main effect on the EXAFS region is simply an overall reduction in the amplitude of the fine structure, which is taken into account by the S_0^2 factor in the EXAFS equation. Even in simpler systems, the quasiparticle self-energy approximations such as the plasmon pole model, are most suitable to construct an effective single-particle Hamiltonian in the EXAFS regime. When calculating XANES they must be augmented, *e.g.*, with the many-pole self-energy model.[26] Another approximation has to do with the description of vibrational and structural distortions. In particular, vibrational effects damp the EXAFS via the Debye–Waller factor $exp(-2k^2\sigma^2)$. In the XANES, the effects of vibrations and disorder are sometimes more related to symmetry breaking, which allows transitions to states that were previously dipole forbidden, resulting in additional peaks in the near edge spectrum.

EXAFS is also less sensitive to the non-spherical details of the potentials thanks to the high photoelectron speeds, and a simple overlapped atomic muffin tin potential is adequate for most practical calculations. On the contrary, near-edge spectra are quite sensitive to details of charge transfer and changes in Fermi level due to solid state effects. Thus, the use of self-consistent potentials and often non-spherical symmetry are essential for accurate calculations of XANES. Finally, calculations of the single particle Fermi golden rule must be treated differently in the near edge region because the path expansion often fails to converge (or converges very slowly) for low energy photoelectrons. This slow convergence is caused by two factors. First, the inelastic mean free path becomes large for low energy electrons so that very long paths must be included in the expansion. Second, the large angle scattering amplitudes are not small at low energies, so that the XANES signal is not dominated by nearly linear scattering paths, and all multiple scattering paths must be taken into account.

XANES can be used to probe a variety of properties, including local electronic and chemical structure, atomic structure, and sometimes vibrational properties. For example, $L_{2,3}$-edges of transition metals can be related to the number of holes in the d-orbitals.[27] Shifts in the onset of the spectrum can also indicate charge transfer to and from a metallic particle, or changes in formal oxidation state. While EXAFS is extremely sensitive to the first shell distances and, to a lesser extent, to coordination numbers, XANES is complementary in the sense that it is more sensitive to the local symmetry around the absorbing atom.[28] For example, the spectrum of an atom with four nearest neighbors in tetrahedral symmetry changes appreciably when the symmetry is changed to square planar. Another effect of local symmetry is symmetry breaking due to vibrational effects.[29]

3.4 XANES Simulations in the Single-electronic Scheme

3.4.1 From multi- to single-electronic

The single-electronic absorption cross-section is given by:

$$\sigma = 4\pi^2 \alpha \hbar \omega S_0^2 \sum_{if} \left| \langle \varphi_f | \hat{O} | \varphi_i \rangle \right|^2 \delta(\varepsilon_f - \varepsilon_i + \Delta E_{scr} - \hbar\omega). \quad (3.6)$$

The sum over f is now not over all the configurations but overall the single-electronic states at a same energy ε_f. More precisely in reciprocal space techniques, it indicates a sum over bands and an integral over crystal momentum; in real space cluster techniques, it is most often a sum over quantum numbers associated with spherical wave expansions. Note that most often S_0^2 and ΔE_{scr} are not explicitly written in the equation and that will be discussed in the rest of this chapter.

In the simplest case with only the electric dipole approximation, one can ignore the $l = l_i - 1$ contribution and only consider the polarization along $z \left(l_o = 1, m_o = 0, \epsilon \cdot r = \sqrt{\frac{4\pi}{3}} r Y_1^0 \right)$. In electric transition, there is no spin-flip during the transition, thus σ and σ_i merge and the geometrical constants $\Gamma_{L_i L_o L}^{\sigma_i \sigma}$ become $\Gamma_{L_i L_o L}$, the values of which are tabulated. They are non-zero for $m_i = m_o + m$, thus $m_i = m$ in the present case. In heavy elements such as platinum, it is often a good approximation to ignore the spin-orbit in the valence state and merge s with σ, and assume that the radial wave functions do not depend on m and thus also the radial integral, R. For simplicity, $a^i_{l,m+\sigma-s,s} = a^i_{i,\sigma}$. Finally, the single-electronic absorption cross-section becomes:

$$\sigma = 4\pi^2\alpha\hbar\omega \sum_{i,\sigma} \frac{4\pi}{3} a^2_{i,\sigma} \Gamma^2_{L_i L_o L} \mathrm{R}^2_{l,1,l_i,\sigma} \sum_f \left| a^f_{l,m_o,\sigma} \right|^2 \delta(\varepsilon_f - \varepsilon_i - \omega). \tag{3.7}$$

At the K-edge, i merges with σ, $a^i_{i,\sigma} = 1$ and $\Gamma = \delta_{m,m_o}/\sqrt{4\pi}$. When the system is not magnetic, the summation on σ can be replaced by a multiplication by 2. Thus the absorption cross-section is mainly the sum of the modulus square of the amplitudes, $a^f_{l,m_o,\sigma}$, times the square is a radial integral, this one only slightly depending on the energy. This means that the spectra shape mainly comes from the amplitudes. These are closely related to the density of state (DOS). Indeed, the projection of the DOS on the atom is given by

$$\rho_{L\sigma} = \sum_S R^{\rho}_{LS\sigma} \sum_f \left| a^f_{l,m+\sigma-s,s} \right|^2, \tag{3.8}$$

where R^{ρ} is the radial integral of the modulus square of the radial wave function b. The absorption cross-section is nearly proportional to a specific projection of the density of state. The proportionality factor is nevertheless not the same for the E1 and E2 transition because R contains an extra r factor in the second case. Because kr is smaller than 1, and the overlap between core valence orbitals decreases with l, this factor is smaller for E2. Thus, at the K-edge of 3d elements, the pre-edges sometimes attributed to 3d levels appear very small in comparison with the main edge due to the 4p band.

3.4.2 Cluster and reciprocal space techniques

Although the formula above looks simpler after the above approximations, the evaluation of the final state amplitudes (the computation of the electronic structure of the material) is still complex and sometimes time-consuming. Two types of technique can be adopted to resolve this issue. The first solves the Schroedinger (or Dyson, or

Dirac) equation in a cluster centered on the absorbing atom. The second uses the 3D periodicity of the material, like band structure calculations. Reciprocal space techniques are efficient, but the excitation of the absorbing atom usually breaks the 3D periodicity. Consequently, such calculations are usually performed with a super-cell so that the excited atoms are sufficiently far from each other. From this point of view, the cluster approach is better because the central atom is different. However, the cluster must be chosen with a sufficiently large radius, typically corresponding to the space probed by the photoelectron. This radius can be important just at the edge, and especially for the semiconductors. In the other cases, even in well-ordered materials, for the XANES energy range, a radius up to about 7 Å is usually sufficient (in the EXAFS range, the photoelectron mean free path is shorter and a smaller radius is sufficient). Anyway, 3D and cluster techniques must give the same result when converged.

All the techniques to calculate electronic states are convenient to obtain absorption cross sections. In principle, all these methods should be equivalent, but in practice, they give different results due to different approximations are employed inside each software. Since expansion in spherical harmonics or in plane waves are limited, interpolations are often used and sometimes some bugs remain.

Many codes have been developed to calculate the absorption cross-section. Some are just an extra output from a large purpose electronic structure code and have the advantage of being more general. In other cases, the code is especially devoted to X-ray absorption spectroscopy.

The band-structure based approaches include (1) linear augmented plane wave (LAPW), Wien2k, by Blaha et al. ;[30] (2) linear combination of atomic orbital (LCAO), StoBe, by Hermann and Pettersson ;[31](3) pseudo potential, Xspectra (Quantum-espresso), by Cabaret *et al.*;[32] (4) The Korringa-Kohn-Rostoker (KKR) Green Function Method, SPRKKR, by Ebert *et al.*;[33] (4) linear muffin-tin orbital (LMTO), by Andersen *et al.*[34]

The cluster approach is more common because often it is more user-friendly. It is also less time-consuming on the computer if using multiple-scattering theory (MST). Some of the cluster-based codes are (1) Continuum, by Natoli *et al.* ,[35] which was the first; (2) FEFF, by

Rehr, Ankudinov *et al.*,[36] which is the most common; (3) FDMNES, by Bun˘au and Joly, [37] which can also use the Finite Difference Method to avoid the muffin-tin approximation; (4) MXAN, by Benfatto and Della Longa;[38] (5) XKDQ, by Kraizman *et al.* [39]

3.4.3 About the potential

No matter what technique of simulation is used, a potential is first needed in the XANES simulation. The local spin density approximation (LSDA) of the density functional theory (DFT) can be utilized for single-electronic simulations. In this way, exchange and correlation are taken into account in an average way by stating that the effect is local. The electron feels the surrounding area through an exchange-correlation potential which depends only on the density of electron on the point where the potential acts on the wave function. In the case of magnetism, the exchange-correlation potential also depends on the local difference between the density of spin-up and spin-down. This potential can take different forms. Presently the most common ones used were developed by Hedin and Lundqvist [40] and the Perdew and Wang.[41] Globally, they give an attractive potential whose amplitude increases with increasing electron density. As suggested by Von Bart and Hedin,[42] the energy dependence of this potential should be considered for XANES simulation. In fact, the magnitude of the exchange-correlation potential decreases as the electron kinetic energy increases. In a relatively narrow energy range around the plasmon energy (usually between 10 and 30 eV), this exchange-correlation potential could change by several eV. Without considering this phenomenon, the resulting structure in the spectrum will change by the same amount. Because the position of the oscillation is related to the distance between atoms, this may lead to incorrect fitting of the corresponding parameters by mistake.

The screening phenomenon of the core-hole is partly taken into account by the convenient choice of the potential. A rule of thumb is to calculate the absorbing atom potential with one hole, in the core state, but with also one more electron in the first non-occupied atomic orbital. From this, self-consistency can eventually be performed. In insulators, better simulations without the extra electron were reported.[43] Sometimes fractional screening is also utilized to improve the agreement with experimental spectrum. Such adding of a phenomenological parameter is nevertheless dangerous, and when the simplest schemes are not sufficient, which is not so often at the *K*-edges, it is better to apply multi-electronic approaches.

To make the calculation simpler, the multiple scattering theory, as linear muffin-tin orbital (LMTO), usually makes an approximation on the potential shape. The most commonly used one is the so-called muffin-tin approximation in which the potential is taken spherically symmetrical in the atoms and constant in the interstitial area between them. Therefore, the radius of the (spherical part) atoms is a technical problem, and each code author has his own recipe. In most cases, small overlaps are allowed, because from a practical point of view, this can improve consistency with the experiment. However, from a mathematical point of view, this technique is unreasonable. Now, using the finite difference method, the pseudo-potential or FLAPW technique, for example, can have a free-form (or complete) potential, but at a higher computational cost. However, when the structure is not dense, there are few symmetric elements or relatively light absorption atoms, due to the muffin-tin approximation, important differences can be observed. [44]A recent study showed that the complete potential energy shape can also be used in the multiple-scattering, although the current multiple-scattering theory always utilizes the muffin-tin approximation.[45]

Some codes like FEFF adopt a self-consistent potential policy. As the Green's function naturally replaces all the orbitals used in standard density functional theory codes, the total electron density can also be calculated using Green's function, *i.e.*, $\rho(\vec{r}) = -\frac{1}{\pi}\int_{-\infty}^{\mu} Im[G(\vec{r},\vec{r},E)]dE$. And the Kohn–Sham Hamiltonian can be solved self-consistently using the Green's function.[46] In the real-space multiple scattering code FEFF, the potential is obtained through a self-consistent approach, which works as follows:

1. Atomic Dirac–Fock eigenfunctions and densities are calculated for each type of atom, and an overlapped atom approximation to the density is used as the initial guess in the self-consistency loop.
2. The density and exchange correlation functional is used to create a Kohn–Sham (local density approximation) potential.
3. The spherical muffin-tin approximation is applied.
4. The Green's function is calculated at each energy using a scattering state basis.

5. A new chemical potential μ is found by requiring that the total number of electrons N_e, is conserved, *i.e.*, $N_e = -\frac{1}{\pi}\int_{-\infty}^{\mu}\int Im[G(\vec{r},\vec{r},E)]d^3rdE$.
6. New densities are calculated from the new Green's function and chemical potential.
7. Steps 2–6 are repeated until the output density is sufficiently close to the input density.

Along with details of the self-consistency algorithm, several other approximations made in FEFF are worth mentioning. First, FEFF allows the user to treat several atoms as identical, even if they are physically different, i.e., two oxygen atoms in different local environments can be constrained to have the same potentials. Second, the integral over space shown in step 5 above is approximated by a sum of integrals about each atom. Each of these integrals spans the volume of a sphere (the Norman sphere) defined such that the total overlapped atomic charge (including the nucleus) is zero.

Self-consistency of the densities and potentials can be very important for XANES calculations. For example, if an initial guess of the Fermi energy is too high, pre-edge peaks are missing. Self-consistency and the shift in the Fermi energy due to solid state effects also play a role in chemical shifts, and are very important in systems with multiple, physically unique absorbing sites, such as nanoparticles on a support.

3.5 The Multiple-scattering Theory

The multiple-scattering theory has been described in details by Natoli and coworkers [47] or Brouder.[48] There are two ways to explain it. The first way uses the Green's function approach, the second uses the scattering wave approach, which remains in the DFT single-electronic frame.

3.5.1 Scattering wave approach

First, a complete wave basis is built just for one atom in the surrounding vacuum. With the constant potential, the solutions of the radial Schroedinger equation are the Bessel, j_l, Neuman and Hankel functions. The phase-shift theory can be utilized to look at how the atom scatters all the Bessel functions with the consideration of the continuity of the wave function and its derivative at the border. For simplicity, in the non-magnetic case, and it can be:

$$a_l b_l(R) Y_l^m = \sqrt{\frac{k}{\pi}}\left(j_l(r) - i t_l h_l^+(R)\right) Y_l^m, \tag{3.9}$$

for the wave function at the muffin-tin radius R. $h_l^+(R)$ is the Hankel outgoing function, t_l is the atomic scattering amplitude, $b_l(R)$ the solution of the radial Schroedinger equation in the atom, k the modulus of the electron wave vector and a_l the amplitude inside the atom. All these electron-kinetic-energy dependent terms are normalized by the square root of the density of states in vacuum $\sqrt{\frac{k}{\pi}}$ where $\frac{\hbar^2}{m}$ is usually omitted. For continuity, the asymptotic completeness theorem of differential equations implies that the state density is included in the atomic amplitude. a_l and t_l can be obtained by using the two equations of continuity (function and derivative).

The atoms are now embedded in the cluster. However, the incident wave is no longer included in the simple Bessel function, but it is included in the superposition of this Bessel function and all the other waves of Hankel type, which are backscattered by other atoms. Therefore, the problem is no longer spherical, and the scattering and atomic amplitude will also depend on the quantum number m (and ultimately on the spin index). All scattering processes from one atom to any other atom must be considered. To do this, a multiple scattering matrix needs to be filled, which contains all single expansions in the spherical harmonic function for all atoms. Its diagonal contains the amplitude of atomic scattering. The off-diagonal part contains the propagation terms which connect the scattering from the (l, m) of an atom a, to the (l', m') of another atom a'. The inversion of this matrix gives the multiple scattering amplitudes, $\tau_{lma}^{l'm'a'}$, by the relation:

$$\tau_{lma}^{l'm'a'} = \left[\frac{T}{1-TH}\right]_{lma}^{l'm'a'}, \tag{3.10}$$

T is a diagonal matrix containing the atomic scattering amplitudes, when no spin-orbit and spherical symmetry inside the atoms. The matrix H contains the propagation terms. Most of the computing time resides in the inversion of this matrix. When the low photo-electron energy range is not of interested, it is possible to perform a Taylor expansion (this is the so-called path expansion):

$$\tau_{lma}^{l'm'a'} \cong [T + THT + (TH)^2T + (TH)^3T + \cdots]_{lma}^{l'm'a'}, \quad (3.11)$$

where the number of paths increases with the power of *TH*. The limitation of this number is one of the key point in the IFEFfit [49] and GNXAFS [50] codes. At the rising edge it is not possible to perform this path expansion, because terms in the denominator are bigger than one and the series never converges.[51] One must then invert the matrix, this is the so-called full multiple scattering approach.

Using $a = a'$, skipping the atom index, but introducing again the spin-orbit in the final states, the optical theorem gives:

$$-\Im\left(\tau_{l,m+\sigma-S,S}^{l',m'+\sigma-S',S'}\right) = \left|\sum_f a_{l,m+\sigma-S,S}^{f^*} a_{l',m'+\sigma-S',S'}^{f}\right|, \quad (3.12)$$

when the potential is real. Introducing this in the XANES equations 4.34 and 4.35, the absorption cross section becomes:

$$\sigma = -4\pi^2\alpha\hbar\omega \sum_{i\sigma} \sum_{LL'} \Im\left(\langle\varphi_i|\hat{o}^*|\bar{b}_{Ls\sigma}Y_l^m\rangle \tau_{l,m+\sigma-S,S}^{l',m'+\sigma-S',S'} \left\langle\bar{b}_{L's'\sigma}Y_{l'}^{m'}\middle|\hat{o}\middle|\varphi_i\right\rangle\right), \quad (3.13)$$

where the spin-flip coming from M1 is skipped. The central term $|\bar{b}_{Ls\sigma}Y_l^m\rangle \tau_{l,m+\sigma-S,S}^{l',m'+\sigma-S',S'} \langle\bar{b}_{L's'\sigma}Y_{l'}^{m'}|$ is the Green's function. The bar in $\bar{b}$ means that they are normalized by the density of state in vacuum. The Kronecker symbol is not necessary because the summation on the final states (*f*) is already performed. As the potential is complex, the expression contains more terms, because the irregular solutions of the radial Schroedinger equation must also be considered.

Sometimes the codes using multiple-scattering theory do not give necessarily exactly the same simulation for the same edge of the same compound. The main reason is often that this theory suffers from two parameters, the muffin-tin radius and the interstitial potential. The ways to calculate them are not, by default, always the same. A secondary reason can come from different choices of the exchange-correlation potential or differences in the self-consistency techniques when this one is used.

3.5.2 Green's function approach – full multiple-scattering

In a condensed matter system, the presence of neighboring atoms around the absorbing atom modifies the final state of the transition. The calculation of the absorption cross-section could be done by a multiple-scattering theory, in which the muffin-tin approximation of the scattering potential is often employed. It assumes the scattering potential as a spherical average of the potential around each atom and a constant interstitial potential in between. Then the absorption cross section can be rewritten using a Green's function:

$$\sigma(E) \propto -\frac{1}{\pi} Im\langle i \mid \vec{\epsilon} \cdot \vec{r}' G(r, r', E) \vec{\epsilon} \cdot \vec{r} \mid i\rangle \theta\left(E - E_f\right), \quad (3.14)$$

Here, $\theta\left(E - E_f\right)$ is the Heaviside step function assuring that the cross-sectionis non-zero only above the Fermi energy E, the function G describes all possible ways for a photoelectron to interact with the surrounding atoms in the presence of the scattering potential. It could be expanded in a series by the Dyson equation in terms of the free electron propagator G_0 and atomic scattering matrix T:

$$G = \frac{G^0}{(1-G^0T)} = G^0 + G^0TG^0 + G^0TG^0TG^0 + G^0TG^0TG^0TG^0 + \cdots, \quad (3.15)$$

G^0 describes how an electron propagates between two points in space, and T describes how a photo-electron scatters from a surrounding atom. In the EXAFS energy region where the X-ray photon has sufficiently high energy above the absorption edge, the formal expansion of G is convergent very quickly. Hence the multiple-scattering series of the propagator as a sum of limited number of terms is very effective, corresponding to the path-by-path theory. However, in the XANES region, the matrix expansion does not converge efficiently. Solving the inversion of thematrix $G^0/(1 - G^0T)$ is the most computationally expensive part in the XANES calculation. This actually is equal to expand the propagator G by considering an infinite number of scattering paths, therefore this method is called full multiple-scattering technique. Full multiple-scattering calculation techniques with different details of implementation have been developed separately by several groups[52, 53]

Although the path expansion method of calculation is not appropriate for the XANES, for any finite cluster of atoms, the expansion to all orders can be summed via matrix inversion instead, $G = \frac{G^0}{1-G^0T}$. The matrices in this equation are represented in a site and angular momentum basis $|L, R\rangle$, so the rank of the matrix to be inverted is $N(l+1)^2$, where N is the number of atoms in the cluster, and l is the maximum angular momentum of an expansion in spherical harmonics. Matrix inversion can be expensive computationally for large matrices, since the time to invert scales like the cube of the rank of the matrix. Therefore, practical calculations must limit the cluster size and truncate the sum over angular momentum states. The cluster size cannot be too large, typically ~1000 atoms or less for most computations. Fortunately, XANES calculations usually do not need such large clusters since the finite core-hole lifetime keeps the inelastic mean free path from becoming too large. Additionally, reasonable accuracy requires that l_{max} is of order $k_{max}R_{mt} + l_{occ}$, where l_{occ} is the largest angular momentum of the occupied states and $k_{max}^2 \sim E_{max}$ in Rydberg units. Thus, in the XANES region one can safely set the maximum angular momentum to 4 at most, while at higher energies in the EXAFS matrix inversion becomes impractical. In the intermediate energy regime, Lanczos techniques can be used.[54] In addition, the Green's function $G(E)$ is parameterized by the photoelectron energy, thus one inversion must be performed for each energy point in the spectrum. This makes the calculations parallelize naturally and can be done very efficiently on parallel computers. [55]

3.6 Many-Body Effects: Quasiparticle Self-Energy Models

Many-body effects also play a large role in the XANES and are responsible for energy dependent shifts and broadenings associated with the quasiparticle self-energy, as well as satellite peaks brought about by multi-electron excitations. While simple models such as the plasmon pole self-energy [56,57] have been quite successful for calculations of EXAFS, more accurate models must be used near the edge. Most current calculations of these effects are based on the *GW* approximation of Hedin, [58] *i.e.*,

$$\Sigma(E) = \frac{i}{2\pi} \int G(E-\omega)W(\omega)e^{-i\eta\omega}d\omega, \qquad (3.16)$$

where G is the electron Green's function, and $W = \frac{v}{\varepsilon}$ is the screened Coulomb interaction. *To retain the efficiency of the plasmon pole model, a many-pole model self-energy was developed which uses multiple poles to represent the loss function $L(\vec{q}, \omega) = -Im[\varepsilon^{-1}(\vec{q}, \omega)]$,* i.e.,

$$L(\vec{q}, \omega) = \frac{\pi}{2} \sum_i g_i \omega_j \delta(\omega - \omega_i(\vec{q})). \tag{3.17}$$

The dispersion of the excitations is approximated by a polynomial $\omega_i(\vec{q})^2 = \omega_i^2 + v_F^2 q^2 + \frac{q^4}{4}$ which retains the high momentum transfer limit. The amplitudes g_i are fixed by matching to the loss function in the zero-momentum transfer limit $q \to 0$, giving $g_i = -[2\Delta\omega/\pi\omega_i]L(0, \omega_i)$. In this way the model is consistent with the inverse and first moments of the spectrum for any number of poles, and thus convergence with number of poles is relatively fast. Given this model of the dielectric function, the total self-energy can be expressed as a sum of single plasmon pole self-energies arising from interactions with "plasmons" at different frequencies, i.e., $\Sigma(E) = \Sigma_x + \sum_{(i)} g_i \Sigma_1(E, \omega_i)$. *This model has been shown to produce reasonably accurate quasiparticle self-energies in comparison with more accurate but much more time-consuming calculations.* The many-pole model self-energy has been applied to a variety of systems and gives improved results for amplitudes and phases in XANES calculations using FEFF. The model can also be applied in a post processing step as a convolution of the spectrum with an energy dependent Lorentzian.[59]

In addition to quasiparticle energies, the inelastic mean free path is related to the imaginary part of the self-energy, i.e., $\lambda_k(E) = k/|Im[\Sigma(E)]|$. The values of inelastic mean free path calculated using self-energy method, show improved agreement with experiment ones over the plasmon pole model. [60]

Multi-electron excitations can also be calculated given the quasiparticle self-energy, via a convolution of the spectrum with an effective spectral function.[61,62] These calculations are based on a quasi-boson model, which also gives an estimate of the amplitude reduction factor S_0^2 . Additionally, the calculations yield the satellites on the spectrum. These satellites are roughly displayed as "repetitive" peaks in the spectrum, shifted by energy of the main bosonic excitations of the system, and are reflected by peaks in the observed loss function.

Most often, at the *K*-edge and at the $L_{2,3}$ edge of heavy elements (say, from Kr), the problem can be greatly simplified, because the perturbation induced on the other electrons by the transition of a core electron up to some empty state, can be treated in a phenomenological way. The core-hole just induces a modification of the potential associated with the broadening effect. And the other electrons are just passive, with an energy shift and a radial contraction of their orbital due to the core hole. In the so-called final-state rule, the final-state energies are measured, and the dynamics of the excitation process are neglected. [63] After an atomic self-consistent calculation is performed with the core-hole and one more electron in the first non-occupied atomic orbital, the excited atomic potential can be obtained, which includes the internal screening of the core-hole by the other electrons of the atom. This approach is slightly better than simulating the excited atom with the so-called Z+1 approximation. In this case, the absorbing atoms are simply replaced in the simulations by the following one in the Mendeleev table.

3.6.1 Intra-atomic core-hole screening

The electrons of the absorbing atoms and their neighbors tend to screen the core-hole. The contraction of their orbital caused by the core-hole, makes the scalar product between their state before and after the core-hole creation less than one. For all the atoms, there is a correction: $S_0 = \prod_j \langle \varphi_{F,j} | \varphi_{I,j} \rangle$, where *j* index all the electrons of the atom but the one making the transition, and $\varphi_{I,j}$ and $\varphi_{F,j}$ are their wave functions before and after the core-hole creation. In practice, the factor S_0^2 used in EXAFS is typically 0.8 showing that the radial contraction of the orbital due to the core-hole is not negligible. With this new factor we can make explicit the connection between the multiple and single-electronic approaches: $\langle \phi_F | \hat{o} | \phi_I \rangle \approx S_0 \sum_i \langle \varphi_{F,i} | \hat{o} | \varphi_{I,i} \rangle$, where *i* indexes the core electrons making the transition, which correspond to the core electron states of the edge. This implies that the transitions from the core states up to the empty final states are supposed to be independent.

Energy shift

The energy conservation implies that $\hbar\omega = \varepsilon_f - \varepsilon_i$. As the core-hole induces a more attractive potential for the electrons of the absorbing atom with the screening effect, it means an energy shift for all of them, the sum of which is written as ΔE_{src}. The energy conservation can thus be simplified in: $\hbar\omega = \varepsilon_f - \varepsilon_i + \Delta E_{src}$ where ε_i and ε_f are just the energies of the electron making the transition. ΔE_{src} is supposed to be a unique constant for one edge of one element. In other words, when comparing with the experiment, an *ad-hoc* shift is allowed to apply, but this shift must always remain the same. The multi-electronic point of view fundamentally describes the splitting in two different edges L_2 and L_3 by a final state effect, because the hole of the final state can be either in $2p_{1/2}$ or in $2p_{3/2}$. Indeed, the energy difference comes from the two different values of ε_f. Nevertheless, in the single-electronic approximation, it is convenient to call the initial state where the electron stands before its transition. Thus the splitting between L_2 and L_3 originates from the difference between the two corresponding ε_i values due to the same spin-orbit interaction.

The single-electronic equation

Since the photoelectron is not in any bound states but stays in the so-called continuum, it can escape from the absorbing atom. Then formally it is convenient to replace the Kronecker symbol by the density of state at the corresponding energy ε_f. Nevertheless, single-electronic codes using Green's functions or scattering theory, use the basic state not normalized to one, but directly by the square root of the density of state in a vacuum. So, in this case, the Kronecker symbol must be kept. Since this scattering theory utilizes complex energy, the bound states can be treated in the same framework as the continuum.

With all this, the single-electronic absorption cross-section is given by:

$$\sigma = 4\pi\alpha^2\hbar\omega \textstyle\sum_f \left|\langle\varphi_f|\hat{o}|\varphi_i\rangle^2\right|\delta(\varepsilon_f - \varepsilon_i + \Delta E_{src} - \omega). \qquad (3.18)$$

Now the sum over *f* is not over all the configurations but overall the single-electronic states at a same energy ε_f. More precisely, in reciprocal space techniques, it indicates a sum over bands and an integral over crystal momentum; in real space cluster techniques, it sums over quantum numbers associated with spherical wave expansions. More often, S_0^2 and ΔE_{src} are not explicitly written in the equation.

3.6.2 The Hubbard correction

The LSDA picture is appropriate for band behavior but misses the physics of local correlated electrons. The DFT method, called LSDA+U, employs functionals constructed from the mean field approximation of the Hubbard model.[64] To some extent, it reconciles the band and the local aspects. Successful calculations of X-ray absorption spectra using it have been reported [65, 66] and it is a possible option in standard codes such as FEFF [67] and FDMNES. [68]

3.6.3 Broadening

As a life-time is associated with the excited state of the atoms, a corresponding energy width of both the core-hole and photoelectron levels, this phenomenon can be taken into account by a Lorentzian broadening. Its width is simply the sum of the core and valence state widths. The core state width is highly dependent on the atomic number of the element and on the quantum numbers of the edge. The valence state widths is determined by the photoelectron kinetic energy, which is low close to the Fermi level, and strongly increasing, typically at the plasmon energy (say, around 10 eV). Typically, the natural width [69] of the 1s core state ranges from some tenth of eV for the low Z elements (C, N...), reaching 1 to 2 eV for the 3d elements. The width can be very important for heavy elements (Z > 55), no matter what the edge is. In experiments, a selected decay to fill the core-hole can be chosen by detecting the fluorescence yield of XANES at high energy resolution, the broadening is reduced to the corresponding width.[70]

3.6.4 Thermal and structural disorder

For complicated systems such as amorphous materials or systems at higher temperatures, the EXAFS can be calculated via Debye–Waller factors which damp each path at high *k*. The XANES is more difficult, since changes in symmetry can allow previously forbidden transitions, resulting in new peaks in the spectrum. Thus, a more accurate method for XANES calculation is to perform averaging over physically unique absorbing sites. This configurational averaging must be performed even when disorder is not present if the system in question has more than one physically unique absorber. For example, magnetite has both tetrahedrally and octahedrally coordinated iron atoms, thus one calculation should be performed for each of these sites, and the final signal obtained from weighted average of the calculated results, where the weighting depends on the site stoichiometry. Doped systems can also require this kind of configurational averaging if the doping concentration is high. Also, in nanoparticles there can be a large number of physically unique sites, based on the proximity to the surface of the particle, interaction with the support, and interaction with adsorbed molecules. All of these effects cause distortions, which ideally should be modeled along with the dynamic, temperature dependent effects. This can be done using, for example, ab initio molecular dynamics to obtain snapshots of the structure of the system.[71, 72, 73] The XANES can be calculated for each snapshot by averaging over unique sites, and an average over snapshots gives the final result. There are several approximation applied in the FEFF code that can affect the outcome of such calculations. First, core level shifts due to charge transfer can be important in the final averaged result, since these cause red or blue shifting of the absorption edge. Second, the chemical potential found by FEFF must be calculated carefully since this also causes shifting of the edge. For example, all calculations performed on a static nanoparticle should result in the (nearly) same chemical potential. This sometimes requires that a unique potential be defined for each type of atom, although in many cases, atoms with similar chemical surroundings can be defined as the same type without much loss of accuracy in the results.

3.7 Multiplet Ligand Field Theory

Multiplet ligand-field theory (MLFT) is one of the earliest quantum chemistry methods developed to describe the electronic structure of transition-metal and rare-earth compounds.[74] MLFT calculations traditionally use parameters fitted to experiments. Nevertheless, over the years, numerous theoretical studies have therefore been devoted to obtaining MLFT parameters *ab initio*. The molecular orbitals can then be simply constructed from Hartree-Fock orbitals.[75] Better results are now found by obtaining these orbitals from LDA simulations on the cluster around the absorbing atoms.[76] More recently Haverkort *et al.* [77] used another DFT approach producing typically the transition metal 3d and ligand 2p orbitals which, taken together, exactly describe the DFT 3d and 2p bands. This method avoids the embedding errors and the parameters for the multiplet are thus calculated in the best way. The atomic multiplet description including the crystal field for XANES simulation is well documented.[78] They are performed in the two steps: the atomic multiplets, and the introduction of the crystal field.

3.7.1 Atomic multiplets

Besides, time-dependent density-functional theory (TDDFT) has also been successfully applied to X-ray absorption spectra at the $L_{2,3}$ edges for the 3d transmission metal series.[79, 80] However,because of the strong electron correlation effects in the L-edge spectra of 3d transmission metals, atomic multiplet theory is more commonly used to interpret the spectra.[81,82] At present a fully quantitative treatment of the near-edge structures is still challenging, due to various complex many-body effects,such as the photoelectron core-hole interaction, multiplet effects, thephotoelectron self-energy and inelastic losses. However, it still finds wide applications in many different fields, often used as a "fingerprint" to identify an unknown sample by comparing it with those of a number of reference compounds with known structures.

A variety of quantitative tools are also available for interpreting XANES data, such as edge positioning, peak fitting, linear combination fitting, and principle components analysis. In combination with theoretical simulations or interpretations, analysis of XANES spectrum could also provide valuable information on the site symmetry, oxidation states, bond length and so forth.

First the atomic wave function is simply expanded in spherical harmonics, the set of radial part $\varphi_l(r)$ usually depends only on l. Then, the Hamiltonian for atomic multiplets can be written as:

$$H = \sum_i \frac{p_i^2}{2m} - \sum_i \frac{Ze^2}{r_i} + \sum_i \frac{e^2}{|r_i - r_j|} + \sum_i L_i \cdot S_i. \quad (3.19)$$

where the sums are over the electrons i and j (usually only the 2p and 3d ones for a $L_{2,3}$ edge). The two first terms are the kinetic energy of the electrons and the potential energy from the electrostatic interaction with the nucleus with atomic number Z. These terms are the same for all electrons in a given atomic configuration. The third term, the electron-electron repulsion can be split between its spherical average and its non-spherical part. This latter term plus the spin-orbit term in the equation above are the only ones contributing to the multiplet splittings. They are expressed in terms of spherical harmonics with a radial part, called the Slater integral, and an angular part. From this, the radial part is moreover divided into direct Coulomb terms, $F^k(l,l';l,l')$ and exchange terms, $G^k(l,l';l',l)$. The corresponding parameters, and the spin-orbit $L_i \cdot S_i$ are calculated. Some of the F and G parameters are then often scaled down to 80% of their original value. For example, for a configuration with the term symbol $^{2S+1}L_J$, the Coulomb energy is given by:

$$\left\langle {}^{2S+1}L_J \left| \frac{e^2}{|r_i - r_j|} \right| {}^{2S+1}L_J \right\rangle = \sum_k f_k F^k + g_k G^k, \tag{3.20}$$

where f_k and g_k are the geometrical terms. They are non-zero for only some values of k depending on l and l'. From this, to get the X-ray absorption spectra, one must consider the transitions from ground or initial states up to final states. In a closed shell system, it is easy because the transition is typically $2p^6 3d^0 \rightarrow 2p^5 3d^1$, the ground state has the single $^1S_0$ symmetry. On the contrary, there are 12 possible final states, but only three term symbols $^1P_1$, $^3P_1$ and $^3D_1$ can be reached from the selection rules. When ignoring the 3d spin-orbit and the Coulomb repulsion, the triplet states have equal energy. One thus just keeps the 2p spin-orbit splitting and gets only two peaks, as in single-electronic simulations, with the ratio 2:1 between $p_{3/2}$ and $p_{1/2}$ (or L_3 and L_2). Spectra can thus present three peaks. In fact, the Coulomb repulsion cannot be ignored. Consequently, a third peak appear. The relative position and amplitude of these peaks depend on the spin-orbit and Coulomb amplitudes.

3.7.2 The crystal field

The hybridization of the absorbing atom orbital with the ligand ones is usually small, so the number of electrons can be the supposed integer. The problem is in this way simplified and can be handled just

by the reduction of the symmetry from O_3 to a lower one. The use of the group theory and the corresponding algebra is then mandatory. All the irreducible representations are projected onto the group of the molecule around the absorbing atom. The branching technique can be found in Butler's book [83] and papers by van der Laan and Thole [84] and uses the atomic structure of Cowan's program.[85] One now gets a series of peaks, for example 7 for the O_h (octahedral) symmetry. A broadening, which can be different for the different peaks, permits then, the comparison with the experiment. A single-electronic simulation, in the low energy part of the spectra, would give only four peaks, the so-called e_g and t_{2g} states for both L_3 and L_2.
For $3d^n$ configurations with $n > 0$, the number of transitions greatly increases but the philosophy remains the same. An important difference is that in this case, the ground state is affected by *dd* interactions and crystal field.

3.8 Current Theoretical Developments

Many attempts and a variety of theoretical frameworks, including atomic multiplets,[86,87] configuration interaction,[88,89] time-dependent density-functional theory (TDDFT), [90,91,92] dynamical mean-field theory (DMFT),[93] multichannel multiple-scattering,[94] and the Bethe-Salpeter equation (BSE) based approaches,[95,96] have been proposed to improve the most difficult cases, unifying the local atomic multiplet approach with extended states. Ideally this would have to be done within a first-principle theory.

None of them is nowadays, at the same time, sufficiently general, precise, with a reasonable computing resource demand, and user-friendly! Nevertheless, they often currently equal or outperform the other available methods in accuracy. The use of such a theory depends not only on its general power but also on the final agreement really needed and so on the ease of handling it. Indeed, the final goal of the XANES users, but also for some theoreticians, is not to exactly reproduce the experimental spectra, but to understand its material.

Among them, TDDFT is particularly convenient in terms of simplicity and computation time. Improvements on the ratio L_3 / L_2 often taken as a key test parameter are found only for the first half of the 3d transition metal elements, typically from Sc to Mn. The TDDFT can include the many body effects within the response function: $\hat{\chi} = \hat{\chi}_0 +$

$\hat{\chi}_0\hat{K}\hat{\chi}$, where $\hat{\chi}_0$ is the ground state susceptibility and $\hat{K}$ a so-called Kernel. The effect of this kernel is to mix all the previously independent transitions from the core states up to the empty states. $\hat{\chi}_0$ is calculated in a previous step and contains typically the multiple scattering amplitudes. All these terms are expressed in the absorbing atom harmonics basis. The dimension of the matrix involved thus takes up only a small amount of computing time, and TDDFT appears as a perturbation from the DFT case. The problem lies in the evaluation of the kernel. Theoretically, high precision could be obtained with a convenient kernel. Up to now nevertheless only local kernels have been used and, for example, no excitonic effects can be reproduced.

The Bethe-Salpeter equation-based approache should be more precise. By using this BSE technique, Vinson and Rehr [97] recently performed *ab initio* calculations of the $L_{2,3}$ edges of several insulating and metallic compounds which covers a range of 3d-electron occupations and got a very good agreement. Their method includes the unified processing of the key elements of the extended state and the multiple effects of atoms, namely the Bloch state, the GW self-energy correction, the self-consistent crystal potential, the spin-orbit term, the ground state magnetism, and the Coulomb interaction between the L_2 and L_3 levels. This method has been implemented in the OCEAN software package,[98] which uses plane-wave pseudo potential wave functions as a basis, and a resolvent formalism for the calculation of Bethe-Salpeter equation. The results are in near quantitative agreement with the experiment, when the anisotropy is not too high, including both the fine structure at the edges and the L_3/L_2 ratios observed for these systems.

Finally, the multi-configuration interaction techniques could be in principle the best. The difficulty is of course, for real molecules, and the corresponding number of electrons, the very large computational effort required. The increasing power of computers nevertheless means that systems bigger than the absorbing atom and its surrounding shell can become tractable in the future. However, fully quantitative calculations remain a challenge.

3.9 Other Related Spectroscopies

There are many other spectroscopies involve similar physics as XANES and an analogous golden rule formalism. XANES-related techniques include electron energy loss (EELS), X-ray emission (XES) and non-resonant inelastic X-ray scattering (NRIXS). The main difference is that they probe different states, either because the transition operator is different, as in EELS and NRIXS, or because the initial and final states are different, as in XES. XES is particularly useful since it probes the occupied states of the system via the dipole operator, and is thus complementary to XANES.[99] EELS and NRIXS both obey a Fermi's golden rule with transition operators $e^{iq \cdot r}$ replacing the dipole operator, where q is the momentum transfer due to the scattering of the electron or photon.[100] This momentum transfer can be controlled by adjusting the angle of detection, and thus the dipole limit can be reached in which case the signal is equivalent to XANES, with $\hat{q}$ replacing the polarization direction. Within the dipole limit, the EELS amplitude contains a factor that damps the high momentum transfer signal, while NRIXS does not contain this factor, making it possible to probe the local unoccupied density of states for all angular momenta.[101] In addition, NRIXS is useful as a bulk probe of low energy edges, since the energy loss is required to match the edge energy, and high energy X-rays can still be used as the probe.[102]

Some resonant spectroscopies are very close to XANES. At the center of resonant elastic X-ray scattering (REXS), there is an absorption of a photon from a core state up to some intermediate state.[103] The difference in real absorption is that photoelectrons are virtual and exist only for the time given by Heisenberg's uncertainty principle. It returns to its initial state coherently, emitting a new photon with the same energy as the incident light. Therefore, the transition is the same, and its overall probability is closely related to the real absorption process. Of course, the measurements are not done in the same way but from a theoretical point of view, it is relatively easy to adapt the simulation technique from one to the other. Many things for XANES can also be directly used for the resonant process. When conducting this type of study, the Bragg peak of the diffraction is recorded, which is often referred to as resonance X-ray diffraction (RXD). The term "diffractive anomaly fine structure" (DAFS) is often used when studying extensions. [104] Similarly, multiwave-length diffraction (MAD) records only some energy points around an absorption edge, instead of spectra. It is extensively utilized in biology to resolve the structure of metalloprotein.[105] Indeed, the sensitivity of the metallic absorbing element is observed to be reinforced in these complex molecules. The resonant scattering amplitude in the single-electronic case is:

$$f' - if'' \approx m\omega^2 \lim_{n \to 0^+} \sum_{f,i} \frac{\langle \varphi_f | \hat{O}_i | \varphi_i \rangle \langle \varphi_i | \hat{O}_S^* | \varphi_f \rangle}{\hbar\omega - (\varepsilon_f - \varepsilon_i) + i\eta}, \tag{3.21}$$

where $\hat{O}_{i,s}$ is an operator depending on polarization and the wave vector at second order. Polarization can be different for the incoming (i) and scattered (s) waves. $\varphi_{i,f}$ are the core and intermediate state wave functions and $\varepsilon_{i,f}$ their energies. m is the electron mass and $\hbar\omega$ the photon energy. The interpretation of this formula can be read from left to right: the incoming photon i is absorbed, promoting a core electron from φ_i to φ_f through $\langle \varphi_f | \hat{O}_i | \varphi_i \rangle$. Then it comes back to its initial state through $\langle \varphi_i | \hat{O}_S^* | \varphi_f \rangle$. When incoming and outgoing waves have the same polarization, the numerator is the same as the central term of the XANES formula showing the close connection between both spectrocopies. To find the diffracted peak intensity, these atomic resonant scattering amplitudes are added to the non-resonant ones. Then, multiplied by the Bragg term for a chosen Bragg peak, with a summation over the atoms of the unit cell, the modulus square gives the intensity of the diffraction peak.

Resonant X-ray emission (RXES) is given by the Kramers–Heisenberg formula, which is similar to Fermi's golden rule, but includes a resonant energy denominator that focuses on a given transition. Although the formula is more complicated than XANES, the RXES spectrum can be approximated by the convolution of an effective XANES signal with the XES spectrum,[106] *i.e.*,

$$\frac{d^2\sigma}{d\Omega d\omega} = \frac{\omega}{\Omega} \frac{\int \mu_{XES}(\omega_1) \bar{\mu}(\Omega, \Omega - \omega - \omega_1 + E_b) d\omega_1}{|\omega - \omega_1 + i\Gamma_b|}, \tag{3.22}$$

where μ_{XES} is the XES spectrum, $\bar{\mu}$ is the effective XANES spectrum, and E_b and Γ_b are the core level energy and broadening corresponding to the initial deep-core excitation, respectively. This formula can be utilized to estimate both core to core resonant inelastic X-ray scattering, core-valence RXES, and high energy resolution fluorescence XANES. Finally, Compton scattering can be calculated within the impulse approximation from the density matrix, $\rho(\vec{r}, \vec{r}') =$

$-\frac{1}{\pi}\int_{-\infty}^{\mu} Im[G(\vec{r},\vec{r}',E)]\, dE$, making multiple scattering a natural formalism for these calculations. [107]

---

[1] A.L. Ankudinov, A.I. Nesviszhskii, & J.J. Rehr, (2003) Dynamic screening effects in X-ray absorption spectra. *Phys. Rev. B* **67,** 115120.

[2] J.A. Soininen, J.J. Rehr, & E.L. Shirley, (2003) Electron self-energy calculation using a general multi-pole approximation. *J. Phys.: Condens. Matter* 15, 2573–86.

[3] L. Hedin, (1999) On correlation effects in electron spectroscopies and the GW approximation. *J. Phys.: Condens. Matter* **11,** R489–528.

[4] M. Petersilka, U.J. Gossmann, & E.K.U. Gross, (1996) Excitation energies from time-dependent density-functional theory. *Phys. Rev. Lett.* **76,** 1212–15.

[5] J.J. Rehr, (2003) Excited state electronic structure and the theory of X-ray spectra *J. Phys.: Condens. Matter* **15,** S647–54.

[6] F. M. F. De Groot, (1993) X-ray absorption of transition metal oxides: an overview of the theoretical approaches. *J. Electron Spectrosc. Relat. Phenom.* 62, 111.

[7] F.M.F. De Groot, (1994) X-ray absorption and dichroism of transition metals and their compounds. *J. Electron Spectrosc. Relat. Phenom.* 67, 529.

[8] J.J. Rehr, & A.L. Ankudinov, (2001) New developments in the theory of X-ray absorption and core photoemission. *J. Electron. Spectrosc. Relat. Phenom.* **114–116,** 1115–21.

[9] L.F. Mattheiss, & R. E. Dietz, (1980) Relativistic tight-binding calculation of core-valence transitions in Pt and Au. *Phys. Rev. B* 22, 1663–1676.

[10] J.J. Rehr, R.C. Albers, & S.I. Zabinsky, (1992) High-order multiple-scattering calculations of X-ray-absorption fine structure. *Phys. Rev. Lett.* 69, 3397–3400.

[11] T.A. Tyson, K.O. Hodgson, C.R. Natoli, & M. Benfatto, (1992) General multiple-scattering scheme for the computation and interpretation of X-ray-absorption fine structure in atomic clusters with applications to SF_6, $GeCl_4$, and Br_2 molecules. *Phys. Rev. B* 46, 5997–6019.

[12] A. Gonis, X. G. Zhang, & D. M. Nicholson, Multiple-scattering green-function method for space-filling cell potentials. *Phys. Rev.* B 40, 947 (1989).

[13] D. E. Ellis, & G. S. Painter, (1970) Discrete variational method for the energy-band problem with general crystal potentials. *Phys. Rev. B* 2, 2887.

[14] D. Ahlers, K. Attenkofer, & G. Schuetz, (1998) Spin-dependent extended X-ray absorption fine structure in magnetic oxides. *J. Appl. Phys.* 83, 7085.

[15] T. Huhne T, T. C. Zecha, H. Ebert, P. H. Dederichs, & R. Zeller, (1998) Full-potential spin-polarized relativistic Korringa-Kohn-Rostoker method implemented and applied to bcc Fe, fcc Co, and fcc Ni. *Phys. Rev. B* 58, 10 236.

[16] T. Huhne, & H. Ebert, (1999) Calculation of X-ray absorption spectra using the full potential spin-polarized relativistic multiple scattering formalism. *Solid State Commun.* 109, 577.
[17] P. Blaha, K. Schwarz, P. Sorantin, & S. Trickey, (1990) Full-potential, linearized augmented plane wave programs for crystalline systems. *Comput. Phys. Commun.* 59, 399.
[18] Y. Joly, (2001) X-ray absorption near-edge structure calculations beyond the muffin-tin approximation. *Phys. Rev. B* 63, 125120-125129.
[19] S. Grenier, & Y. Joly, (2014) Basics of resonant elastic X-ray scattering theory. *J. Phys.: Conf. Ser.*, 519, 012001.
[20] C. Noguera, D. Spanjaard & J. Friedel. (1979) Dynamic screening of a core hole: i. semiclassical model, *J. Phys. F: Metal Phys.*, 9, 1189.
[21] P. Kruger, & C.R. Natoli, (2005) Multi-channel multiple scattering theory for XAFS application to $L_{2,3}$ edges of Ca, *Phys. Scripta*, T115, 146.
[22] J.A. Soininen, & E.L. Shirley, (2001) Scheme to calculate core hole: electron interactions in solids, *Phys. Rev. B*, 64, 165112.
[23] A.L. Ankudinov, A.I. Nesvizhskii, & J.J. Rehr, (2003) Dynamic screening effects in X-ray absorption spectra, *Phys. Rev. B* 67, 115120.
[24] M. Roemelt, D. Manganas, S. DeBeer, & F. Neese, (2013) A combined DTF and restricted open-shell configuration interaction method including spin-orbit coupling: application to transition metal L-edge X-ray absorption spectroscopy, *J. Chem. Phys.*, 138, 204101.
[25] A. Bianconi, A. Marcelli, H. Dexpert, R. Karnatak, A. Kotani, T. Jo, & J. Petiau, (1987) Specific intermediate-valence state of insulating 4f compounds detected by L_3 X-ray absorption. *Phys. Rev. B* 35, 806.
[26] J.J. Kas, A.P. Sorini, M.P. Prange, L.W. Cambell, J.A. Soininen, & J.J. Rehr, (2007) Many-pole model of inelastic losses in X-ray absorption spectra. *Phys Rev B* 76, 195116.
[27] A.I. Nesvizhskii, A.L. Ankudinov, & J.J. Rehr, (2001) Normalization and convergence of X-ray absorption sum rules. *Phys. Rev. B* 63, 094412.
[28] F. Farges, G.E. Brown, & J.J. Rehr, (1996) Coordination chemistry of Ti(IV) in silicate glasses and melts: 1. XAFS study of titanium coordination in oxide model compounds. *Geochim. Cosmochim. Acta.* 60, 3023.
[29] A.L. Ankudinov, & J.J. Rehr, (2005) Nonspherical potential, vibronic and local field effects in X-ray absorption. *Phys. Scr.* T115, 24.
[30] G.K.H. Madsen, P. Blaha, K. Schwarz, E. Sjöstedt, & L. Nordström (2001) Efficient linearization of the augmented plane-wave method, *Phys. Rev. B* 64, 195134.
[31] http://www.fhi-berlin.mpg.de/KHsoftware/StoBe/
[32] D. Cabaret, E. Gaudry, M. Taillefumier, P. Sainctavit, & F. Mauri, (2005) XANES calculation with an efficient "non muffin-tin" method: application to the angular dependence of the Al K-edge in corundum, *Phys. Scripta*, T115, 131.

33 H. Ebert, D. Koedderitzsch, & J. Minar, (2011) Calculating condensed matter properties using the KKR-Green's function method—recent developments and applications, *Rep. Prog. Phys.* 74, 096501.

34 O.K. Andersen, O. Jepsen, & M. Sob, (1987) in Lecture Notes in Physics: *Electronic Band Structure and Its Applications* (ed. M. Yussouff), Springer-Verlag, Berlin.

35 C.R. Natoli, D.K. Misemer, S. Doniach, & F.W. Kutzler, (1980) First-principles calculation of X-ray absorption-edge structure in molecular clusters, *Phys. Rev. A*, 22, 1104.

36 A.L. Ankudinov, B. Ravel, J.J. Rehr, & S.D. Conradson, (1998) Real-space multiple-scattering calculation and interpretation of X-ray absorption near-edge structure, *Phys. Rev. B* 58, 7565. http://leonardo.phys.washington.edu/feff/

37 O. Bunau, & Y. Joly, (2009) Self-consistent aspects of X-ray absorption calculations, *J. Phys.: Condens. Matter*, 21, 345501. http://www.neel.cnrs.fr/fdmnes.

38 M. Benfatto, & S. Della Long, (2001) Geometrical fitting of experimental XANES spectra by a full multiplescattering procedure, *J. Synchrotron Rad.*, 8, 1087.

39 R.V. Vedrinskii, V.L. Kraizman, A.A. Novakovich, & V. Sh. Machavariani, (1992) Theory of X-ray scattering in the anomalous dispersion region and the problem of atomic-ternary-correlation-function determination in amorphous media, *J. Phys. Condens. Matter,* 4, 6155.

40 L. Hedin, & S. Lundqvist, (1971) Explicit local exchange-correlation potentials, *J. Phys. C: Solid State Phys.*, 4, 2064.

41 J.P. Perdew, & Y. Wang, (1992) Accurate and simple analytic representation of the electron-gas correlation energy, *Phys. Rev. B* 45, 13244.

42 I.U. Von Barth, & L. Hedin, (1972) A local exchange-correlation potential for the spin polarized case, *J. Phys. C: Solid State Phys.*, 5, 1629.

43 M. Taillefumier, D. Cabaret, A.-M. Flank, & F. Mauri, (2002) X-ray absorption near-edge structure calculations with the pseudopotentials: application to the K edge in diamond and α-quartz, *Phys. Rev. B* 66, 195107.

44 Y. Joly, (2001) X-ray absorption near edge structure calculations beyond the muffin-tin approximation, *Phys. Rev. B* 63, 125120.

45 K. Hatada, K. Hayakawa, M. Benfatto, & C.R. Natoli, (2007) Full-potential multiple scattering for X-ray spectroscopies, *Phys. Rev. B* 76, 060102.

46 A.L. Ankudinov, B. Ravel, J.J. Rehr, & S.D. Conradson, (1998) Real-space multiple-scattering calculation and interpretation of X-ray-absorption near-edge structure. *Phys. Rev. B* 58, 7565.

47 C.R. Natoli, D.K. Misemer, S. Doniach, & F.W. Kutzler, (1980) First-principles calculation of X-ray absorption-edge structure in molecular clusters, *Phys. Rev. A*, 22, 1104.

48 Ch. Brouder, & J.-P. Kapler, (1997) Prolegomena to magnetic circular dichroism in X-ray absorption spectroscopy, in Magnetism and Synchrotron

Radiation (eds E. Beaurepaire, B. Carriere, and J.-P. Kappler), pp. 33–54, *Les Editions de Physiques*, Paris.

49 http://cars9.uchicago.edu/ifeffit/

50 http://gnxas.unicam.it/

51 C.R. Natoli, & M. Benfatto, (1986) A unifying scheme of interpretation of X-ray absorption spectra based on multiple scattering theory, *J. de Phys. Colloques*, 47, C8–11.

52 C. R. Natoli, D. K. Misemer, S. Doniach, & F. W. Kutzler, (1980) First-principles calculation of X-ray absorption-edge structure in molecular clusters. *Phys. Rev. A* 22, 1104.

53 T. Fujikawa, (1981) Theory of photoionization cross sections by dynamical theory in the X-ray region. *J. Phys. Soc. Jpn.* 50, 1321.

54 K. Jorissen, & J.J. Rehr, (2010) Calculations of electron energy loss and X-ray absorption spectra in periodic systems without a supercell. *Phys. Rev. B* 81, 245124.

55 A.L. Ankudinov, C.E. Bouldin, J.J. Rehr, J. Sims, & H. Hung, (2002) Parallel calculation of electron multiple scattering using Lanczos algorithms. *Phys. Rev. B* 65, 104107.

56 B.I. Lundqvist, (1967) Single-particle spectrum of degenerate electron gas: 1. Structure of spectral weight function. *Phys Konden Mater* 6, 193.

57 B.I. Lundqvist, (1967) Single-particle spectrum of degenerate electron gas: 2. Numerical results for electrons coupled to plasmons. *Phys Konden Mater* 6, 206.

58 L. Hedin, & S. Lundqvist, (1970) Effects of electron-electron and electron-phonon interactions on the one-electron states of solids. *Solid State Phys* 23, 1.

59 J.J. Kas, J. Vinson, N. Trcera, D. Cabaret, E.L. Shirley, & J.J. Rehr, (2009) Many-pole model of inelastic losses applied to calculations of XANES. *J Phys Conf Ser* 190, 012009.

60 A.P. Sorini, J.J. Kas, J.J. Rehr, M.P. Prange, & Z.H. Levine, (2006) Ab initio calculations of electron inelastic mean free paths and stopping powers. *Phys. Rev. B* 74, 165111.

61 L.W. Campbell, L. Hedin, J.J. Rehr, & W. Bardyszewski, (2002) Calculations of inelastic losses and interference terms in X-ray absorption. *Phys. Rev. B* 65, 064107.

62 J.J. Kas, A.P. Sorini, M.P. Prange, L.W. Cambell, J.A. Soininen, & J. J. Rehr. (2007) Many-pole model of inelastic losses in X-ray absorption spectra. *Phys. Rev. B* 76, 195116.

63 U. von Barth, & G. Grossman, (1982) Dynamical effects in X-ray spectra and the final-state rule, *Phys. Rev. B* 25, 5150.

64 V.I. Anisimov, J. Zaanen, & O.K. Andersen, (1991) Band theory and Mott insulators: Hubbard U instead of stoner I, *Phys. Rev. B* 44, 943.

65 N. Binggeli & M. Altarelli, (2004) Orbital ordering, Jahn-Teller distortion, and resonant X-ray scattering in $KCuF_3$, *Phys. Rev. B* 70, 085117.

[66] M.W. Haverkort, Z. Hu, A. Tanaka, W. Reichelt, S. V. Streltsov, M. A. Korotin, V. I. Anisimov, H. H. Hsieh, H.-J. Lin, C. T. Chen, D. I. Khomskii, & L. H. Tjeng, (2005) Orbital-assisted metal-insulator transition in VO_2, *Phys. Rev. Lett.* 95, 196404.

[67] K. Mathew, C. Zheng, D. Winston, C. Chen, A. Dozier, J. J. Rehr, S.P. Ong & K. A. Persson, (2018) High-throughput computational X-ray absorption spectroscopy. *Sci. Data* 5, 180151.

[68] S. Guda, A.A. Guda, M.A. Soldatov, K.A. Lomachenko, A.L. Bugaev, C. Lamberti, W. Gawelda, Ch. Bressler, G. Smolentsev, A. V. Soldatov, & Y. Joly, (2015) Optimized finite difference method for the full-potential XANES simulations: Application to molecular adsorption geometries in MOFs and metal-ligand intersystem crossing transients. *J. Chem. Theory Comput.* 11, 4512-4521.

[69] M.O. Krause, & J.H. Oliver, (1979) Natural widths of atomic K and L Levels, $K\alpha$ X-ray lines and several KLL Auger Lines, *J. Phys. Chem. Ref. Data*, 8, 2.

[70] P. Glatzel, J. Singh, K.O. Kvashnina, & J.A. van Bokhovn, (2010) In-situ characterization of the 5d density of states of pt nanoparticles upon adsorption of CO, *J. Am. Chem. Soc.*, 132, 2555.

[71] F.D. Vila, J.J. Rehr, J.J. Kas, R.G. Nuzzo, & A.I. Frenkel, (2008) Dynamic structure in supported Pt nanoclusters: real-time density functional theory and X-ray spectroscopy simulations. *Phys. Rev. B* 78, 121404.

[72] J.L. Fulton, E.J. Bylaska, S. Bogatko, M. Balasubramanian, E. Cauët, G. K. Schenter, & J.H. Weare, (2012) Near-quantitative agreement of model-free DFT-MD predictions with XAFS observations of the hydration structure of highly charged transition-metal ions. *J Phys Chem Lett* 3, 2588.

[73] F.D. Vila, J.J. Rehr, S.D. Kelly, & S.R. Bare, (2013) Operando effects on the structure and dynamics of PtnSnm/γ-Al_2O_3 from ab initio molecular dynamics and X-ray absorption spectra. *J. Phys. Chem. C* 117, 12446.

[74] J. Ballhausen, (1962) *Introduction to Ligand Field Theory*, McGraw-Hill, New York.

[75] S. Sugano, & R.G. Shulman, (1963) Covalency Effects in $KNiF_3$. III. Theoretical Studies, *Phys. Rev.*, 130, 517.

[76] N.E. Brener, & J. Callaway, (1987) Density-functional study of the NiO_6^{10-} cluster, *Phys. Rev. B*, 35, 4001.

[77] M.W. Haverkort, M. Zwierzycki, & O.K. Andersen, (2012) Multiplet ligand-field theory using Wannier orbitals *Phys. Rev. B*, 85, 165113.

[78] F. de Groot, (2005) Multiplet effects in X-ray spectroscopy, *Coord. Chem. Rev.*, 249, 31–63.

[79] J. Schwitalla, & H. Ebert, (1998) Electron core-hole interaction in the X-Ray absorption spectroscopy of transition metals. *Phys. Rev. Lett.* 80, 4586.

[80] A.L. Ankudinov, A.I. Nesvizhskii, & J.J. Rehr, (2003) Dynamic screening effects in X-ray absorption spectra. *Phys. Rev. B* 67, 115120.

[81] F.M.F. Degroot, J.C. Fuggle, B.T. Thole, & G.A. Sawatzky, (1990) 2p X-ray absorption of 3d transition-metal compounds: An atomic multiplet description including the crystal field. Phys. Rev. B 42, 5459.

[82] F. de Groot, (2005) Multiplet effects in X-ray spectroscopy. *Coord. Chem. Rev.* 249, 31.

[83] P.H. Butler, (1981) *Point Group Symmetry: Applications, Methods and Tables*, Plenum, New York.

[84] G. van der Laan, & B.T. Thole, (1997) Anisotropic branching ratio in X-ray absorption spectra, *J. Electron Spectrosc. Relat. Phenom.*, 86, 57.

[85] R.D. Cowan, (1981) The Theory of Atomic Structure and Spectra, University of California Press, Berkeley.

[86] M.W. Haverkort, M. Zwierzycki, & O.K. Andersen, (2012) Multiplet ligand-field theory using Wannier orbitals *Phys. Rev. B* 85, 165113.

[87] F. de Groot, (2005) Multiplet effects in X-ray spectroscopy, *Coord. Chem. Rev.*, 249, 31–63.

[88] P.S. Miedema, H. Ikeno, and F.M.F. de Groot, (2011) First principles multiplet calculations of the calcium $L_{2,3}$ X-ray absorption spectra of CaO and CaF_2, *J. Phys.: Condens. Matter*, 23, 145501.

[89] H. Ikeno, T. Mizoguchi, & I. Tanaka, (2011) Ab initio charge transfer multiplet calculations on the $L_{2,3}$ XANES and ELNES of 3d transition metal oxides, *Phys. Rev. B* 83, 155107.

[90] J. Schwitalla, & H. Ebert, (1998) Electron core-hole interaction in the X-ray absorption spectroscopy of 3d transition metals, *Phys. Rev. Lett.*, 80, 4586.

[91] A.L. Ankudinov, Y. Takimoto, & J.J. Rehr, (2005) Combined Bethe-Saltpeter equations and time-dependent density-functional theory approach for X-ray absorption calculations, *Phys. Rev. B* 71, 165110.

[92] O. Bunau, & Y. Joly, (2012) Time-dependent density functional theory applied to X-ray absorption spectroscopy, *Phys. Rev. B*, 85, 155121.

[93] O. Sipr, J. Minar, A. Scherz, H. Wende, & H. Ebert, (2011) Many-body effects in X-ray absorption and magnetic circular dichroism spectra within the LSDA + DMFT framework, *Phys. Rev. B* 84, 115102.

[94] P. Kruger, (2010) Multichannel multiple scattering calculation of $L_{2,3}$-edge spectra of TiO_2 and $SrTiO_3$: importance of multiplet coupling and band structure, *Phys. Rev. B* 81, 125121.

[95] E.L. Shirley, (2006) Local screening of a core hole: a real-space approach applied to hafnium oxide, *Ultramicroscopy*, 106, 986.

[96] R. Laskowski, & P. Blaha, (2010) Understanding the $L_{2,3}$ X-ray absorption spectra of early 3d transition elements, *Phys. Rev. B* 82, 205104.

[97] J. Vinson, & J.J. Rehr, (2012) Ab initio Bethe-Salpeter calculations of the X-ray absorption spectra of transition metals at the L-shell edges, *Phys. Rev. B* 86, 195135.

[98] J. Vinson, J.J. Rehr, J.J. Kas, & E.L. Shirley, (2011) Bethe-Salpeter equation calculations of core excitation spectra, *Phys. Rev. B* 83, 115106.

[99] J.J. Rehr, & A.L. Ankudinov, (2004) Solid state effects on X-ray absorption, emission and scattering processes. *Radiat Phys Chem* 70, 453.

[100] K. Jorissen, & J.J. Rehr, (2010) Calculations of electron energy loss and X-ray absorption spectra in periodic systems without a supercell. *Phys. Rev. B* 81, 245124.

[101] T.T. Fister, F.D. Vila, G.T. Seidler, L. Svec, J.C. Linehan, & J.O. Cross, (2008) Local electronic structure of dicarba-closo-dodecarboranes $C_2B_{10}H_{12}$. *J. Am. Chem. Soc.* 130, 925.

[102] R.A. Gordon, G.T. Seidler, T.T. Fister, & K.P. Nagle, (2011) Studying low-energy core-valence transitions with bulk sensitivity using q-dependent NIXS. *J Electron Spectrosc Relat Phenom* 184, 220.

[103] Y. Joly, S.D. Matteo, & O. Bunau, (2012) Resonant X-ray diffraction: basic theoretical principles, in resonant elastic X-ray scattering in condensed matter, *Eur. Phys. J. Special Topics*, 208, 21.

[104] I.J. Pickering, M. Sansone, J. Marsch, & G.N. George, (1993) Diffraction anomalous fine-structure: a newtechnique for probing local atomic environment, *J. Am. Chem. Soc.*, 115, 6302.

[105] W.A. Hendrickson, (1991) Determination of macromolecular structures from anomalous diffraction of synchrotron radiation, *Science*, 254, 51.

[106] J.J. Kas, J.J. Rehr, J.A. Soininen, & P. Glatzel, (2011) Real-space Green's function approach to resonant inelastic X-ray scattering. *Phys. Rev. B* 83, 235114.

[107] B.A. Mattern, G.T. Seidler, J.J. Kas, J.I. Pacold, & J.J. Rehr, (2012) Real-space Green's function calculations of Compton profiles. *Phys. Rev. B* 85, 115135.

Chapter 4

Synchrotron X-ray Sources

4.1 Introduction

Synchrotron radiation is the electromagnetic wave emitted from charged particles traveling at relativistic speeds, toward their running-direction when their trajectory is curved by a magnetic field. The kind of light commonly come from either circular motion (bending magnets) or sinusoidal motion (periodic magnetic structures) of electron beam when they are subject to an acceleration perpendicular to their velocity ($\mathbf{a} \perp \mathbf{v}$).[1] Strong magnetic fields force electrons to propagate in closed paths. The emission is called cyclotron radiation if the electrons are non-relativistic ($v << c$). When the electron moves at a relativistic speed ($v \sim c$), due to the Doppler effect, the relativistic speed changes the observed frequency by γ times. The relativistic Lorentz contraction boasts the frequency by another factor γ. As a consequence, the gigahertz frequency of the resonant cavity which accelerates electrons is multiplied into the X-ray range. The frequency of the produced light can cover the entire electromagnetic spectrum, which is also called continuum radiation. Also due to the relativistic effect, the isotropic dipole radiation pattern of the non-relativistic electrons is dramatically squeezed into a radiation cone pointing extremely forward, which makes synchrotron radiation the most brilliant X-ray source. The relativistic Lamor formula governs the radiated power, and the Abraham-Lorentz-Dirac force acts on the emitting electrons. The planar acceleration geometry of electron storage ring makes the synchrotron radiation linearly polarized when viewed in the orbital plane and circularly polarized when observed at a small angle relative to the plane. However, the amplitude and frequency are concentrated on the extreme ecliptic. Thus, synchrotron light source is characterized by many useful features such as high brightness, high directionality, tuneable frequency, and variable polarization.

Synchrotron radiation was discovered for the first time on a synchrotron particle accelerator by General Electric Company of the United States in 1947. It occurs in accelerators as a nuisance, causing undesired energy loss in high energy particle physics contexts, but in the

1960s it was then recognized as excellent light source with exceptional properties that overcame the shortcomings of X-ray tubes. The first-generation synchrotron light experiments were carried out in a parasitical mode by using the storage ring of an electron synchrotron or positron collider as the light source. The light-emitting element is a bending magnet, and a lead-out pipe is connected to the outside of the vacuum outdoor near it, and the synchrotron light is emitted along the orbit. The second-generation synchrotron uses a dedicated electron storage ring as a light source, and employs bending magnets to generate synchrotron light. Wigglers, undulators and other components are inserted into the linear section of the storage ring to improve the brightness of the light, which can reach 10^{15} to 10^{16} photons / (sec · mm^2 · Milliradians^2) × 10^{-3} bandwidth. As accelerator synchrotron radiation became more intense and its applications more promising, special arrays of magnets called insertion devices, which cause the electrons to wiggle, are built into existing rings to create even more intense and tuneable beams of light. These led to the construction of the "third-generation" synchrotrons, such as European Synchrotron Radiation Facility (6 GeV) in the late 1980s, and shortly thereafter the Advanced Photon Source (7 GeV) in the United States and SPring-8 (8 GeV) in Japan in early 1990s. The third generation is based on the second generation, further improving the quality of the electron beam in the storage ring and arranging inserts with better modulation performance, so that the brightness of the synchronization light reaches 10^{17} to 10^{19} photons / (second · mm^2 · mrad^2) × 10^{-3} bandwidth. Fourth-generation sources are likely to be free electron lasers, which include different concepts for producing ultrabrilliant, pulsed time-structured X-rays and offer even more advanced capabilities for extremely demanding and also probably yet-to-be-conceived experiments. A radiation source with an emissivity that can reach the diffraction limit (0.01 nm radians), due to the key performance indicators that are much higher than the third-generation synchrotron radiation source currently in operation, the diffraction-limited light source is also known as the "fourth-generation synchrotron radiation device". Accelerator physics, electronics and computing, as well as magnet and vacuum technology continue to make impressive advances, with the goal of producing brighter and more stable X-ray beams with lower emittance.

Many synchrotron radiation facilities have been built across the world. Now a days, synchrotron radiation is widely useful for crystal structure analysis, and chemical composition analysis and atomic structure research. Synchrotron radiation has significantly impacted human's understanding of this world and instantly revolutionized the fields of physics, chemistry, materials, biology, geology, medicine and so on.

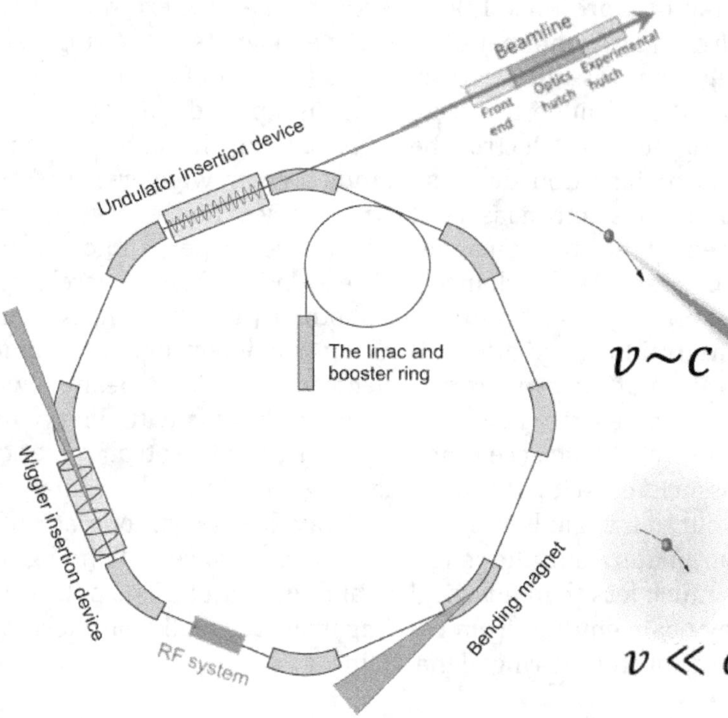

Figure 4.1 Essential elements of a third-generation synchrotron radiation facility.

4.2 Synchrotron Radiation Facility Structure

As a generator of electromagnetic radiation for scientific and technical purposes, synchrotron light source is usually produced by a storage ring. The electrons reach very high energy through a series of pre-acceleration stages before being injected into the storage ring. Here, taking the Advanced Photon Source (APS) at the Argonne National Laboratory as an example, the electron beam is generated by a hot cathode heated to about 1100°C. In a linear accelerator, the electrons get accelerated to the relativistic speed (99.999 +% of the speed of light) with an energy of 450 MeV. From the linear accelerator, the electrons are injected into the booster synchrotron where the electrons are further accelerated to 7 GeV around an oval racetrack of electromagnets just within one-half second. Upon reaching that energy, the electrons are injected and stored in an ultrahigh vacuum auxiliary magnetic storage ring, a 1,104-metre (3,622 ft) circumference ring of more than 1,000 electromagnets where the electrons may circle a large number of times. The magnets in the ring also repeatedly compress the electron beam against Coulomb (space charge) forces, as the Coulomb forces often disrupt the electron bunches. When high-energy electron beam is passing through the bending magnets or insertion devices (undulators or wigglers) in storage rings and free electron lasers, the trajectory of high energy electrons is altered by the strong magnetic field which is perpendicular to the electron beam. As the change in direction is a form of acceleration, the lost energy was emitted as photons (also known as magnetobremsstrahlung radiation). Bending magnets provide for angular acceleration of the synchrotron electron or positron beam,[2] which can let the accelerated electrons to emit photons with energy up to the GeV level. As the acceleration generates synchrotron radiation, a port associated with the bending magnet allows the radiation to travel out to a beam line. The total X-ray flux is characteristic of the ring parameters and the magnetic field. Brilliance of bending magnets is much less than for wigglers and undulators. The polarization of X-ray beam emitted from bending magnet usually parallels to the synchrotron storage ring plane, is linear one.

In order to accelerate electrons to extremely high energy and then make them change direction periodically, a synchrotron machine (Figure 4.1) composes many parts such as electron gun, linear accelerator, booster synchrotron, storage ring, beamlines and control system, and all these components have to work cooperatively as an integrated system day and night, with periodic shutdowns for maintenance and upgrade. The functions of each parts are as follows.

Electron gun

The electron beam used to provide the synchrotron light is first produced by an electron gun. A high voltage cathode is heated under vacuum, giving the electrons in the metal material sufficient thermal energy to "evaporate" from the surface and escape through a process called thermionic emission. These liberated electrons are accelerated to produce a stream of electrons with an energy of nearly 100 keV by a potential applied across the gun and make their way into the linear accelerator.

Linear accelerator

The linear accelerator (or linac) uses a series of RF cavities, operating at a GHz frequency, to accelerate the electron beam to an energy of 100 MeV subjecting them to a series of oscillating electric potentials along a linear beamline. Due to the nature of this acceleration, the beam must be separated into discrete packets, or 'bunches'. The bunching process is done at the start of the linac, using several 'bunching' cavities. The linac can accelerate a beam once every second. Further along the linac quadrupole magnets are used to help focus the electron beam.

Booster synchrotron

A linac long enough to accelerate electrons to the energy needed by a synchrotron radiation facility would not fit inside the building. Instead, a circular booster synchrotron is used in which the electrons receive a "boost" from an accelerating chamber (RF cavity) each time they go around. In less than one second inside the booster, the electrons make million revolutions, attain their target energy, and are ready to be transported to the storage ring. As the second particle accelerator, booster synchrotron usually consists of two straight sections joined together with semicircular curves, forming "athletics track" shaped trajectory. Dipole bending magnets are used to curve the electrons around the bends. A radio frequency voltage source is used to accelerate the electrons in the straight sections. In order to maintain the orbital path of the electrons, the magnetic fields produced by the bending and focusing magnets increase in synchronization with the rf field as the electron energy rises up to a final extraction energy before they are transferred into the storage ring.

Storage ring

As a type of circular particle accelerator, storage ring can keep a continuous or pulsed electron beam circulating inside typically for many hours. The storage ring is the final destination for the accelerated electrons. It consists of many straight sections angled together to form a closed loop in circumference. Each sector consists of a straight section and an arc, with the arcs containing two dipole 'bending' magnets each. Large electromagnets (called dipole magnets or bending magnets) are used to curve the electron beam between adjacent straight sections. Each dipole magnet is a potential source of synchrotron light and most straight sections can also host an insertion device, giving the possibility of 30+ beamlines at the third-generation synchrotrons.

Synchrotron light is emitted when the electrons travel round the storage ring. However, energy doesn't come from nowhere, and the electrons lose energy and momentum as they give out synchrotron light. The lighter electrons start to take a different path through the storage ring until they eventually spiral into the walls of the accelerator vacuum vessel and are lost. However, the RF cavities in some of the straight sections provide a boost to the electrons every time they complete a turn of the storage ring, replacing the energy and the momentum lost last time around so that they follow the correct path.

There are many types of magnets used in the storage ring of a synchrotron facility. The larger dipole magnet (bending magnet) is employed to bend the trajectory of electron beam and generate the synchrotron radiation. However, dipole magnets alone can only provide weak focusing, a storage ring composed of only bending magnets would result in a relatively large electron beam size. The storage ring also contains a large number of quadrupole and sextupole magnets for beam focusing and chromaticity correction. When the electrons follow this curved path, they generate synchrotron light. Dipole and quadrupole magnets deflect electrons with different energy by different amounts, which is called chromaticity by analogy with physical optics. Therefore, the energy spread inherent in any stored electron beam will cause the spread of transverse and longitudinal focusing, thus, causing various electron beam instabilities. The sextupole magnet and the quadrupole magnets are utilized for focusing and to maintain chromaticity, respectively. Sextupole and higher order multipole magnets are needed in chromaticity correction, but this leads to non-linear motion, which is one of the major problems faced by storage ring designers. A suitable strong focusing system which can give a smaller beam size, can be achieved by interleaving dipole magnets with properly arranged quadrupole and sextupole magnets.

In order to minimize electrons scattering off air molecules, the entire storage ring is maintained under high vacuum. Electrons complete the circuit in approximately two millionths of a second, equivalent to travelling around the earth 7.5 times in a single second. As the electrons circulate in the storage ring, they collide with each other and with the few gas molecules that remain in the vacuum, leading to their loss. To make up for this loss, new electrons are added to the ring every 10 minutes in a process known as "top-up". The advantage of top-up, compared to the previous method of injecting electrons only twice per day, is that the light beams delivered to the beamlines are more stable and remain at maximum intensity at all times. The ring is designed to hold a few hundred mA of stored current with a beam lifetime of over a few tens of hours. Depending on the application of the storage ring, new electrons can be injected into the storage ring in various ways. The simplest method is to use one or more pulse deflection dipole magnets (injection kicker magnets) to direct a beam of incident electrons to the stored electron beam path. Before the train of the newly stored electrons returns to the injection point, the kicker magnets are turned off, thereby fresh electrons are injected into a storage ring. This is single-turn injection. Multi-turn injection can accumulate many incoming trains of electrons to achieve a large stored current. If there is no significant beam damping, each pulse is injected onto a particular point in the transverse or longitudinal phase space of the stored beam, a careful arrangement of beam deflection and coherent oscillations in the stored beam is made to avoid ejecting previously injected trains. If significant beam damping exists, such as radiation damping of electrons, a pulse may be injected at the edge of phase space and then left to damp into the stored beam in transverse phase space before the next injection. Many pulses can be accumulated per second thanks to the short damping time of synchrotron radiation (typically tens of milliseconds).

Vacuum systems

As mentioned above, if the electrons were to travel through air as they orbited the storage ring, they would quickly collide with air molecules and be lost from the electron beam. This gradual loss of electrons in the storage ring is called beam lifetime and means that a fresh complement of electrons must be periodically injected into storage rings. Additionally, the electron collisions with air molecules also increase the size of the bunch and the energy spread. Therefore, the higher vacuum, the better beam dynamics. To minimize these losses, the electron beam is kept within a very high vacuum at all times during the acceleration process and within the storage ring. This vacuum is necessary as any beam collisions with gas molecules will quickly degrade the beam quality and reduce the lifetime of the beam. The vacuum is achieved by enclosing the beam in a stainless-steel pipe system, with numerous vacuum pump systems continually working to keep the vacuum quality high. Pressure within the storage ring is typically around 10^{-13} bar (10 nPa). That's only 10 times greater than the pressure on the moon.

Beamlines

Beamline captures the radiation tangent to the electron storage ring. These beamlines may start at bending magnets at the corners of the storage ring; or insertion devices in the straight sections of the storage ring. The two types of devices usually have different spectrum and energy of X-rays. Although the breadth of the research purpose requires an extensive probe kit, the beamline share some similarities in the basic components. All beamline is made of X-ray optical devices including slits, crystal monochromators, mirrors, and attenuators. These optical devices are installed to control the bandwidth, photon flux, beam sizes, focus, and collimation of the rays. For instance, in order to achieve a high photon flux in a small area, the mirrors with curves or toroidal shapes are utilized to focus the beam. The beamline is designed according to its applications. The beamline guides radiation to specimens in the experimental station, where the resulting diffraction, scattering or secondary radiation are measured by various detectors.

Control system

Control system is the brain of a synchrotron radiation facility. It controls the console, networking and data acquisition. Each digital and analogue I/O channel is associated with a database entry in a customized distributed database system (like EPICS, Experimental Physics and Industrial Control System). The condition of the system is monitored and controlled by connecting specialized GUIs to the specified database entries (also known as process variables, most of which relate to the physical I/O). Some high-level control of the physics-related parameters of beam is provided through certain software (like MATLAB) which also provides data analysis tools and an interface with a computerized model of the accelerator. Personnel and equipment protection are achieved through the use of PLC-based systems, which also transfer data to EPICS. The Beamlines also use EPICS as the basis for their control.

4.3 Bending Magnet and Insertion Devices

In a real synchrotron facility, there are many different undulators and wigglers. There are, for example, sophisticated magnet arrays that force the electrons to move along helices, thus producing elliptically polarized radiation. In addition to undulators and wigglers, a storage ring includes the dipole bending magnets that deflect the electrons to keep them in a closed-loop trajectory. The emission of bending-magnet synchrotron radiation is a by-product of the deflection function.

In a simple picture, the path of an electron interacting with a bending magnet can be considered a portion of a "cyclotron" circular trajectory. This is the plane trajectory of an electron moving at speed V perpendicular to a magnetic field B. The Lorentz force of magnitude eVB determines the motion and in particular the angular speed ω_c, the so-called "cyclotron frequency". The central wavelength emitted by the electron is, *in the electron reference frame*, $2\pi c/\omega_c$, where ω_c equals to eB/m_o and $\gamma eB/m_o$, respectively, in classic and relativity cases. In fact, the magnetic field of the laboratory frame corresponds in the electron frame to an electric field of strength γVB: the force magnitude is $e\gamma VB$ rather than eVB. The central emitted wavelength in the electron reference frame is $2\pi cm_o/(\gamma eB)\omega_c$. Additionally, the Doppler shift, i.e., the factor $1/(2\gamma)$ must be taken into account when detecting the emission in the laboratory frame. Therefore, the central emitted wavelength thus becomes $\lambda_c = \frac{2\pi cm_o}{2\gamma^2 eB}$. The $2\gamma^2$ factor in the denominator that pushes the wavelengths to short X-ray values. This factor arises from a combination of two relativistic phenomena: the Doppler shift and the effect of relativity on the emission in the electron frame. Furthermore, stronger bending magnets also yield shorter wavelengths. Traditionally, the spectral peak of bending magnet emission is characterized using not the central wavelength but the so-called "critical wavelength". Its definition is that equal amounts of power are emitted at shorter and longer wavelengths. This central wavelength is the peak of a broadband emission with bandwidth $\Delta\lambda$. The bending-magnet emission is a bell-shaped curve peaked at a short wavelength λ_c and with a bandwidth $\approx\lambda_c$. Likewise, the spectrum as a function of the photon energy is a bell-shaped curve. The critical wavelength is approximately $\frac{4}{3}\lambda_c$.

To quantitatively understand the characteristics of bending magnet and insertion device emission, we adopt the Liénard–Wiechert potentials to describe the classical electromagnetic effect of a moving electron in terms of a vector potential and a scalar potential in the Lorenz gauge. The Liénard–Wiechert potentials are built directly from Maxwell's equations, which can describe the relativistically corrected and time-varying electromagnetic field for a point charge in arbitrary motion, however, are not corrected for quantum-mechanical effects. Electromagnetic radiation in the form of waves can be obtained from these potentials. For a source point charge q at position $\mathbf{r}_s$ traveling with velocity $\mathbf{v}_s$, the Liénard–Wiechert potentials φ (scalar potential field) and $\mathbf{A}$ (vector potential field) can be written as:

$$\varphi(\mathbf{r},t) = \frac{1}{4\pi\epsilon_0}\left(\frac{q}{(1-\mathbf{n}_s\cdot\boldsymbol{\beta}_s)|\mathbf{r}-\mathbf{r}_s|}\right)_{t_r} \text{ and}$$

$$\boldsymbol{A}(\mathbf{r},t) = \frac{\mu_0 c}{4\pi}\left(\frac{q\boldsymbol{\beta}_s}{(1-\mathbf{n}_s\cdot\boldsymbol{\beta}_s)|\mathbf{r}-\mathbf{r}_s|}\right)_{t_r} = \frac{\boldsymbol{\beta}_s(t_r)}{c}\varphi(\mathbf{r},t), \quad (4.1)$$

where $\boldsymbol{\beta}_s(t) = \frac{\mathbf{v}_s(t)}{c}$ is the velocity of the source expressed as a fraction of the speed of light, $|\mathbf{r}-\mathbf{r}_s|$ is the distance from the source and $\mathbf{n}_s = \frac{\mathbf{r}-\mathbf{r}_s}{|\mathbf{r}-\mathbf{r}_s|}$ is the unit vector pointing in the direction from the source.

Thus, the electric and magnetic fields directly from the potentials can be calculated using the definitions: $\boldsymbol{E} = -\nabla\varphi - \frac{\partial\mathbf{A}}{\partial t}$ and $\boldsymbol{B} = \nabla\times\mathbf{A}$.

After nontrivial and multi-step calculation, the Liénard–Wiechert expressions for the electric and magnetic fields can be obtained in non-covariant form:[3]

$$\mathbf{B}(\mathbf{r},t) = -\frac{\mu_0 q}{4\pi}\left[\frac{c\hat{n}\times\vec{\beta}}{\gamma^2R^2(1-\vec{\beta}\cdot\hat{n})^3} + \frac{\hat{n}\times\left[\dot{\vec{\beta}}+\hat{n}\times\left(\vec{\beta}\times\dot{\vec{\beta}}\right)\right]}{R(1-\vec{\beta}\cdot\hat{n})^3}\right]_{\text{retarded}}, \quad (4.2a)$$

$$\mathbf{E}(\mathbf{r},t) = \frac{q}{4\pi\varepsilon_0}\left[\frac{\hat{n}-}{\gamma^2R^2(1-\cdot\hat{n})^3} + \frac{\hat{n}\times[(\hat{n}-)\times\ ]}{cR(1-\cdot\hat{n})^3}\right]_{\text{retarded}}. \quad (4.2b)$$

where $R(t') = |\mathbf{R}(t')|=|\mathbf{r} - \mathbf{r}_0(t')|$, the unit vector between the observer and the electron position at the retarded time, $\hat{n}(t') = \mathbf{R}(t')/R(t')$, and t′ is the retarded time. The first terms for **B** and **E** (called the generalized Coulomb field or velocity field) represent the electron static field effect, which is a function of the component of its motion that has zero or **constant velocity**, as seen by a distant observer at **r**. By contrast, the second terms (called the *acceleration field* or *radiation field*) represent components of field due to the charge's **acceleration** (changing velocity), and these second terms fall off as the inverse *first* power of the distance from the source, representing E and B which are emitted as electromagnetic radiation from the particle to an observer at **r**.

By ignoring the *velocity* field in order to find the power of emitted EM radiation only, the radial component of Poynting's vector resulting from the Liénard–Wiechert fields can be calculated to be

$$[S \cdot \hat{n}] = \frac{q^2}{16\pi^2 \varepsilon_0 c}\left\{\frac{1}{R^2}\left|\frac{\hat{n}\times[(\hat{n}-\vec{\beta})\times\dot{\vec{\beta}}]}{(1-\vec{\beta}\cdot\hat{n})^3}\right|^2\right\}_{\text{retarded}} \tag{4.3}$$

Note that the spatial relationship between $\vec{\beta}$ and $\dot{\vec{\beta}}$ determines the detailed angular power distribution; the relativistic effect of transforming from the rest frame of the particle to the observer's frame manifests itself by the presence of the factors $(1 - \vec{\beta} \cdot \hat{n})$ in the denominator of Eq. 4.3; the latter effect dominates the whole angular distribution for relativistic electrons. The energy radiated into per solid angle during a finite period of acceleration from $t' = T_1$ to $t' = T_2$ can be expressed as

$$\frac{dP}{d\Omega} = R(t')^2[\mathbf{S}(t')\cdot\hat{n}(t')]\frac{\mathrm{d}t}{\mathrm{d}t'} = R(t')^2\mathbf{S}(t')\cdot\hat{n}(t')[1-\vec{\beta}(t')\cdot\hat{n}(t')]$$

$$= \frac{q^2}{16\pi^2\varepsilon_0 c}\frac{|\hat{n}(t')\times\{[\hat{n}(t')-\vec{\beta}(t')]\times\dot{\vec{\beta}}(t')\}|^2}{[1-\vec{\beta}(t')\cdot\vec{\mathbf{n}}(t')]^5} \tag{4.4}$$

After integration of Eq. 4.4 over the all solid angles, the relativistic generalization of Larmor's formula becomes

$$P = \frac{q^2}{6\pi\varepsilon_0 c}\gamma^6[|\dot{\vec{\beta}}|^2 - |\vec{\beta}\times\dot{\vec{\beta}}|^2]$$

However, this also can be derived by relativistic transformation of the 4-acceleration in Larmor's formula.

When the charge is in instantaneous circular motion, its acceleration $\dot{\vec{\beta}}$ is perpendicular to its velocity $\vec{\beta}$. Choosing a coordinate system such that instantaneously $\vec{\beta}$ is in the z direction and $\dot{\vec{\beta}}$ is in the x direction, with the polar and azimuth angles θ and φ defining the direction of observation, the general formula Eq. 4.4 reduces to

$$\frac{dP}{d\Omega} = \frac{q^2}{16\pi^2\epsilon_0 c}\frac{|\dot{\beta}|^2}{(1-\beta\cos\,\theta)^3}\left[1 - \frac{\sin^2\theta\cos^2\phi}{\gamma^2(1-\beta\cos\,\theta)^2}\right] \tag{4.5}$$

With the relativistic limit, the angular distribution can be approximated as

$$\frac{dP}{d\Omega} \simeq \frac{2}{\pi}\frac{q^2}{c^3}\gamma^6\frac{|\dot{v}|^2}{(1+\gamma^2\theta^2)^3}\left[1 - \frac{4\gamma^2\theta^2\cos^2\phi}{(1+\gamma^2\theta^2)^2}\right] \tag{4.6}$$

The factors $(1 - \beta\cos\theta)$ in the denominators tip the angular distribution forward into a narrow cone, just like the beam of the headlight is directed in front of the electron. The angular distribution ($dP/d\Omega$ vs. $\gamma\theta$) shows a sharp peak around $\theta = 0$. Integration over the whole solid angle yields the total power radiated by one electron:

$$P = \frac{q^2}{6\pi\epsilon_0 c}\left|\dot{\vec{\beta}}\right|^2\gamma^4 = \frac{q^2 c}{6\pi\epsilon_0}\frac{\beta^4\gamma^4}{\rho^2} = \frac{q^4}{6\pi\epsilon_0 m^4 c^5}B^2E^2\beta^2 = \frac{q^4}{6\pi\epsilon_0 m^4 c^5}B^2(E^2 - m^2c^4)$$

where E is the electron energy, B is the magnetic field, and ρ is the radius of curvature of the track in the field. So, the radiated power is inversely proportional to m^4, ρ^2, and $1/B^2$. Using $B = \frac{E\beta}{qr\sin(\alpha)}$ where α is the angle between the velocity and magnetic field and r is the radius of the circular acceleration, the power emitted is:

$$P = \frac{q^2}{6\pi\epsilon_0 m^4 c^5 r^2\sin^2(\alpha)}E^4\beta^4 = \frac{q^2}{6\pi\epsilon_0 m^4 c^5 r^2\sin^2(\alpha)}(E^2 - m^2c^4)^2 \tag{4.7}$$

Thus, the power emitted scales to 4th power of energy and decreases with the square of radius and 4th power of particle mass.

Radiation integral

The energy received by an observer (per unit solid angle at the source) is

$$\frac{d^2W}{d\Omega} = \int_{-\infty}^{\infty} \frac{d^2P}{d\Omega} dt = c\varepsilon_0 \int_{-\infty}^{\infty} |R\vec{E}(t)|^2 \, dt \tag{4.8}$$

Using the Fourier transformation, we move to the frequency space

$$\frac{d^2W}{d\Omega} = 2c\varepsilon_0 \int_0^{\infty} |R\vec{E}(\omega)|^2 \, d\omega$$

Considering only the radiation field, the angular and frequency distribution of the energy received by an observer can be expressed as

$$\frac{d^3W}{d\Omega d\omega} = 2c\varepsilon_0 R^2 |\vec{E}(\omega)|^2 = \frac{q^2}{4\pi\varepsilon_0 4\pi^2 c} \left| \int_{-\infty}^{\infty} \frac{\hat{n}\times[(\hat{n}-\vec{\beta})\times\dot{\vec{\beta}}]}{(1-\hat{n}\cdot\vec{\beta})^2} e^{i\omega(t-\hat{n}\cdot\vec{r}(t)/c)} dt \right|^2 \tag{4.9}$$

Therefore, given the particle's motion, cross products term, and phase factor, the radiation integral could be calculated through require Airy function or the modified Bessel functions.

4.3.1 Bending magnet

As dipole magnets, bending magnets curve the electron beam around the storage ring, and also produce synchrotron light with a broad spectrum including infrared, visible, UV and X-ray. They are very important for spectroscopy experiments on synchrotrons. For the bending magnet, the trajectory of the arc of circumference can be expressed as

$$\vec{r}(t) = \left(\rho \sin\frac{\beta c}{\rho} t, \rho\left(1 - \cos\frac{\beta c}{\rho} t\right), 0\right) \tag{4.10}$$

In the limit of small angles, $\hat{n} \times (\hat{n} \times \vec{\beta}) = \beta[-\vec{\varepsilon}_{\parallel} \sin\left(\frac{\beta ct}{\rho}\right) + \vec{\varepsilon}_{\perp} \cos\left(\frac{\beta ct}{\rho}\right) \sin\theta]$ and $\omega(t - \frac{\hat{n}\cdot\vec{r}(t)}{c}) = \omega[t - \frac{\rho}{c}\sin\left(\frac{\beta ct}{\rho}\right)\cos\theta]$.

Substituting into the radiation integral and introducing $= \frac{\rho\omega}{3c\gamma^3}(1 + \gamma^2\theta^2)^{3/2}$

$$\frac{d^3W}{d\Omega d\omega} = \frac{3q^2}{16\pi^3\varepsilon_0 c}\left(\frac{2\omega\rho}{3c\gamma^2}\right)^2 (1 + \gamma^2\theta^2)^2 [K_{2/3}^2(\xi) + \frac{\gamma^2\theta^2}{1+\gamma^2\theta^2} K_{1/3}^2(\xi)], \tag{4.11}$$

where the function K_n is a modified Bessel function of the second kind.

Frequency distribution of radiated energy

From Eq. 4.11, the radiation intensity is negligible for $\omega_c = \frac{3}{2}\frac{c}{\rho}\gamma^3$. *Critical frequency* is the frequency at $\xi = 0.5$ and $\theta = 0$. So $\omega_c = \frac{3}{2}\frac{c}{\rho}\gamma^3$ and *critical angle* is the one for which $\xi(\theta_c) \simeq \xi(0) + 1$ and is approximately $\theta_c \simeq \frac{1}{\gamma}\left(\frac{2\omega_c}{\omega}\right)^{1/3}$.[4] The synchrotron radiation emission is negligible when frequencies much larger than the critical frequency and angles much larger than the critical angle. After integration on all angles, the frequency distribution of the energy radiated can be written as

$$\frac{dW}{d\omega} = \oint \frac{d^3W}{d\omega d\Omega} d\Omega = \frac{\sqrt{3}e^2}{4\pi\varepsilon_0 c}\gamma\frac{\omega}{\omega_c}\int_{\omega/\omega_c}^{\infty} K_{5/3}(x)\, dx \tag{4.12}$$

By defining $S(y) \equiv \frac{9\sqrt{3}}{8\pi} y \int_y^{\infty} K_{\frac{5}{3}}(x)\, dx \int_0^{\infty} S(y)\, dy = 1$, where $y = \omega/\omega_c$, it can be written as

$$\frac{dW}{d\omega} = \frac{2q^2\gamma}{9\varepsilon_0 c} S(y)$$

Note that $\frac{dW}{d\omega} \sim \frac{q^2}{4\pi\varepsilon_0 c}\left(\frac{\omega\rho}{c}\right)^{\frac{1}{3}}$, if $\omega \ll \omega_c$; and $\frac{dW}{d\omega} \approx \sqrt{\frac{3\pi}{2}\frac{q^2}{4\pi\varepsilon_0 c}}\gamma\left(\frac{\omega}{\omega_c}\right)^{\frac{1}{2}} e^{-\frac{\omega}{\omega_c}}$, if $\omega \gg \omega_c$.

The above formula for spectral distribution of synchrotron radiation can be expressed in terms of a rapidly converging integral with no special functions involved [5] by means of the relation:

$$\int_{\xi}^{\infty} K_{5/3}(x)dx = \frac{1}{\sqrt{3}}\int_0^{\infty} \frac{9+36x^2+16x^4}{(3+4x^2)\sqrt{1+x^2/3}} \exp\left[-\xi\left(1+\frac{4x^2}{3}\right)\sqrt{1+\frac{x^2}{3}}\right] dx \tag{4.13}$$

Synchrotron radiation emission as a function of the beam energy

The critical photon energy is defined as $\varepsilon_c = \hbar\omega_c = \frac{3}{2}\frac{\hbar c}{\rho}\gamma^3$. The relationship between radiated power and photon energy shows that the higher the critical energy, the more photons with high energies are generated, but no dependence on the energy at longer wavelength.

Polarization of synchrotron radiation

In Eq. 4.11, the first term $K_{2/3}^2(\xi)$ is the radiation power with polarization in the orbit plane, and the second term $\frac{\gamma^2\theta^2}{1+\gamma^2\theta^2}K_{1/3}^2(\xi)$ is the polarization orthogonal to the orbit plane. The polarization is purely horizontal in the orbit plane $\theta = 0$. Integration on all frequencies leads to the angular distribution of the radiated energy

$$\frac{d^2W}{d\omega} = \int_0^\infty \frac{d^3W}{d\omega d\Omega} d\omega = \frac{7q^2\gamma^5}{64\pi\varepsilon_0\rho}\frac{1}{(1+\gamma^2\theta^2)^{5/2}}[1 + \frac{5}{7}\frac{\gamma^2\theta^2}{1+\gamma^2\theta^2}] \quad (4.14)$$

After integrating on all the angles, it can be found that parallel polarized radiation has seven times the energy of vertically polarized radiation. The radiation generated by the relativistic moving charge is very strong in the plane of motion, but it is not fully polarized.

Current (third-generation) synchrotron radiation sources often rely on insertion devices to produce more intense, tuneable light. As suggested by its name, insertion device is a set of magnets installed into the straight sections of the storage ring. The periodic magnetic structures force the electrons into a sinusoidal or helical path. Therefore, instead of a single bend, tens or hundreds of "swings" at precisely calculated positions add or multiply the total intensity of the beam. There are two main types of insertion devices: wigglers and undulators. The main difference between an undulator and a wiggler is their magnetic field strength and the magnitude of deviation from the straight path of the electron. A wiggler has a broader spectrum of radiation than an undulator. In a wiggler or an undulator, electrons travel through a periodic magnetic array. Considering the case where the magnetic field B varies sinusoidally and is in the vertical direction: $B(x) = B_0 \sin(2\pi x/\lambda_u)$, where x is the distance along the wiggler axis, B_0 the peak magnetic strength and the λ_u magnetic period. The electronic motion is also sinusoidal and lies in the horizontal plane. The important parameter that characterizes the motion of electrons is the deflection parameter K, given by $K = \frac{eB_0\lambda_u}{2\pi mc}$. In terms of K, the maximum angular deflection of the orbit is $\delta = K/\gamma$. For $K \leq 1$, since the angular deviation of the electrons is within the nominal radiation cone, the radiation from each period will show strong interference; in this case, the structure is called an undulator. In the case $K \gg 1$, interference effect is not important, and the structure is called a wiggler.

4.3.2 Wiggler

A wiggler is a series of magnets designed to periodically laterally wiggle electron beam in the straight section of the storage ring. Passing through such alternating magnetic field structures, electrons oscillate perpendicularly to the direction of their motion and therefore emit synchrotron radiation during each individual wiggle. In fact, a wiggler is essentially equivalent to a series of bending magnets. The spectral distribution of a wiggler is thus similar to that of the bending magnet regime. These deflections cause changes in acceleration, which in turn produce wide synchrotron radiation tangent to the curve, very similar to that of a bending magnet, but due to the contribution of many magnetic dipoles in the wiggler, the intensity is higher. In addition, as the wavelength (λ) decreases, this means that the radiation frequency (*f*) increases. Therefore, the wiggler can produce higher energy photons, so it is also called "wave shifter".

Typically, the magnets in a wiggler are arranged in a Halbach array, which bend the electron beam back and forth, returning the original beam to the same direction at the end of the device. They are thus inserted into straight sections of a storage ring. A *wiggler* is realized through large magnetic fields and long period lengths ($K = 0.934\lambda_0 B_0$ [cm·T], $K > 1$). In this case, the emitted radiation interferes incoherently and therefore the spectrum is similar to that of a bending magnet. However, the central brightness of the wiggler within the opening angle 2δ is increased by about two times the number of the wiggler periods N. Intensity and brilliance are related to the number of wiggles, so the wiggler is essentially the sum of bending magnets. Its intensity is one to two orders of magnitude higher than that of bending magnet.

4.3.3 Undulators

Undulators are wiggler-like devices whose magnet spacing and gap are adjusted to cause interference in synchrotron radiation.[6] In an "undulator", a series of dipole magnets of period L are placed in a straight section of the electron accelerator. The magnetic field applies to the electrons a periodic Lorentz force that causes them to oscillate. Since the Lorentz force does not do any work (being perpendicular to the velocity), it cannot change the magnitude of the velocity. Thus, the transverse oscillations correspond to a slower longitudinal motion. An undulator employs smaller magnetic fields (K<1). This results in a coherent superposition of the emitted radiation. Due to this interference effect the undulator spectrum consists of sharp peaks at characteristic wavelengths (fundamental and higher harmonics), and the central brightness of the undulator radiation is up to N times larger when compared with that of a bending magnet. Therefore, the emitted radiation is not as continuous as a wiggler or bending magnet, but peaks within a few specific energy ranges. So the undulator produces more intense X-rays than the wiggler. Its intensities can be three orders of magnitude higher than that of a bending magnet. The emission of radiation caused by the oscillations tends to decrease the electron energy. In a closed-loop storage ring, this energy loss is compensated for by radiofrequency cavities that apply an accelerating electric field synchronized with the passage of electrons.

The properties of the undulator emission can be understood from the point of view of a relativistic electron. The Lorentz transformation from the laboratory frame to the electron frame changes the static periodic magnetic field of maximum strength B into a combination of periodic magnetic and electric fields, both transverse, perpendicular to each other and traveling at speed -V. The E'/B' ratio is $V \approx c$ since maxima of the electric field strength and magnetic field are $E' = \gamma B$, $B' = \gamma B$, respectively.

All these characteristics are reminiscent of an electromagnetic wave of wavelength equal to the field period, which in the laboratory frame is L corrected for the Lorentz contraction, giving L/γ. The electron backscatters this "wave" and this is the mechanism that produces synchrotron radiation.

The emitted radiation has the angular collimation, polarization, brightness, and coherence discussed above. In the laboratory reference frame the emission comes from the moving electron, the Doppler effect shifts the wavelength by a factor of $1/(2\gamma)$; thus, the electron-frame wavelength L/γ is Doppler-shifted to $\lambda_u \approx \frac{L}{2\gamma^2}$, a very short wavelength due to the large value of γ. Relativity takes the macroscopic period L of a technically feasible undulator (from mm to cm) and transforms it into ultraviolet and X-ray wavelengths.

Most applications of synchrotron radiation require a tunable wavelength. Tuning λ by changing L is not feasible. Change γ is also problematic, since the corresponding change in the electron energy would affect all the other beamlines in the synchrotron facility. Fortunately, the undulator-induced transverse oscillations slow down the longitudinal motion – and therefore the longitudinal value of γ that determines the Lorentz contraction and the Doppler shift. The longitudinal γ-value and the corresponding emitted wavelength can thus be tuned by changing the amplitude of the transverse oscillations. In turn, this amplitude can be modified by changing the strength of the undulator magnetic field. This introduces a dependence on magnetic field: $\lambda_u \approx \frac{L}{2\gamma^2}\left(1+\frac{K^2}{2}\right)$, where K is the so-called "undulator parameter", proportional to B: $K = \frac{eBL}{2\pi m_0 c}$, ($m_o$ = electron rest mass). Practically speaking, the tuning of the wavelength is achieved by changing *magnetic field* through using electromagnets or changing the gap size of the permanent magnets.

The above discussion implicitly adopted several approximations that merit a more detailed analysis. The first approximation was considering only the wavelengths related to L. In reality, an undulator also emits harmonics of the fundamental value; if K is not much greater than 1, this is limited to odd harmonics, so that $\lambda_u \approx \frac{L}{2n\gamma^2}, n = 1,3,5,\cdots$.

The second approximation was to consider only the emission in the forward (longitudinal) direction, for which the Doppler shift factor is $1/(2\gamma)$. This is not unreasonable since the emission is confined to a narrow angular range. Actually, for an undulator, this range is even narrower than $1/\gamma$.

Consider an electron moving through an undulator with N periods. Its emitted wave is a pulse composed of N oscillations. For an infinitely long undulator with an infinite N, the pulse would correspond to only one wavelength. But a finite-length undulator corresponds to a pulse with a finite number of oscillations, whose bandwidth $\Delta\lambda$ is not zero but has $\frac{\Delta\lambda}{\lambda} \approx \frac{1}{N}$, according to the Fourier transform theory. This result impacts the angular collimation of the undulator emission. For the emission at an angle θ (in the laboratory frame) from the longitudinal direction, the Doppler factor is not $1/(2\gamma)$ but $\gamma(1 - \beta\cos\theta')$. For small values of θ and $\beta \approx 1$, $(1 - \beta\cos\theta') \approx \frac{1}{2\gamma}(1 + \gamma^2\theta'^2)$; so the wavelength emitted at θ' is shifted by $\Delta\beta = \lambda_u(\gamma^2\theta'^2)$ with respect to the longitudinal value λ_u. Since $\lambda_u\gamma^2\theta'^2 < \frac{\lambda_u}{N}$, and $\theta' < \frac{1}{\sqrt{N}\gamma}$, the undulator angular spread is reduced by $\sqrt{N}$ with respect to the "torchlight" collimation $\approx 1/\gamma$. This makes it difficult to obtain different wavelengths from an undulator by collecting the off-axis emission, because the intensity drops very rapidly as the angle increases.

Solution of equation of motion and undulator equation

A periodic array of magnets in an undulator provides a sinusoidal magnetic field, $\vec{B} = (0, B_0\sin(k_u z), 0)$, which cause the electrons to oscillate and thus radiate. These magnetic fields also slow the electrons' axial (z) velocity, thereby reducing both the Lorentz contraction and the Doppler shift, so the observed radiation wavelength will not be that short. The force equation for an electron is $\frac{dP}{dt} = \frac{d(\gamma mV)}{dt} = e(\boldsymbol{E} + \boldsymbol{V} \times \boldsymbol{B})$. The solution of this motion equation is $\vec{r}(t) = \frac{\lambda_u K}{2\pi\gamma}\sin\omega_u t \cdot \hat{x} + \left(\bar{\beta}_z ct + \frac{\lambda_u K^2}{16\pi\gamma^2}\cos(2\omega_u t)\right) \cdot \hat{z}$, where $K = \frac{qB_0\lambda_u}{2\pi mc}$, and $\bar{\beta}_z = 1 - \frac{1}{2\gamma^2}\left(1 + \frac{K^2}{2}\right)$, and the parameter $d = \frac{\lambda_u}{\bar{\beta}} - \lambda_u\cos\theta = n\lambda$ is called the *undulator parameter*. Condition for the constructive interference of radiation emitted at different poles is Expanding $\cos\theta \approx 1 - \frac{\theta^2}{2}$ and neglecting the terms in the resulting equation, one obtains

$$\lambda_n = \frac{\lambda_u}{2\gamma^2 n}\left(\frac{1 + \frac{K^2}{2} + \gamma^2\theta^2}{\bar{\beta}}\right) \tag{4.15}$$

For $\bar{\beta} \to 1$, one finally gets $\lambda_n = \frac{\lambda_u}{2\gamma^2 n}\left(1 + \frac{K^2}{2} + \gamma^2\theta^2\right)$. This equation is called the *undulator equation*.

Radiation from the undulator

Radiation integral is $\frac{d^3W}{d\Omega d\omega}=\frac{q^2}{4\pi\varepsilon_0 4\pi^2 c}\left|\int_{-\infty}^{\infty}\frac{\hat{n}\times[(\hat{n}-\vec{\beta})\times\vec{\beta}]}{(1-\hat{n}\cdot\vec{\beta})^2}e^{i\omega(t-\hat{n}\cdot\vec{r}(t)/c)}dt\right|^2$. By using the periodicity of the trajectory, the radiation integral can be split into a sum over N terms, where $2N$ is the total number of bending magnets of the undulator.

$$\frac{d^3W}{d\Omega d\omega}=\frac{q^2\omega^2}{4\pi\varepsilon_0 4\pi^2 c}\left|\int_{-\lambda_u/2\overline{\beta}c}^{\lambda_u/2\overline{\beta}c}\hat{n}\times(\hat{n}\times\vec{\beta})e^{i\omega(t-\hat{n}\cdot\vec{r}(t)/c)}dt\right|^2\left|1+e^{i\delta}+e^{2i\delta}+\cdots+e^{i(N_u-1)\delta}\right|^2$$

where , $\overline{\beta}=\beta\left(1-\frac{K^2}{4\gamma^2}\right)$ and $\delta=\frac{2\pi\omega}{\omega_{\text{res}}(\theta)}$.

With $\omega_{res}(\theta)=\frac{2\pi c}{\lambda_{\text{res}}(\theta)}$ and $\lambda_{\text{res}}(\theta)=\frac{\lambda_u}{2\gamma^2}\left(1+\frac{K^2}{2}+\gamma^2\theta^2\right)$, radiation integral in an undulator can be written as $\frac{d^3W}{d\Omega d\omega}=\frac{q^2\gamma^2N^2}{4\pi\varepsilon_0 c}L\left(N\frac{\Delta\omega_n}{\omega_{\text{res}}(\theta)}\right)F_n(K,\theta,\phi)$ where $\Delta\omega_n=\omega-n\omega_{\text{res}}$ is the frequency difference to the n-th harmonic. The sum of δ generates a series of sharp peaks in the frequency spectrum harmonics of fundamental wavelength $L(N\frac{\Delta\omega_n}{\omega_{\text{res}}(\theta)})=\frac{\sin^2(N\pi\Delta\omega_n/\omega_{\text{res}}(\theta))}{N^2(\pi\Delta\omega_n/\omega_{\text{res}}(\theta))^2}$, and F_n depends on the angles of observations and K,

$$F_n(K,\theta,\phi)\propto\left|\int_{-\lambda_u/2\overline{\beta}c}^{\lambda_u/2\overline{\beta}c}\hat{n}\times(\hat{n}\times\vec{\beta})e^{i\omega(t-\hat{n}\cdot\vec{r}(t)/c)}dt\right|^2$$

On the axis (θ = 0, φ = 0), the radiation integral becomes

$$\frac{d^3W}{d\Omega d\omega}=\frac{e^2\gamma^2N^2}{4\pi\varepsilon_0 c}L\left(N\frac{\Delta\omega_n}{\omega_{\text{res}}(0)}\right)F_n(K,0{,}0)$$

and

$$F_n(K,0{,}0)=\frac{n^2K^2}{1+K^2/2}[J_{\frac{n+1}{2}}(Z)-J_{\frac{n-1}{2}}(Z)]^2$$

where $=\frac{nK^2}{4(1+K^2/2)}$. Note that only odd harmonics are radiated on-axis, and as K increases, higher harmonic becomes stronger.

4.4 The 4th Generation Light Sources: FEL and DLSR

As a fourth-generation synchrotron radiation, free-electron laser (FEL) makes use of the physics of the undulator. The modern FEL concept was proposed by John Madey in 1970. Based on his calculation he claimed in his PhD thesis that an undulator can amplify resonant radiation, and "finite gain is available from the far-infrared through the visible region raising the possibility of continuously tunable amplifiers and oscillators at these frequencies with the further possibility of partially coherent radiation sources in the ultraviolet and X-ray regions to beyond 10 keV". Free electron lasers are tunable and have the widest frequency range of any laser type. Currently the wavelength ranges from microwave to terahertz radiation and infrared, to visible spectrum, ultraviolet and X-ray.[7] Since 1990s, the progress of FELs boosted the brightness to very high values with a pulsed time structure, take FLASH (Free Electron Laser in Hamburg) as example, its peak brilliance has reached up to 10^{28} - 10^{31} photons/sec/mm^2/mrad^2/0.1%, and the peak values of brightness and flux still have a wide margin for improvement.

FELs are relativistic machines. To generate FEL, a electron beam is accelerated to near the speed of light. Free electron lasers need to use electron accelerators with related shielding functions, because if the electrons are not properly absorbed, the accelerating electrons may cause radiation hazards. These accelerators are usually powered by klystrons. The light beam passes through a periodic magnet array with alternating polarities through the beam path, thereby generating a side to side magnetic field. Free electron lasers use long wave undulators (K <1), where the Lorentz force of the field forces the electrons in the beam to swing transversely and travel along a sinusoidal path around the axis of the undulator. The direction of the beam is called the longitudinal direction, and the direction through the beam path is called the transverse direction.

When the electron bunch enters the undulator, the transverse acceleration of the electrons across this path causes the release of photons (synchrotron radiation), which are monochromatic but still incoherent, because the electromagnetic waves from randomly distributed electrons interfere constructively and destructively in time. The resulting radiation power is linearly proportional to the number of electrons.

Initially, all electrons are evenly distributed, and they only emit incoherent spontaneous radiation. Such radiation waves travel together with the electron bunch, and their transverse electromagnetic field interacts with each electron. The wave magnetic field B_W together with the transverse oscillation velocity v_T can cause a longitudinal Lorentz force that pushes the electrons towards a "node" of the wave. This tends to accumulate the electrons in one-half of the nodes. Through the interaction of radiation and electronic oscillations, they progressively drift into and form thin sheets ("microbunches"), and the distance between them is equal to one wavelength of the radiation, like an "electron crystal". Through this interaction, all electrons begin to emit coherent radiation. All emitted radiation is perfectly self-enhancing, so the crests and troughs always superimposed best with each other. The increased radiation intensity causes additional micro-aggregation of electrons, and the electrons continue to radiate in phase with each other.[8] This causes the emitted radiation power to increase exponentially. This process continues until the electrons are fully microbunched, and the saturation power reached by the radiation is orders of magnitude higher than the saturation power radiated by the undulator, which results in high beam intensity and laser-like characteristics. The microbunching causes the strong pulses generated by X-ray lasers, this FEL principle is called the self-amplifying spontaneous emission (SASE).

However, SASE FEL lacks temporal consistency due to the noisy startup process. To avoid this, a laser adjusted to FEL resonance can be used to "fill" the FEL. Such temporally coherent seeds can be generated in a more conventional manner, for example by high-harmonic generation (HHG) using optical laser pulses. This leads to coherent amplification of the input signal. In fact, the quality of the output laser is characterized by seeds. For infrared and visible FEL, mirrors can be added at both ends of the undulator to create an optical cavity, so that the radiation forms a standing wave, or provide an external excitation laser. The synchrotron radiation becomes strong enough that the transverse electric field of the radiation beam interacts with the transverse electron flow generated by the sinusoidal oscillating motion, causing some electrons to gain and others to lose energy to the optical field through the ponderomotive force. This energy modulation evolved into an electron density (current) modulation with a light wavelength period. Therefore, electrons are grouped into microbeams in the longitudinal direction, separated by a wavelength of light along the axis. A single undulator causes electrons to radiate independently (incoherently), the radiation emitted by the bunched electrons is in phase, and the electric field adds coherently.

Due to the lack of materials for mirrors that can reflect extreme ultraviolet and X-rays, FEL at these frequencies cannot use a resonant cavity to reflect radiation like other optical lasers, so it will pass through the undulator many times. Additionally, although the high-harmonic generation seeds can boost coherent amplification of the input signal but is available with wavelength as low as extreme ultraviolet light, the lack of conventional X-ray lasers makes seeding at X-ray wavelengths impossible. Consequently, in an X-ray FEL (XFEL) the output radiation is produced by a single pass of a much better-quality electron beam in terms of density and emittance through a very long undulator, so that a single pass can provide enough amplification until saturation to produce an adequately bright X-ray beam. By adjusting the energy of the electron beam or/and the magnetic field strength of the undulator, the wavelength of the emitted radiation can be easily tuned.

The peak brightness and flux of FEL are likely to increase by many orders of magnitude, thanks to a radical change in the emission mechanism. Although the properties of synchrotron sources– high flux, extreme angular collimation, high brightness – are reminiscent of those of lasers, a synchrotron source is not a laser. A laser for infrared or visible light is characterized by several elements: (1) optical amplification, *i.e.* the progressive increase of the initial emission along its path, (2) an active medium – a solid, a gas or a liquid for this amplification, (3) the quantum "stimulated emission" based amplification mechanism, (4) the inversion of a particle energy distribution with respect to thermal equilibrium for the amplification by stimulated emission, (5) an optical cavity, typically formed by two mirrors, that increases the photon path along which amplification occurs for the enhanced effectiveness. For a synchrotron X-ray laser, stimulated emission and population inversion are not feasible, nor is an optical cavity (since X-ray reflection is very ineffective except at grazing incidence). But optical amplification is possible in a special kind of active medium: electrons interacting with an undulator. This hypothetic layered electron crystal structure consists of periodic thin planes of electrons separated by a distance equal to an electromagnetic wave of wavelength. The wave forces the electrons in the planes to oscillate and emit additional waves of the same wavelength. All these emissions combine constructively, producing a more intense wave – hence the optical amplification.

In addition to X-ray free electron lasers, another fourth-generation light source is a diffraction-limited storage ring (DLSR) or ultra-low emissivity storage ring, where the emittance of the electron beams in the storage ring is less than or equal to that of the X-ray photon beam they generate at the end of the insertion device. These devices can operate in the soft to hard X-ray range (100eV–100keV) and have very high brightness ($\sim 10^{21}$–10^{22} photons/s/mm/mrad^2/0.1%BW). Brilliance and coherent fraction are maximized for smaller electron beam emittances until the diffraction limit is reached. DLSRs optimize for extended physical and chemical characterizations at the nano-scale and enable the highest possible spatial resolution coupled to broad temporal sensitivity and X-ray contrast mechanisms – lattice structure and strain, atomic and chemical species, magnetic structure. New and upgraded facilities will have very low electron beam emittance, which will greatly improve many areas of X-ray science, especially for experiments that directly require transversely coherent X-ray wavefronts, such as nanoprobe spectroscopy, coherent scattering and imaging, and interferometric techniques.

Currently most development work for future X-ray sources is on FELs, including more simultaneous users to reduce beam-time costs (European XFEL: 3 undulators to support simultaneous users); shorter wavelength to open new science possibilities (LANL MaRIE ~0.25Å); narrower linewidth to improve monocromator throughput (LCLS operates at ~0.2% bandwidth); seeding expected to reduce bandwidth X10-X100 for future FELs; shorter pulse duration for better time resolution measurements(LCLS: <10fs); higher repetition rate to reduce experiment time (European XFEL: 30 KHz (10 Hz X 3000 Bunches)); higher peak power for nonlinear and single-shot imaging experiments (LCLS maximum is ~70GW).

Compact synchrotron light sources

Due to the usefulness of tuneable collimated coherent X-ray radiation, efforts have been made to make smaller more economical sources of the light produced by synchrotrons. The aim is to make such sources available within a research laboratory for cost and convenience reasons; at present, researchers have to travel to a facility to perform experiments. One method of making a compact light source is to utilize the energy shift from Compton scattering near-visible laser photons from electrons stored at relatively low energies of tens of megaelectronvolts. However, a relatively low cross-section of collision can be obtained in this manner, and the repetition rate of the lasers is limited to a few hertz rather than the megahertz repetition rates naturally arising in normal storage ring emission. Another method is to use plasma acceleration to reduce the distance required to accelerate electrons from rest to the energies required for UV or X-ray emission within magnetic devices. The types and spectrum feature of synchrotron radiation is displayed in Figure 4.2.

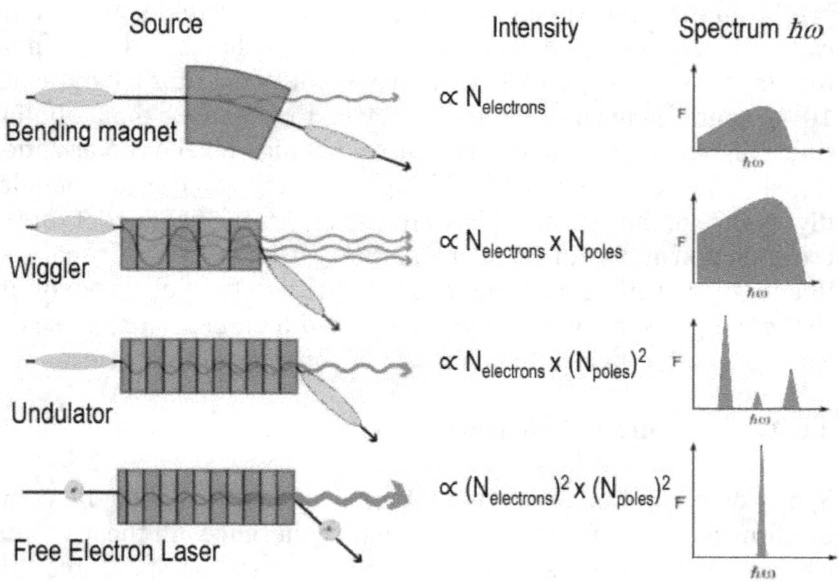

Figure 4.2 types of synchrotron radiation and spectrum features.

4.5 Properties of Synchrotron Radiation Sources

When evaluating X-ray sources, an important parameter of measuring the quality of the source is called brightness. It describes the brightness and angular spread of the beam. It takes into account the number of photons generated per second, the angular divergence of the photons, or the diffusion speed of the beam, the cross-sectional area of the beam, and the photons falling within a bandwidth (BW) of 0.1% of the central wavelength or frequency. In most X-ray literature, the units for brilliance appear as:

$$\text{brilliance} = \frac{\text{photons}}{\text{second}\cdot\text{mrad}^2\cdot\text{mm}^2\cdot 0.1\%\text{BW}}.$$

The greater the brightness, the more the photons of a given wavelength and direction will be concentrated at one point per unit time.

Thanks to the use of insertion devices and bending magnets, compared with other light source, synchrotron radiation is notable for its (1) high brilliance, many orders of magnitude more than with X-rays produced in conventional X-ray tubes, 3rd generation sources typically have a brilliance larger than 10^{18} photons/s/mm^2/mrad^2/0.1%BW, where 0.1%BW denotes a bandwidth $10^{-3}\omega$ centered around the frequency ω, that is more than a billion times higher than a laboratory source; (2) high level of polarization (linear, elliptical or circular); (3) high collimation, *i.e.*, small angular divergence of the beam; (4) low emittance, *i.e.*, the product of source cross section and solid angle of emission is small; (5) wide tunability in energy/wavelength by monochromatization (sub-electronvolt up to the megaelectronvolt range); (6) pulsed light emission, pulse durations are at or below one nanosecond, or a billionth of a second.

4.6 Applications of Synchrotron Radiation

Synchrotron radiation features tunable spectrum, high flux and brightness, fine time structure and high coherence. All these unique characteristics make synchrotron radiation an indispensable and ideal tool for many types of research in the fields such as materials science, physics, biology, geoscience, metrology, and chemistry, medicine and pharmaceutical sciences. It has been widely used by academic, industrial, and government researchers. The high intensity, adjustable frequency, collimation, and selectable polarization of the synchrotron radiation make many specific types of experiments feasible. For instance, the high intensity and penetrating power of synchrotron light enables experiments to be performed inside specimen cells designed for specific environments. The tuneable wavelengths of synchrotron X-ray from the atomic level to biological cells provide keen probes for advanced research in various areas. The characteristics of synchrotron radiation especially well fit the research needs of nanoscience. Most of current synchrotron facilities also provide various in-situ specimen treatment capabilities such as heated, cooled, or exposed to gas, liquid, or high-pressure environments. These in-situ experiments can characterize phenomena from the atomic to nanoscale that most other characterization tools cannot reach. In operation, the measurement is designed to be as close as possible to the actual working conditions of the material.[9]

The major categories of the synchrotron experimental measurement techniques are spectroscopy, scattering, imaging. These experiments correspond to measuring the fundamental parameters such as energy, momentum, and position, respectively. By utilizing the short pulse length radiated by the synchrotron, each technique can be performed in a timed manner, it is especially suitable for time-dependent dynamics. Synchrotron radiation enable us to obtain all these fundamental parameters for perceiving the physical world.

4.6.1 Diffraction and scattering

Since the 1970s, the scientific community has increasingly realized the advantages of using synchrotron radiation for spectroscopy and scattering. The tuneable wavelength makes X-ray diffraction (XRD) suitable for various structural analysis of crystalline and amorphous materials in any kind form including powders, single crystals or thin films. The high resolution and intensity of the synchrotron beam make it feasible to measure the scattering of the dilute phase or analyze the residual stress. A diamond anvil box can be employed to investigate materials at high temperature and pressure to simulate extreme geological environments or create peculiar forms of matter. X-ray crystallographic analysis of proteins and other large molecules is routinely performed. The structure of ribosomes has been solved by using synchrotron-based crystallography experiments, [10] this work has been recognized by the 2009 Nobel Prize in Chemistry.

The size and shape of the nanoparticles can be characterized using small angle X-ray scattering (SAXS). nano-sized features on the surface can be measured by using grazing incident small-angle X-ray scattering (GISAXS). [11] In this and other methods, surface sensitivity can be achieved by placing the crystal surface at a small angle relative to the crystal surface. The incident beam can realize total external reflection and minimize the penetration of X-rays to the material. Techniques such as X-ray reflectivity (XRR) and crystal truncation rod (CTR) analysis can be used to characterize the atomic to nanoscale details of surfaces, interfaces, and thin films.[12] X-ray standing wave (XSW) measurements can also be used to measure the position of atoms at or near the surface; these measurements require high-resolution optics to resolve dynamic diffraction phenomena.

By adjusting the beam energy passing through the absorption edge of a particular element of interest, the scattering from atoms of that element will be modified. These so-called resonance anomalous X-ray scattering methods can help solve the scattering contribution of specific elements in the sample. Other scattering techniques include energy-dispersive X-ray diffraction, resonance inelastic X-ray scattering and magnetic scattering.

4.6.2 Spectroscopy

Broad and tuneable light source strengthen spectroscopy. Synchrotron-based X-ray absorption spectroscopy (XAS) can provide information about the coordination structure of atoms in any form of materials and molecules. By scanning the energy of the synchrotron beam across the absorption edge of the element of interest, and the modulation in absorption coefficient is measured. Photoelectron transitions cause modulation near the absorption edge. Analysis of the X-ray absorption near-edge structure (XANES) reveals information about the chemical state and local symmetry of the element. At a much higher incident beam energy than the absorption edge, photoelectron scattering causes "rippling" modulation, called extended X-ray absorption fine structure (EXAFS). Fourier analysis of the EXAFS scheme can obtain local structural information of the absorbed atoms, such as bond length and number of ligands. Therefore, it is useful for studying liquid and amorphous materials,[13] and sparse substances (such as impurities). A related technique, X-ray magnetic circular dichroism (XMCD), uses circularly polarized X-rays to measure the magnetic properties of elements.

X-ray photoelectron spectroscopy (XPS) can be performed on a beamline equipped with a photoelectron analyzer. The synchrotron's light intensity is high, and XPS measurements can be performed on the surface under gas near ambient pressure. The ambient pressure XPS can be used to measure chemical phenomena under simulated catalytic or liquid conditions. [14]The use of high-energy photons produces high-kinetic energy photoelectrons, whose inelastic mean free path is much longer than the mean free path generated on laboratory XPS instruments. Therefore, the detection depth of the synchrotron high-energy XPS can be extended to a few nanometers, so that the buried interface can be studied.[15] Angle-resolved optical emission spectroscopy (ARPES) is a direct experimental technique for observing the distribution of the density of single-particle electronic excitations in the reciprocal space of solids. By studying the photoemission of electrons from a specimen usually achieved by soft X-ray irradiation, one can obtain information about the direction, speed, and scattering process of valence electrons in the solid surface. The detailed information on band dispersion and Fermi surface can be gained from ARPES data analysis which reveals both the energy and momentum of an electron. Through the band mapping process, one can determine several fundamental physical properties of a solid such as kinetic energy of the electron(s), elasticity, electrical/magnetic properties, and even optical properties.

X-ray fluorescence (XRF) can be used to quantitatively analyze the material composition. XRF detection is also used in several other technologies, such as X-ray absorption spectroscopy and X-ray standing waves, in which the absorption changes of specific elements must be measured. Other spectroscopy techniques include soft X-ray emission spectroscopy and nuclear magnetic resonance vibration spectroscopy, which are related to Mössbauer spectroscopy.

4.6.3 Imaging

The X-ray Ångström scale wavelength enables imaging at a much higher resolution, which is far below the diffraction limit of visible light. Synchrotron radiation can be used as light source for traditional X-ray imaging, phase-contrast X-ray imaging and tomography. With the advance in X-ray focusing devices, the smallest spatial resolution obtained so far reaches about 8 nm.[16] The X-ray nanoprobe source has been utilized in scanning transmission X-ray microscopy (STXM). One can use imaging in combination with spectroscopic techniques such as X-ray fluorescence spectroscopy or X-ray absorption spectroscopy to map the chemical composition or oxidation state of a sample with submicron resolution.[17] Other imaging techniques like coherent diffraction imaging are also under development.

Synchrotron beams can also be utilized as part of the LIGA process in MEMS structure fabrications, synchrotron-based lithography is becoming a hot topic. When electron beams circulate in a magnetic field with high energy, the synchrotron radiation will cause the self-polarization of the electrons in the beam, this phenomenon is called Sokolov–Ternov effect,[18] which has been utilized to generate highly polarized electron beams for various experiments. Synchrotron radiation sets the electron beam size in the electron storage ring (determined by the electron beam emittance) through the effects of radiation attenuation and quantum excitation.

---

[1] D. Attwood, (1999) *Soft X-Rays and Extreme Ultraviolet Radiation: Principles and Applications* (Cambridge Univ. Press, Cambridge); see especially Chaps. 5 and 8.

[2] K.-J. Kim, (1986) Brightness, coherence and propagation characteristics of synchrotron radiation. *Nucl. Instrum. Methods Phys. Res. A* 246, 71.

[3] E. Wiechert, (1901) Elektrodynamische Elementargesetze. *Annalen der Physik.* 309 (4),667–689.

[4] J.D. Jackson, (1999) *Classical Electrodynamics* (3rd ed.). Chichester: Wiley. p. 680. ISBN 978-0-471-30932-1.

[5] M. Kh. Khokonov, (2004) Cascade processes of energy loss by emission of hard phonons. *J. Exp. Theor. Phys.* 99 (4): 690–707.

[6] D. F. Alferov, Yu. Bashmakov, & E. G. Bessonov, (1974) *Sov. Phys. Tech. Phys.* 18, 1336.

[7] Z. Huang, & K. J. Kim, (2007) Review of X-ray free-electron laser theory. *Phys. Rev. Spec. Top.-AC* 10(3), 034801.

[8] J. Feldhaus, J. Arthur, & J. B. Hastings, (2005) X-ray free-electron lasers. *J. Phys. B-At. Mol. Opt.* 38 (9), S799.

[9] J. Nelson, S. Misra, Y. Yang, A. Jackson, Y. Liu, H. Wang, H. Dai, J.C. Andrews, Y. Cui, & M.F. Toney, (2012) In Operando X-ray Diffraction and Transmission X-ray Microscopy of Lithium Sulfur Batteries. *J. Am. Chem. Soc.* 134 (14): 6337–6343.

[10] G. Renaud, R. Lazzari, & F. Leroy, (2009) Probing surface and interface morphology with Grazing Incidence Small Angle X-Ray Scattering. *Surf. Sci. Rep.* 64 (8): 255–380.

[11] I .K. Robinson, & D.J. Tweet, (1992) Surface X-ray diffraction. *Rep. Prog. Phys.* 55 (5), 599–651.

[12] J.A. Golovchenko, J.R. Patel, D.R. Kaplan, P.L. Cowan, & M.J. Bedzyk, (1982) Solution to the Surface Registration Problem Using X-Ray Standing Waves. *Phys. Rev. Lett.* 49 (8), 560–563.

[13] D.E. Sayers, E.A. Stern, & F.W. Lytle, (1971) New Technique for Investigating Noncrystalline Structures: Fourier Analysis of the Extended X-Ray—Absorption Fine Structure. *Phys. Rev. Lett.* 27 (18), 1204–1207.

[14] H. Bluhm, M. Hävecker, A. Knop-Gericke, M. Kiskinova, R. Schlögl, & M. Salmeron, (2007) In Situ X-Ray Photoelectron Spectroscopy Studies of Gas-Solid Interfaces at Near-Ambient Conditions. *MRS Bull.* 32 (12), 1022–1030.

[15] M. Sing, G. Berner, K. Goß, A. Müller, A. Ruff, A. Wetscherek, S. Thiel, J. Mannhart, S.A. Pauli, C.W. Schneider, P.R. Willmott, M. Gorgoi, F. Schäfers, & R. Claessen, (2009) Profiling the Interface Electron Gas of $LaAlO_3/SrTiO_3$ Heterostructures with Hard X-Ray Photoelectron Spectroscopy. *Phys. Rev. Lett.* 102 (17), 176805.

[16] J. Deng, C. Preissner, J.A. Klug1, S. Mashrafi, C. Roehrig, Y. Jiang, Y. Yao, M. Wojcik, M.D. Wyman, D. Vine, K. Yue, S. Chen, T. Mooney, M. Wang, Z. Feng, D. Jin, Z. Cai, B. Lai, & S. Vogt, (2019) The Velociprobe: an ultrafast X-ray nanoprobe for high-resolution ptychographic imaging. *Rev. Sci. Instrum.* 90, 083701.

[17] A. M.Beale, S.D.M. Jacques, & B.M. Weckhuysen, (2010) Chemical imaging of catalytic solids with synchrotron radiation. *Chem. Soc. Rev.* 39 (12), 4656–4672.

[18] A. A. Sokolov, & I. M. Ternov, (1986) *Radiation from Relativistic Electrons.* New York: American Institute of Physics Translation Series. Edited by C. W. Kilmister. ISBN 978-0-88318-507-0.

Chapter 5

XAS Experimental Setups and Measurement Modes

5.1 Introduction

A circulating beam of relativistic electrons in a storage ring can produce electromagnetic waves. They usually span a wide frequency range from infrared to the hard X-ray. The emitted light is characterized by high brightness, many orders of magnitude brilliance than conventional sources, high polarization, tunable wavelength, high collimation (consisting of almost parallel rays), etc. These synchrotron facilities provide high quality light source for a wide range of applications, which range from condensed matter physics and structural biology to chemical engineering and cultural heritage. Now more than 50 dedicated second and third-generation synchrotron facilities are available in the world. Among them, most third-generation sources work with electron beams in the mid-energy range (2-4 GeV). In order to generate high-intensity X-ray in the range of 10 to 50 keV, only a handful of third-generation synchrotrons (ESRF, APS, Spring8) operate in the high-energy range of the electron beam (6–8 GeV).

The different synchrotron facilities share quite similar experimental configurations. Storage ring act as a type of circular particle accelerator and can keep a continuous or pulsed electron beam circulating inside for many hours. As the final destination for the accelerated electrons, the storage ring consists of many straight sections angled together to form a closed loop in circumference. Each sector consists of a straight section and an arc. Large electromagnets curve the electron beam between adjacent straight sections. Each dipole magnet also serves as a potential source of synchrotron light, and the straightest part can also accommodate an insertion device, allowing the third-generation synchrotron to have more than 30 beamlines. Beamlines are built tangentially to the electron beam orbit of the storage ring and capture the radiation emitted from bending magnets, wigglers or undulators. Beamline captures the radiation tangent to the electron storage ring. Although the breadth of the research purpose requires an extensive probe kit, all beamlines consist of X-ray optical devices e.g., slits, crystal monochromators, mirrors and attenuators. These optical devices are installed in order to achieve proper bandwidth, photon flux, beam sizes, focus, and collimation of the rays. The beamline guides radiation to the specimens in the experimental station, where scientific researchers set up their experiments and collect data.

The precise missions of the beamlines are likely to evolve according to their scientific goals, and their capabilities will be continuously improved through collaborative interaction with the user community. The following pages briefly introduce the instruments available on different beam lines according to the working energy and technology. Generally, the essential beamline components described in this section include monochromators, focusing optics, and detectors.

5.2 Elements of Beamline Optics

In order to be used for experiments, the radiation emitted by an undulator, a wiggler or a bending magnet must be optically processed along a beamline and then delivered to an experimental system. Depending on the spectral range and on the experimental technique, such a system may or may not require to be under vacuum. The wide variety of applications of synchrotron radiation requires many different specialized beamline elements. However, some basic components are present in most cases: (1) *A series of valves* that separate the beamline from the storage ring as well as the beamline components from each other. Note that the long-lived circulation of electrons in the ring requires an ultrahigh vacuum of 10^{-10} torr or better. Actions that raise the pressure in the beamline require isolation from the ring. To prevent accidents, there is often a fast-acting valve at the beamline front-end. (2) *A first optical element* is typically a focusing mirror. This element is different with respect to other optical components since it must handle the entire power of the emitted synchrotron radiation. This is very high and can easily cause instabilities and damage. The front-end element requires sophisticated technical solutions to deal with such problems. (3) One or more *focusing elements* like Kirkpatrick-Baez (K-B) mirror or Fresnel zone plate are installed in the beam line. (4) *A monochromator* is employed to select the desired wavelengths. Some of the most common synchrotron monochromators are discussed below. (4) *A photon detector system* is positioned to measure the resulting diffraction, scattering or secondary radiation signals from the specimens.

5.2.1 Focusing devices

Focusing X-ray is quite challenging since both refraction and reflection are weak in their wavelength range. Particularly, the conventional refractive lenses for visible light do not work at X-ray wavelengths due to the fact that the refractive indices are lower than but very close to 1 (exceptions close to some resonance frequencies) in the X-ray regime.

The reflection of X-ray depends on the angle of incidence and is negligible except for very small incidence (grazing) angles. Classical optics show that the critical angle for total reflection is $\sqrt{2\delta}$, where δ is the difference between unity and the real part of the index of refraction. Typical δ-values for X-ray in the 10^{-4} range give critical angles of tens of milliradians. There is practically no reflection at larger incident angles. Therefore, reflecting mirrors for X-ray can only work at grazing incidence. The corresponding technical problems are severe: the illuminated area is large, making it difficult to meet the stringent surface accuracy requirements.

A grazing-incidence X-ray mirror can be produced by accurately bending a flat surface. But when the bending is only in one direction, focusing is only achieved in the plane of curvature. In order to obtain focused X-rays in both directions, Kirkpatrick and Baez proposed the X-ray focusing optical system ("Kirkpatrick-Baez" configuration) in 1948, which consists of two vertically aligned total reflection elliptical mirrors (Figure 5.1a). Using two of elliptical mirrors or one elliptical and one toroidal mirror in a Kirkpatrick-Baez (K-B) design can allow one to have a system almost free of aberrations. Compared with other methods, the focusing optical system can achieve more effective X-ray collection, which can significantly improve the performance of the X-ray microscope. In addition, even if the wavelength of the X-ray is shifted, this method can maintain the focused state with the same optical arrangement. However, in order to achieve an X-ray beam with an ideal focal length, high efficiency, and no background noise around the main peak, it is necessary to prepare an elliptical mirror with high pattern accuracy (peak and valley about 2 nm) and surface smoothness (about 0.2 nm root mean square) position on the entire reflection area and accurately position the two mirrors.

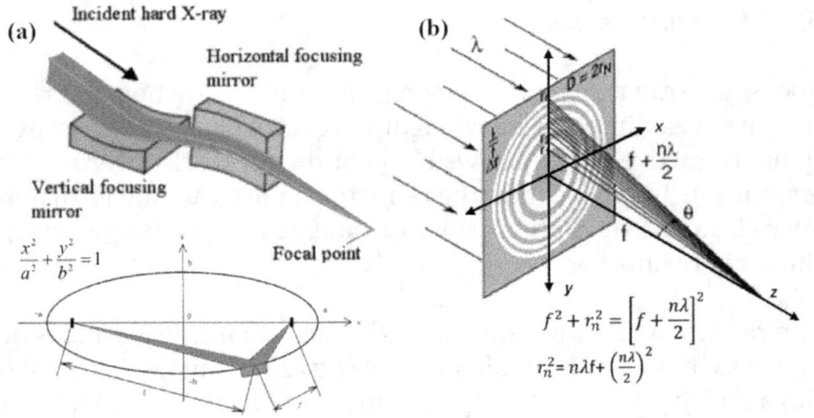

Figure 5.1 (a) Kirkpatrick-Baez optics. It consists of two elliptical mirrors aligned perpendicularly to each other. X-ray from a light source placed at one focus are reflected on the mirror surface and always reach the second focus. (b) A Fresnel zone plate lens with plane wave illumination, showing only the convergent (+1st) order of diffraction. Sequential zones of radius r_n are specified such that the incremental path length to the focal point is nl/2. Alternate zones are opaque in the simple transmission zone plate.

Another popular X-ray focusing lens is the Fresnel zone plate (FZP). The zone plate is a set of circular concentric ridges. As they extend from the middle to the edges, they gradually become narrower and closer. The light passing through the ridge produces a diffraction pattern that constructively interferes, focusing the light onto a point. For X-rays, the effect is enhanced by using heavy elements in the ridges. But these elements absorb X-ray energy and will melt under strong light. X-ray FZP is a short-wavelength counterpart of devices widely used for visible radiation. They consist of alternating circular concentric absorbing and transmitting zones, the latter of radius r_n, $n = 1, 2, 3, \ldots$ The focusing conditions (Figure 5.1b) occur when the rays passing through the transmitting zones interfere constructively. This gives a first-order focal length f if: $\sqrt{f^2 + r_n^2} - f = n\lambda$, which for small wavelengths (and therefore small r_n values) give $r_n \approx \sqrt{2n\lambda f}$. The width Δ_n of the n-th zone is thus given by $\Delta_n = (\sqrt{2n\lambda f} - \sqrt{2(n-1)\lambda f})/2$, which for large n-values becomes $\Delta_n \approx \sqrt{\frac{\lambda f}{2n}}/2 = \frac{\lambda f}{2r_n}$, a result with important technological implications. To increase the FZP diameter, corresponding to the maximum value of r_n, a large number of zones are needed. However, the width of the outermost zone gradually decreases with n until it is not technically feasible. In fact, due to weak X-ray absorption, the absorption zone must be very thick. Therefore, the outer absorption zone is thicker but narrower in width – such a high "aspect ratio" may cause mechanical instability. In fact, the aspect ratio cannot exceed 10–20, and the FZP diameter is still small. This limitation also affects the spatial resolution of the FZP lens, which can prove to be proportional to the width of the outermost zone.[1]

In recent years, the "composite refractive lens" (CRL) has made tremendous progress.[2] Contrary to visible light lenses which are convex for a focusing effect, the index of refraction $1 - \delta$ for X-ray is <1, refraction focusing requires that these lenses must be concave to achieve focusing. Since refractive indices at X-ray wavelengths are so close to 1, the focal lengths of normal lenses get impractically long. A single cylindrical cavity in a metal block would produce focusing, but only weakly because of the small value of δ. To overcome this lenses with very small radii of curvature are used, radii of curvature are typically less than a millimeter making the usable X-ray beam width at most about 1 mm.[3] In a CRL, the effect is enhanced by drilling a series of cylindrical cavities and they are stacked in long rows so that the combined focusing power gets appreciable.[4] In order to reduce the absorption of X-rays in these stacks, it is common to use host metals with very low atomic numbers, such as beryllium or lithium, to reduce their cohesion. Since the refractive index depends largely on the X-ray wavelength, these lenses are highly chromatic, and the focal length must be considered as a function of wavelength in any application.

A polycapillary lens can also be utilized to focus X-rays. An array of small hollow glass tubes guides X-rays and generates many total external reflections inside the tube.[5] The array is tapered so that one large end of the capillaries faces at the X-ray source and the small end points at the sample. Polycapillary optics are achromatic and thus suitable for scanning fluorescence imaging and other applications where a broad X-ray spectrum is useful. They collect X-ray efficiently for photon energies of 0.1 to 30 keV and can achieve gains of 100 to 10000 in flux over using a pinhole at 100 mm from the X-ray source. Since only X-rays entering the capillary at a very narrow angle are totally internally reflected, only X-rays from a small spot can be transmitted through the optical system. Polycapillary optics cannot image more than one point to another, so they are used to illuminate and collect X-rays.

5.2.2 Monochromators

Monochromator can transmit mechanically selectable monochromatic light selected from the wider wavelength range available at the input. For X-ray, grating monochromators and crystal monochromators are the two main classes used for soft and hard X-ray regimes, respectively.[6] Like grating monochromators for visible light, X-ray grating monochromators operate based on controlling the relative positions of the grating and of the entrance and exit slits. Since the early 1970s, different approaches were adopted to solve the problems of reflection efficiency, aberrations and astigmatism, and achieve good performances in terms of flux and spectral resolution.

In the soft X-ray regime, a spherical grating monochromator can combine the spectral filtering and focusing functions. Obviously, they are easier to manufacture than non-spherical counterparts but they tend to be less effective. Spherical grating monochromator is often based on the "Rowland circle" geometry in which a theoretical circle of radius $\rho/2$ – where ρ is the spherical grating radius – is tangent to the grating surface. If the incident light hits the grating through an entrance slit on the Rowland circle, then the monochromatized radiation is focused on another point on the same circle. Thus, the desired wavelength can be selected by moving a small-area detector (or an exit slit) along the Rowland circle. Initially, this geometry was implemented either with a Rowland circle rail or with more sophisticated mechanical solutions such as the Brown-Lien-Pruett "Grasshopper" monochromator.[7] Later, the implementation relied on computer-controlled motions – a solution pioneered by the "Extended Range Grasshopper" design.[8]

Besides Rowland circle geometry, the Wadsworth geometry [9] and the famous Chen-Sette "Dragon" design [10] were also effectively implemented for spherical grating monochromators. Compared to the spherical gratings, the holographically produced toroidal gratings [11] have strong technical advantages for the combined monochromatizing and focusing functions. Despite its high price tag, toroidal grating has resulted in a variety of designs. With the advancements in the fabrication techniques, variable line space (VLS) gratings can solve problems such as the aberrations.[12]

The hard X-ray crystal monochromators work based on Bragg diffraction by the atomic planes of a high-quality single crystal. Alternate approaches use Laue diffraction, also from excellent single crystal– mainly silicon crystals. As in the case of grating monochromators, technical problems such as geometry and focusing, are required to be solved in a sophisticated design in order to provide excellent resolution and yield. Two of the most popular solutions are the "channel-cut" double-crystal monochromators [13] and the single-exit double-crystal monochromators (Figure 5.2a).[14,15] As each photon energy corresponds to a given pair of incidence and Bragg reflection directions, thus the photon energy that is Bragg-diffracted can be changed by rotating the crystal.

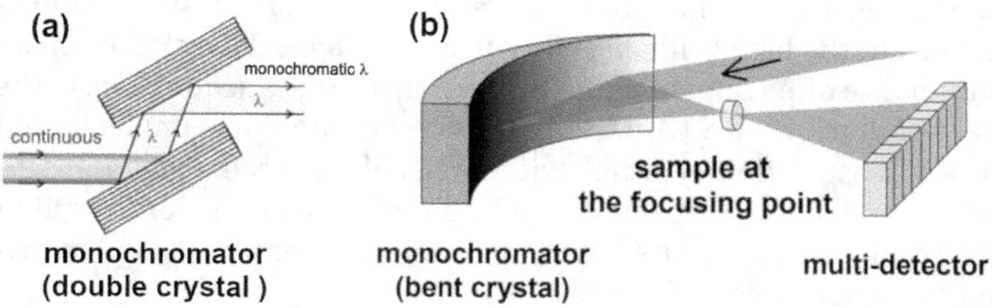

Figure 5.2 (a) typical double crystal monochromator, and (b) bent crystal monochromator for energy dispersive XAFS.

As monochromatized beams exit the system in different directions depending on the photon energy, to solve this problem, a two parallel crystal geometry can be utilized to keep all the exiting photon energies along the same direction as the incident beam. Additionally, the spectral resolution also get enhanced with two Bragg reflections. It necessitates keeping the two crystals parallel to each other with great accuracy, a popular and simple solution is to use only one, channel-cut single crystal instead of two crystals. However, the specific trajectory of the exiting beam – and its exit point from the monochromator – still change with the photon energy. It can be eliminated using more sophisticated solutions based on two linear motions of the second crystals, plus the common rotation of the two crystals. These combined motions in fact keep the trajectory of the exit beam within a fixed exit, many different fixed-exit solutions of this kind have been implemented in the past decades.[16] Since a given Bragg reflection occurs not only for the fundamental photon energy but also for its integer multiples, the Bragg monochromators are also affected by a higher-harmonic problem. One solution is to use the X-ray reflection cut-off to reject higher harmonics with suitable mirrors.

Motived by the requests of fast spectroscopy, [17,18] many new types of crystal monochromators [19] emerged out in recent decades. Their development relies on using additional motions and/or additional crystals [20,21] and crystal types, plus other beamline components such as focusing mirrors. Fast spectrum acquisition is desirable for many reasons, such as following the evolution of dynamic systems and avoiding instrumental instabilities. The spectral scanning time of the monochromator dominates the time per spectrum, whereas the intrinsic high signal-to-noise ratio allows much faster scanning thanks to the high brilliance of synchrotron light. Two classes of solutions are often implemented to speed up the data acquisition. In the case of quick EXAFS (Q-EXAFS), the crystal rotation is implemented as a fast oscillation, typically within tens of milliseconds. In the case of dispersive EXAFS, beams of different photon energy diffracted in different directions are simultaneously processed using an extended-area detector (Figure 5.2b).

5.2.3 Detectors

One of the important elements in the experimental station is the X-ray detector. In many X-ray detectors for synchrotron radiation, X-ray are first converted by a suitable system into secondary particles, such as electrons, visible photons, electron-hole pairs in a semiconductor or ion pairs in a gas, and then the secondary particles are detected with or without spatial resolution. Detectors based on this general principle have developed over decades from the initial devices such as the photomultipliers and the early ionization chambers [22] to a variety of instruments in different classes, such as channel multipliers and channel plates,[23] diodes and in particular avalanche photodiodes,[24,25] and the increasingly popular silicon drift detectors.[26] Sophisticated ionization chambers are commonly used by hard X-ray beamlines. X-ray detectors can be either photon counting or integrating. Photon-counting detectors measure each individual

X-ray photon separately, while integrating detectors measure the total amount of energy deposited in the active region of the detector. In general, the detectors can be classified into three categories: (a) point detectors or 1D detectors which only offer information on the X-ray flux, (b) position sensitive detectors which provide spatial information, and (c) energy dispersive detectors which give information on the energies of the detected X-ray. The field of X-ray detection is large and merits a chapter of its own. Various types of X-ray detectors exist, and the best choice is a trade-off between the nature of the experiment and the merits of different technological approaches.

In diffraction and scattering experiments, spatial information (the scattering angle of X-ray) is very important. Therefore, position-sensitive detectors are often used because they provide a large amount of space and intensity data in a relatively short time. An example of a position-sensitive detector is a direct detection pixel-based solid-state detector. In these detectors, X-ray (or electrons) directly hit the semiconductor, producing free electrons and holes, which are detected by electronic devices. The readout circuit is designed to have low noise, high dynamic range and fast readout. The X-ray pixel detector built into the hybrid design includes a radiation sensor, a reverse-biased pn diode (usually silicon) with a segmented contact structure on the p-side, and a readout ASIC chip with the same pixel structure. The sensor and readout chip are connected by tiny conductive indium balls (called bump bonding). X-rays are detected in the sensor and the resulting signal is amplified in the pixel chip. When the detected charge reaches a certain threshold, photons are counted. Low-energy photons can also be excluded by adjusting the threshold, which can reduce the background caused by fluorescence. Since the silicon sensor is used to directly detect X-rays, the spatial resolution of the detector is very good. The detection quantum efficiency of silicon sensors is 80% at 12 keV and even 100% at 8 keV, which is obtained by X-ray absorption of silicon.

Other types of experiments may require the detection of various energies of X-ray photons, such as fluorescence measurements. In this case, a silicon drift detector (SDD) is the preferred energy dispersion detector. Silicon drift detectors are basically discs of n-type silicon which measure the energy of an incoming photon by the amount of ionization it produces in the detector material. This varying ionization produces varying charge, which the detector electronics measure for each incoming photon. On the other side of the part exposed to X-ray, a transverse electric field is generated by a series of ring electrodes. When X-ray radiation is absorbed, electrons are generated in the silicon. The electric field forces the electrons to drift laterally towards a small anode. The number of electrons reaching the anode is proportional to the energy of the absorbed photons. The small size of the anode and the integration of the first FET of the amplifier circuits in the anode greatly reduces the capacitance of the detector, thereby providing higher energy resolution and a higher photon count rate. The energy resolution is expressed in full-width at half maximum (FWHM) and is typically 125–150 eV for Mn Kα wavelength (~2.1 Å).

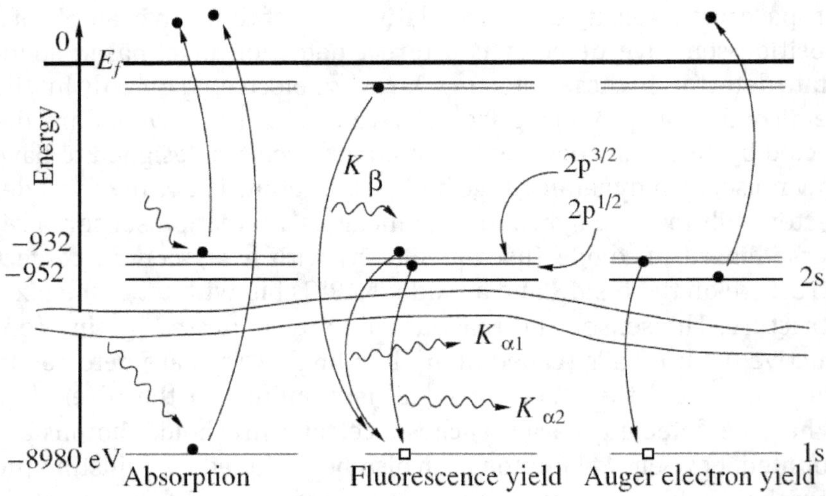

Figure 5.3 Diagram of relaxation of an atomic state exited by the absorbed X-ray photon.

The purpose of the detector has gradually changed from pure X-ray intensity measurement to studying one-dimensional or two-dimensional intensity distribution. There are two main reasons triggering this development. First, advanced X-ray imaging has become one of the main fields of synchrotron research. Second, band (extended area) detectors can be used with crystal or grating monochromators to analyze different wavelengths in parallel. Some of these devices can also be used for two-dimensional inspection. However, the real revolution in this direction is the introduction of charge-coupled devices (CCD).[27] Recently, a big step forward is to integrate the two steps of the process (conversion into secondary particles and detection of such secondary particles) into the same "pixel" in a charge-coupled device. This eliminates some of the inherent problems of the two-step method, thereby improving efficiency and spatial resolution.

5.3 XAFS Measurement Methods

The absorption of an X-ray photon by an atom leads to excitation, which is relaxed either through emission of individual photons (fluorescence) or ejection of Auger electrons with definite energies corresponding to the levels of the atomic system. Figure 5.3 shows the diagram of relaxation of the exited atomic state.[28] The probability of various decay channels (Auger process and fluorescence) depends on the charge of the nucleus. The level widths determine the experimental resolution necessary for their separation.

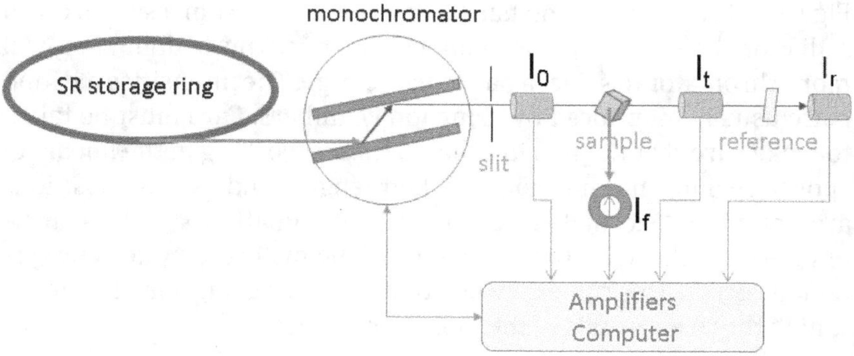

Figure 5.4 Typical XAFS setup diagram. The five parts of the experiment are (1) the X-ray source, SR, (2) the optics, monochromator, (3) sample, (4) detectors and (5) the electronic and computer. For transmission mode, $\mu(E)=\ln(I_0/I_t)$, while for fluorescence mode, $\mu(E)=I_f/I_0$.

XAFS experiments can be conducted in transmission mode (for specimens with high element concentrations), fluorescence-yield mode (for studying elements at low concentrations in a sample) or electron yield mode (for surface measurement, especially light element. Figure 5.4 shows the typical setup needed for conducting both transmission and fluorescence-yield experiments on a hard X-ray beamline.

Critical variables of this experimental arrangement are the vertical spacing of slits in the monochromator chamber and in front of the I_0 or first ion chamber, the type of monochromator crystals employed (typically the (111), (220), and (400) cuts of silicon), the types of gas used in the ion chambers, and the type and configuration of the sample.

5.3.1 Transmission mode

The transmission mode is the most convenient method for collecting XAFS data. It relies on the measurement of monochromatic photons intensities incident on (I_0) and transmitted by (I_t) a specimen of thickness *d*. The incident and transmitted beam intensities I_0 and It are related by $I_t = I_t e^{-\mu(\omega)t}$, then the absorption coefficient is calculated as

$$\mu(\omega)d = ln(I_0/I_t). \tag{5.1}$$

A typical transmission experimental setup is schematically shown in Figure 5.4. The wide-band hard X-ray comes out from a synchrotron radiation beamline and is monochromatized by a double-crystal monochromator to select a particular energy. The intensities of I_0 and I_t are usually monitored by using ion chambers. Care must be taken to make sure that the two ion chambers are working in the linear response region, and the choice of their lengths and gas compositions are essential. The higher-harmonic contamination should also be minimized, otherwise the spectrum will be distorted by serious systematic error. This can easily be done by either using a mirror or detuning the two crystals of the monochromator.

XAFS measurement in transmission mode is the most widely used and has the best data quality. But it is mostly suitable for concentrated specimens in which the element of interest is the major component (>10 atm%). A crucial requirement for the sample is that it must be uniform and pinhole-free. Typically, the sample thickness *d* should be optimized so that the edge jump $\Delta\mu d \approx 1$. Under this condition, the "thickness effects" could be avoided and the data could have a good signal/noise ratio.[29] This requires that the grain sizes for powder samples be smaller than or comparable to the absorption length of the material. For a single substance, the absorption coefficient μ is related to the absorption cross σ by $\mu = \rho\sigma$, where ρ is the density. If the unit of ρ is g/cm^3, the cross-section is in cm2/g and μ is in cm^{-1}. For a compound consisting of different elements, the total absorption coefficient μ is given by:

$$\mu \approx \sum_i \rho_i \sigma_i = \rho_M \sum_i \frac{m_i}{M} \sigma_i = \rho_N \sum_i \frac{n_i}{N} \sigma_i , \tag{5.2}$$

where ρ_M is the mass density of the material as a whole, ρ_N is the number density of the material as a whole, and $\frac{m_i}{M}$ and $\frac{n_i}{N}$ are the mass fraction and number fraction of element *i*, respectively. Sources of the energy-dependent cross-section data can be found in literature.[30,31]

Most XAFS experiments can be carried out in the transmission mode. The best S/N ratio is achieved on a uniform sample when the absorbance by I_0 detector is 0.245 and that by the sample is 2.55. Under this condition the S/N ratio is described as $S/N=0.556(\Delta\mu_A/\mu_T)\sqrt{I_0}$, where $\Delta\mu_A/\mu_T$ is the ratio of the XAFS oscillation magnitude and the total absorbance by the sample and I_0 is the photon flux at the first detector. For the proper amount to place in the cell we can refer to the following formula:

$$Wd = \frac{\ln {}^{I_0}\!/_{I_t}}{\sum\left(\frac{\mu_i}{\rho_i} w_i\right)}, \tag{5.3}$$

where W_d is the superficial density of the sample in g/cm^2, μ_i is the absorption coefficient at the given energy, ρ_i is the density of the element and w_i is the mass fraction of the element. The term $\ln(I_0/I_t)$ is the value that is plotted versus the energy and for a good quality measurement has a value between 1 and 2.5.

5.3.2 Fluorescence yield mode

In a transmission experiment for a dilute specimen, like the lightly doped specimens such as diluted magnetic semiconductors and thin films grown on substrates, the thickness of the sample has to be adjusted so that the edge jump $\Delta\mu d \approx 1$. Under such circumstance, the background absorption due to the matrix or substrate would be very large, and transmitted beam intensity is exponentially attenuated, resulting in very bad quality of the data. To reduce the background absorption, selectively recording the atomic-specific signals is a good choice. Among them, monitoring the X-ray fluorescence is the most preferred, since the fluorescence yield is proportional to the absorption efficient under proper conditions. Fluorescence detection is often employed when the concentration of the atom of interest is very low and a reasonable S/N ratio cannot be realized by the transmission mode measurement.

To make good fluorescence XAFS measurements, one desires to collect the available signal as much as possible. In a fluorescence XAFS measurement, the X-ray emitted from the sample will include the fluorescence line of interest, fluorescence lines from other elements in the sample, and both elastic and inelastic (Compton) scattering X-ray. Although in general the fluorescence is emitted isotropically, the scatter is not emitted isotropically because the X-ray from a synchrotron are polarized in the plane of the synchrotron. This means that elastic scatter is greatly suppressed at 90◦ to the incident beam in the horizontal plane. Therefore, fluorescence detectors are normally placed in the horizontal plane at a right angle to the incident beam, and the sample is placed at 45° to both the incident beam and the fluorescence detector, as shown schematically in Figure 5.4. In this geometry, the fluorescence intensity $I_f(\omega)$ accepted by a detector with a solid angle of Ω/4π is:

$$I_f(\omega) = I_0\left(\frac{\Omega}{4\pi}\right)\gamma\frac{\mu_A(\omega)}{\mu_{tot}(\omega)+\mu(\omega_f)}\left\{1 - e^{-\left(\mu_{tot}(\omega)+\mu(\omega_f)\right)d}\right\}, \quad (5.4)$$

where γ is the fluorescence efficiency, d is the sample thickness, ω is the energy of incident X-ray and ω_f is the energy of the fluorescent photons, $\mu_A(\omega)$ is the absorption coefficient of the element of interest, and $\mu_{tot}(\omega)$ is the total absorption coefficient that is the sum of the contributions $\mu_A(\omega)$ from the element of interest and $\mu_B(\omega)$ from the background absorption coefficient of all the other elements in the matrix, and Ω is the solid angle covered by the detector. Because $\mu_{tot}(\omega)$ depends on $\mu_A(\omega)$, the measured fluorescence intensity is in general not proportional to the absorption coefficient $\mu_A(\omega)$ that one wants to measure. However, there are two limiting cases where this proportion holds: (1) Thin specimens (typically below 300 nm), in which the series expansion of the exponential in Eq. (5.4) can be truncated at the term linear in *d*, *i.e.*, $1 - e^{-\left(\mu_{tot}(\omega)+\mu(\omega_f)\right)d} \approx \left(\mu_{tot}(\omega) + \mu(\omega_f)\right) d$ and thus

$$I_f(\omega) = I_0\left(\frac{\Omega}{4\pi}\right)\gamma\mu_A(\omega)\, d \propto I_0\mu_A(\omega). \tag{5.4a}$$

(2) Thick samples with a dilute concentration of the element of interest, in which $\mu_{tot}(\omega) \gg \mu_A(\omega)$ and $1 - e^{-\left(\mu_{tot}(\omega)+\mu(\omega_f)\right)d} \approx 1$. Then Eq. (5.1) reduces to

$$I_f(\omega) = I_0\left(\frac{\Omega}{4\pi}\right)\gamma\frac{\mu_A(\omega)}{\mu_{tot}(\omega)} \propto I_0\mu_A(\omega). \tag{5.4b}$$

If neither of the above two conditions could be met, a strong suppression of features in the spectrum could be observed. Hence, use of a thick, concentrated sample results in severely distorted EXAFS amplitudes and XANES spectra. Such a phenomenon, known as the "self-absorption" effect of the fluorescence data, should be corrected before further processing, and numerous correction procedures have been proposed.[32] In the fluorescence mode one desires to collect only the fluorescence line of interest, and completely suppress the scatter peak and other fluorescence lines. This would greatly decrease the background intensity and increase the signal/noise ratio. One can reduce the intensity of the scattering by an order of magnitude by inserting an X-ray filter that has the absorption edge between the fluorescence line of interest and the elastic scattering peak. A thin metal foil or a homogeneous powder containing the Z−1 element (Z is the atomic number of the element of interest) is conventionally used as a filter. For example, to measure Co *K*-edge fluorescence XAFS data, a 3- or 6-μm thick Fe foil is usually used. Besides, to avoid re-radiation from the filter itself, a set of slits (Soller slits) is often used to preferentially collect emission from the sample and block the emission from the filter.

To collect fluorescence X-ray, an ionization chamber–Lytle detector [33] could be used, but it cannot discriminate the fluorescence of interest from the background. To suppress the background intensity, detectors with energy-discriminating ability, such as high-purity Ge or Si solid-state detector (SSD) and silicon drift diode detector, have been developed and routinely used. The energy resolution of these detector is typically in the range of 150–200 eV at 6 keV, as a result of a balance between the counting rate and background removal. The maximum counting rate of these detectors is limited by the detector electronics, which can be enhanced by an order of magnitude by assembling multiple detector elements with a close packing geometry. The dilute limit of the fluorescence XAFS measurements is determined by the detector counting rate and incident beam intensity. A state-of-the-art Ge pixel array detector (PAD) with100 segments has been developed to gain high throughput measurement of fluorescence yield by Oyanagi,[34] who describes more details about the design of the PAD in Section 5.3 of this book. Because both the incident and detected signals in the fluorescence mode are penetrating X-ray, fluorescence XAFS is the most suitable choice of study bulk specimens. It could also be a surface-sensitive method by using a grazing-incidence geometry. Nowadays, samples with interested element concentration of several to tens of ppm or thin films with thickness equivalent to 0.1~1 monolayers are measurable by XAFS with adequate signal/noise ratio. [35]

As XAFS is oscillation of μ_A, fluorescence XAFS can be used without correction only in the following extreme cases: a thin, concentrated sample, or a thick, dilute one. The correction for the self-absorption was proposed as following by C. Booth[36]

$$\chi = \frac{-\left[\gamma\left(\alpha - \bar{\mu}_a\left(\chi_{\exp} + 1\right)\right) + \beta\right] + \sqrt{\left[\gamma\left(\alpha - \bar{\mu}_a\left(\chi_{\exp} + 1\right)\right) + \beta\right]^2 + 4\alpha\beta\gamma\chi_{\exp}}}{2\beta},$$

with

$$\beta = \frac{\bar{\mu}_a d\alpha}{\sin\phi} e^{-\frac{\alpha d}{\sin\phi}}, \gamma = 1 - e^{-\frac{\alpha d}{\sin\phi}}, \alpha \equiv \bar{\mu}_T + g\mu_f \tag{5.5}$$

In the thick limit d→ infinity, it becomes

$$\chi = \frac{\chi_{\exp}}{1 - \frac{\bar{\mu}_a}{\alpha}\chi_{\exp} - \frac{\bar{\mu}_a}{\alpha}} \tag{5.6}$$

5.3.3 Electron yield detection mode

The total/partial electron yield method is often used in the soft X-ray region since the absorption is too high in transmission mode. When the core level electron absorbs an X-ray photon with energy above the absorption edge, the core hole left behind could decay by emitting fluorescence and Auger electron. Analogous to the fluorescence mode, detection of the Auger electrons provides another way to collect the XAFS signal. Depending on which electrons are detected, there are various electron detection yield detection modes. Among them, the total electron yield (TEY) detection mode is experimentally the simplest, which measures the drain current using an electrometer under ultra-high vacuum (Fig. 6.7). [37] In this mode, the sample should be made electrically conductive. The TEY mode is especially important for collecting the XAFS data in the soft X-ray region, like O or N *K*-edge and 3d transition metal $L_{2,3}$-edge spectra.[38] A variation of TEY mode is the partial electron yield (PEY) mode in which only electrons of a given energy range are collected with retarding grids and electron detectors. There is also the conceptually simplest Auger electron yield (AEY) detection mode, in which the number of the Auger electrons is recorded by using an electron analyzer operating in ultrahigh vacuum. The short escape-length of the electrons (tens to hundreds of angstroms, depending on the energy of the original Auger electrons) makes the electron yield mode sensitive to near-surface regions or interfaces, in contrast to the bulk-sensitivity of the fluorescence mode. This method works efficiently for thick and concentrated specimens that cannot be measured by other XAFS modes. It also suffers less from the "self-absorption" effect occurring in the fluorescence mode, because the electron absorption has much higher efficiency than the photon absorption.

Saturation effects lead to a recorded signal no longer proportional to the photoabsorption cross-section as the photon energy is varied. The saturation effects arise if the electron yield escape depth is larger or comparable to the X-ray penetration depth.[39] The electrons number (mostly Auger electrons), which reach the surface $dY_e(z)$, decay exponentially as a function of depth: $dY_e(z) = dY_{e,0}(z)e^{-z/\lambda_e}$, where z is the depth from the surface and $dY_{e,0}(z)$ the number of electrons produced in a thickness dz at the depth z which itself is given by

$dY_{e,0}(z) = I_0 e^{-\mu z/\cos\theta} \frac{\mu dz}{\cos\theta}$, where I_0 is the number of incident photons and θ the angle from the normal. Then the electron yield intensity $Y_e(E)$ can be obtained by combining the above two equations (i denotes the element),

$$Y_e(E) \sim \int_0^d \frac{\mu_i(E,z)}{\cos\theta} e^{-\int_0^z \left(\frac{\mu_i(E,z')}{\cos\theta} + \frac{1}{\lambda_e(z')}\right)dz'} dz \quad . \tag{5.7}$$

For a semi-infinite thick film, by integrating over the depth z from z=0 to z =∞, the measured yield $Y_e(E)$ intensity for the total electron yield detection is then $Y_e(E) \sim \frac{\mu(E)}{\mu(E) + \frac{1}{\lambda_e}\cos\theta}$, where $\mu(E)$ is the absorption coefficient, λ_e the electron escape depth and θ the angle from the surface normal.

---

[1] S.R. Wu, C.H. Lin, Y.S. Chen, Y.Y. Chen, Y. Hwu, Y.S. Chu, & G. Margaritondo, (2013) At the frontiers of high-resolution hard-X-ray microscopy: an international programe. *J. Phys. D* 46, 494005.

[2] F. Cerrina, B. Lai, C. Gong, & A. (1989) Ray-Chaudhuri. MAXIMUM: Status of the scanning photoelectron microscope. Rev. Sci. Instrum., 60, 2249.

[3] F. Seiboth, M. Scholz, J. Patommel, R. Hoppe, F. Wittwer, J. Reinhardt, J. Seidel, M. Knaut, A. Jahn, K. Richter, J. Bartha, G. Falkenberg, & Ch. Schroer, (2014) Hard X-ray nanofocusing by refractive lenses of constant thickness. Applied Physics Letters. 105. 10.1063.

[4] A. Snigirev, (1998) Focusing high-energy x rays by compound refractive lenses. *Appl. Opt.* 37 (4), 653–62.

[5] C.A. MacDonald, (2010) Focusing Polycapillary Optics and Their Applications. *X-Ray Optics and Instrumentation*. 2010, 1–17.

[6] D.M. Mills, (2002) X-ray optics for third-generation synchrotron radiation sources, in *Third-Generation Hard X-Ray Synchrotron Radiation Sources: Source Properties, Optics, and Experimental Techniques 1st Edition* (ed. D.M. Mills), John Wiley & Sons, Inc., New York. ISBN-13: 978-0471314332

[7] F.C. Brown, (1980) Improved high resolution grasshopper monochromator. *Nucl. Instr. Meth.*, 172, 100.

[8] F. Cerrina, H. Hoechst, R. Rosenberg, *et al.* (1988) The General Motors-University of Wisconsin extended range grasshopper beam line, *Nucl. Instr. Meth.*, A266, 208.

[9] G. Margaritondo, (1988) *Introduction to Synchrotron Radiation*, Oxford University Press, New York. ISBN 0-19-504524-6.
[10] C.T. Chen, F. Sette, & N.V. Smith, (1990) A double headed dragon monochromator for soft-X-ray circular dichroism studies, *Appl. Optics,* 29, 4535.
[11] W.B. Peatman, (1997) *Gratings, Mirrors and Slits: Beamline Design for Synchrotron Radiation Sources,* CRC Press, Amsterdam. ISBN-13: 978-9056990282.
[12] L. Poletto, & G. Tondello, (2000) Design of a high-throughput grazing-incidence flat-field spectrometer, *Appl. Optics*, 39, 5671.
[13] C. Giannini, & L. Tapfer, (1996) A high-resolution multiple-crystal monochromator for X-ray diffraction studies, J. *Appl. Cryst.*, 29, 230.
[14] S. Bernstoff, E. Busetto, A. Savoia, & M. Colapietro, (1995) The beamlines of ELETTRA and their application to structural biology, *Rev. Sci. Instrum.*, 66, 2065.
[15] B.X. Yang, F.H. Middleton, B.G. Olsson, & G.M. Bancroft, (1992) Double-crystal monochromator beam line on the Aladdin 1-GeV storage ring, *Rev. Sci. Instrum.*, 63, 1355.
[16] P. Suortti, & C. Schulze, (1995) Fixed-exit monochromators for high-energy synchrotron radiation. *J. Synchrotron Rad.*, 2, 6.
[17] M. Nachtegaal, C. Konig, O. Muller, & R. Frahm, (2015) QEXAFS: techniques and scientific applications for time-resolved XAS, in *XAS and XES: Theory and Applications*, vol. I (eds J.A. van Bokhoven and C. Lamberti), John Wiley & Sons, Ltd., pp. 155–184.
[18] O. Mathon, I. Kantor, & S. Pascarelli, (2015) Time-resolved XAS using an energy dispersive spectrometer: techniques and applications, in *XAS and XES: Theory and Applications*, vol. I (eds J.A. van Bokhoven and C. Lamberti), John Wiley & Sons, Ltd., pp. 185–212.
[19] C. Malgrange, (1992) X-ray optics for synchrotron radiation. *Acta Physica Polonica*, 82, 13.
[20] W.J. Bartels, (1983) Characterization of thin layers on perfect crystals with a multipurpose high resolution x-ray diffractometer. *J. Vac. Sci. Technol.*, B1, 338.
[21] S.M. Heald, (1988) A versatile two/four crystal monochromator for X-ray absorption spectroscopy. *Nucl. Instrum. Meth.*, A266, 457.
[22] F. Sauli, (2000) GEM: gas detectors development, *Nucl. Instrum. Methods*, A419, 26.
[23] J. Wiza, (1979) Microchannel plate detectors. *Nucl. Instrum. Methods*, 162, 587.
[24] E.M. Gullikson, E. Gramsch, & M. Szawlowski, (1995) Large-area avalanche photodiodes for the detection of soft x rays, *Appl. Optics*, 34, 4662.
[25] M. Moszynski, M. Kapusta, M. Balcerzyk, M. Szawlowski, & D. Wolski, (2000) Large Area Avalanche Photodiodes in X-ray and scintillation detection. *Nucl. Instr. Meth.*, A442, 230.
[26] E. Gatti, & P. Rehak, (2005) Review of semiconductor drift detectors. *Nucl. Instrum. Methods*, A541, 47.

[27] R. Clarke, W.P. Lowe, R.A. MacHarrie, *et al.*, (1992) Synchrotron radiation applications of charge coupled device detectors, *Rev. Sci. Instrum.*, 63, 784.

[28] A. Fontane, (1997) *Proceedings of School of the Use of Synchrotron Radiation in Science and Technology*, Miramare Trieste, Italy.

[29] E. A. Stern, & K. Kim, (1981) Thickness effect on the extended-X-ray-absorption-fine-structure amplitude. *Phys. Rev. B* 23, 3781.

[30] B. L. Henke, E. M. Gullikson, & J. C. Davis, (1993) X-Ray Interactions: Photoabsorption, Scattering, Transmission, and Reflection at E = 50-30,000 eV, Z = 1-92. *At. Data Nucl. Data Tables* 54, 181.

[31] C. T. Chantler, (1995) Theoretical Form Factor, Attenuation, and Scattering Tabulation for Z=1–92 from E=1–10 eV to E=0.4–1.0 MeV. *J. Phys. Chem. Ref. Data* 24, 71.

[32] L. Troger, D. Arvanitis, K. Baberschke, H. Michaelis, U. Grimm, & E. Zschech, (1992) Full correction of the self-absorption in soft-fluorescence extended X-ray-absorption fine structure. *Phys. Rev. B* 46, 3283.

[33] F.W. Lytle, R. B. Greegor, D. R. Sandstrom, E. C. Marques, J.Wong, C. L. Spiro, G. P. Huffman, & F. E. Huggins, (1984)*Nuclear Instruments & Methods in Physics Research Section a-Accelerators Spectrometers Detectors and Associated Equipment* 226, 542.

[34] H. Oyanagi, K. Sakamoto, R. Shioda, Y. Kuwahara, & K. Haga, (1995) Ge overlayers on Si (001) studied by surface-extended X-ray-absorption fine structure. *Phys. Rev. B* 52, 5824.

[35] Y. Takeda, H. Ofuchi, H. Kyouzu, R. Takahashi, & M. Tabuchi, (2005) Lowest limit for detection of impurity concentration in semiconductors by fluorescence XAFS: resonant Raman scattering and angle dependence. *J. Synchrotron Rad.* 12, 494.

[36] C. H. Booth, & F. Bridges, (2005) Improved self-absorption correction for fluorescence measurements of extended X-ray absorption fine-structure. *Physica Scripta*, T115, 202.

[37] W. Gudat, & C. Kunz, (1972) Close similarity between photoelectric yield and photoabsorption spectra in the soft-X-ray range. *Phys. Rev. Lett.* 29, 169.

[38] J. Stoehr, (1992) *NEXAFS Spectroscopy*: Springer Series in *Surface Sciences* (Springer, Berlin Heidelberg).

[39] R. Nakajima, J. Stöhr, & Y. U. Idzerda, (1999) Electron-yield saturation effects in L-edge X-ray magnetic circular dichroism spectra of Fe, Co, and Ni. *Phys. Rev. B* 59, 6421–6429.

Chapter 6

Time-Resolved and Space-Resolved XAFS

6.1 Introduction

X-ray absorption fine-structure (XAFS) techniques are an excellent analytical choice for resolving the local structure around a selected kind of element in the probed specimen by using the intense X-ray beams. This technique entails both XANES (X-ray absorption near-edge structure) and EXAFS (extended X-ray absorption fine-structure) spectroscopy, providing local information about the chemical state and near-range coordination geometry of the absorbing element, respectively.[1] Performing such measurements on chemically heterogeneous specimens that vary in time and space is very important for the understanding of the dynamics and heterogeneity of the systems. These in-situ and operational XAFS analyses usually requires high temporal and spatial resolution. A consistent scientific background as well as the applicability of various configurations of XAFS was thoroughly discussed in a wonderful review by Bordiga et al. [2] on in situ XAFS study of surface species in heterogeneous catalysts. Many in situ XAFS studies were performed using monochromatic radiation with energy scans.[3,4,5,6]

Nowadays, time-resolved XAFS studies of condensed matter at the atomic level are essential for with an in-depth understanding of the complex mechanisms of dynamic processes in physics and chemistry. Most of these processes occur on the surface, such as layer growth and corrosion phenomena or all catalytic applications. Therefore, XAFS is well-suited for studying short-range systems, such as very thin layers or nanoparticles, which is particularly important for catalytic applications. Adding time resolution to XAFS can access a wide range of new applications. The prerequisite is the high photon flux provided by the third-generation modern synchrotron radiation facility. Obtaining an XAFS spectrum with a signal-to-noise ratio >10^3 in less than a second is statistically feasible and provides a valuable opportunity. The most obvious one is (1) the feasibility to gain deeper structural insights in time-dependent phenomena in physics, chemistry and biology. In addition, for economic reasons, (2) fast XAFS scanning is advantageous, because within the framework of scientific projects, users can conduct more specimen research on the beamline within a limited time. (3) Even in the case of static samples, it is better to measure many spectra, rather than measuring one spectrum with the same total integration time. (4) It is always unavoidable that the sample structure changes unexpectedly within about 20 minutes in the beam. For example, the penetration of the beam causes radiation damage, which is especially true in the case of biological samples. If the sample is not very densely packed powder or exists in a liquid state, the sample will also change due to vibration, so stirring and bubbles may affect the measurement. In addition, (5) random distortion may appear in the spectrum due to spontaneous electrical noise, sudden vibration of the sample, and even sudden beam loss in the storage ring. Measuring many spectra can be controlled, and in many cases these accidents can be filtered out. Although the time-resolved EXAFS method has all these advantages, the experimental technique and the processing of the generated data are much more complicated than conventional EXAFS measurements.

In the last decades several attempts have been made to improve the time resolution of EXAFS measurements, such as replacing the time-consuming stepwise variation of the monochromator by quick-scanning X-ray absorption spectroscopy (QEXAFS).[7] The prerequisite for QEXAFS spectroscopy is a monochromator with a perfect crystal that can only reflect a specific energy determined by the angle of incidence. In the recent QEXAFS setup, the polychromatic beam at the synchrotron radiation source is monochromated by the continuously oscillating channel-cut crystal before the beam passes through the sample.[8] The developed QEXAFS monochromators enable scan rates of above 80 spectra per second, each one covering a 1 keV energy range. Recently by using mechanics to rapidly oscillate the monochromator crystals, time-resolutions of down to a few milliseconds for one spectrum are achievable.

The other experimental device used for time-resolved XANES measurement is called "dispersive XAFS".[9,10] A curved polychromator focuses the "white" beam onto a sample with a width of a few microns. Behind the focal point, the beam diverges again, allowing the entire transmission spectrum to be collected simultaneously by position sensitive detectors or CCD cameras in a single shot.[11,12] Instead of directly exposing the CCD camera to the X-ray beam, the scintillation screen is coupled to the active part of CCD through the lens system.

By replacing the scintillating screen, the energy and time resolution can be easily adjusted to the experimental requirements. A more sophisticated version of dispersive-XAFS on ESRF includes collecting the necessary horizontal divergence obtained from the curved light source of bending magnet. This can be achieved by using multiple optical elements until the radiation interacts with the sample, making this optical scheme more suitable for microfocus / high-brightness applications.[13] Dispersive-XAFS devices are well established at many synchrotron facilities, such as the European Synchrotron Radiation Facility (ESRF), Soleil, Diamond Light Source and SPring-8.[14,15]

The visualization of the heterogeneous structure of solid materials and devices is one of the interesting topics of modern XAFS analysis, because the function of solid materials is highly related to their structure. The conventional XAFS measurement is performed by micron-level X-ray beam, which can provide the average structure information of the target sample in the beam spot. For a powder specimen, the information of local coordination is averaged for all powders with heterogeneous morphology, composition, surface structures, etc. in a beam spot of X-rays. The local coordination of inhomogeneous assemblies cannot be separated by the conventional XAFS measurement. Recently, X-ray beam focusing technology is being developed, and it is possible to prepare micro- to nano-sized X-ray beams in synchrotron facilities. Kirkpatrick-Baez (KB) reflectors and Fresnel zone plates are practical optics that focus X-ray beams to micrometer or even nanometer sizes. The use of this focused X-ray beam can visualize the heterogeneity of solid materials. Micro-XAFS and nano-XAFS have been developed as simple two-dimensional spatial resolution XAFS technology and have been applied to solid samples (such as heterogeneous catalysts).

This chapter presents the work principles of typical setups for time-resolved and space-resolved EXAFS measurement and their applications. For the time–resolved EXAFS setups, the advantages and weakness of both energy dispersive XAFS and quick EXAFS are compared. Micro/nano-XAFS based on focusing optics such as KB mirrors and Fresnel zone plate are discussed. A new concept that comprises both time- and lateral-resolved XAFS information simultaneously in a single shot is introduced.

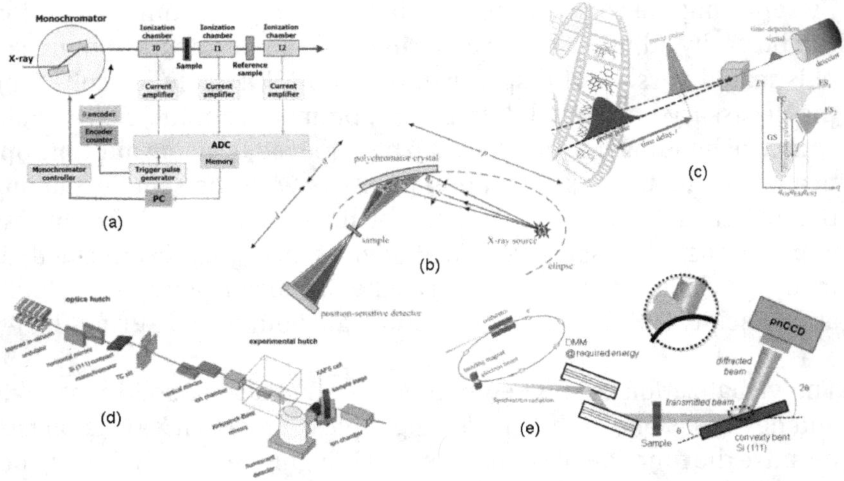

Figure 6.1 (a) Schematic of arrangement and control system for transmission mode QXAFS measurements. (b) Principle of operation of the energy-dispersive XAS spectrometer. The crystal is curved to an elliptical shape. The radius R of the closest circle is given by: $\frac{1}{p} + \frac{1}{q} = \frac{2}{R\sin\theta_0}$, where p, q, and θ_0 are respectively the source–crystal and crystal–sample distances and central Bragg angle. L is the horizontal dimensions of the beam intercepted by the polychromator, and $\frac{L}{\sin\theta_0}$ is the footprint of the beam on the crystal surface. (c) Schematic representation of the pump-and-probe set-up for ultrafast characterization of photo-induced processes. (d) A typical optical setup of nano-XAFS using KB mirrors. (e) Sketch of the set-up for single-shot XANES. The inset shows the detail of the convex shape of the crystal.

6.2 Experimental Set-up

6.2.1 Time-resolved XAFS: QXAFS

QXAFS speeds up conventional step-scan XAFS simply by removing unnecessary operations. Conventionally, in the step scan XAFS measurement, the X-ray absorption of a sample is measured at each energy point using a monochromatic X-ray selected by a double-crystal monochromator (Figure 6.1a). The double-crystal monochromator moves to the target energy and stops, and the X-ray intensities and Bragg angle of the monochromator crystal are then measured. This operation is repeated point by point at each XAFS measurement energy point; for a 200–1000-point measurement, it could loss more than several minutes plus the net X-ray exposure time. In the QXAFS method, the monochromator moves continuously from the start point to the end point during XAFS spectrum scan. The output signal of the X-ray detector and the Bragg angle of the monochromator (*i.e.*, X-ray energy) are separately stored in a high-speed data storage system (*e.g.*, memory) during continuous monochromator scanning. The measurement timing is synchronized by trigger signals sent from the computer through a pulse generation device. The trigger signal interval (*i.e.*, the data sampling time) is usually constant. Therefore, there is no time loss except for the actual X-ray exposure time. QXAFS is also called time-on-the-fly scan mode XAFS. The beamline arrangement and optics of QXAFS can be basically the same as those of conventional step-scan XAFS, thus, conventional XAFS beamlines can adopt an adapted measurement system and control software utilized in the QXAFS method, along with using one double-crystal monochromator for both the step-scan XAFS and QXAFS measurements. Most newly built XAFS beamlines have been installed in QXAFS mode as a basic measurement method.

6.2.2 Dispersive spectrometer for time-resolved experiments

In energy-dispersive XAFS (EDXAS), a curved crystal disperses and focuses a polychromatic X-ray beam onto the specimen (Figure 6.1b). Similar to the monochromator used on the energy scanning spectrometer, the crystal is called a polychromator. The Bragg incidence angle continuously changes from one side of the crystal to the other. Therefore, the energy of the X-ray beam diffracted by the crystal also continuously changes from one side of the crystal to the other, thereby introducing a correlation between the propagation direction of the X-ray beam and its energy. The light beam transmitted by the sample is then detected by a position-sensitive detector, where the energy direction correlation introduced by the polychromator is converted into energy position correlation. To order to collect a XAFS spectrum, the incident and transmitted intensities as a function of photon energy $I_0(E)$ and $I_t(E)$ are measured from the spatial distribution of photons on the detector without and with the specimen respectively. All energy points of the absorption spectrum are acquired simultaneously (in parallel). Thanks to the parallel acquisition scheme, an EDXAS spectrometer operates intrinsically in a time resolved manner. The time resolution is ultimately defined by the performance of the detector (minimum response time and maximum frame rate) and the characteristics of the X-ray source (number of photons per detector pixel). In addition to the acquisition speed, another major advantage of EDXAS is its inherent stability, because there are no moving components during the data acquisition. This leads to a reduction in system noise and a high stability of energy scale and focus position. The combination of speed and stability brings unprecedented sensitivity to the detection of atomic displacement through measurement in differential mode.[16] If a high brilliance source is utilized, focal spots of the order of a few microns can be achieved,[17] and EDXAS becomes an ideal playground for high pressure research [18] or hyperspectral mapping on heterogeneous specimens.[19] EDXAS has also been applied as a chemical and orbital selective probe of magnetism.[20,21]

As compared with energy scanning spectrometers, EDXAS spectrometers mainly offer the following advantages: (1) the high speed of data acquisition, because all energy points are collected in parallel on position-sensitive detectors; (2) a small focal spot size together with a high flux, and (3) the stability of the energy scale and focal spot position, because there is no moving component during XAFS data acquisition. By abandoning parallel detection in the entire energy range and using the sequential acquisition mode, the energy dispersive spectrometer proved its capability for fast-time resolved XAFS studies in fluorescence mode.[22]

However, there exist some limitations that are intrinsic to EDXAS: (1) one need to maintain the energy-direction correlation created by the polychromator throughout the whole path from the crystal to the detector. This poses severe constraints on the morphology and microstructure of the specimen: systems which produce strong small angle scattering are not suited for EDXAS (see for example Ref. [23]). (2) Detection using de-excitation processes (fluorescence, total electron yield, etc.) is impossible, since all energies impinge on the specimen simultaneously. (3) The EXAFS *k*-range is limited at low energies because the energy bandwidth of the polychromatic fan scales with the cotangent of the Bragg angle. (4) Finally, the incident and transmitted intensity (I_0 and I_t) are not measured simultaneously, which may represent a drawback especially in case of incident intensity fluctuations with time (X-ray beam instabilities). These limitations are certainly critical in the development of EDXAS. Although this technique has spread quite rapidly, the EDXAS user community remains a minor portion of global XAFS community. In addition to these intrinsic limitations, the technical challenges to build and use an EDXAS spectrometer are important: for example, dealing with stability issues, designing efficient polychromator crystals and benders, or developing dedicated detectors.

In the field of dynamical studies, EDXAS beamlines have been historically employed for the investigations of homogeneous and heterogeneous catalysis.[24] The typical timescales of this kind of experiments ranges from minutes down to milliseconds. The new development of fast energy scanning spectrometers (QEXAFS)[25] enable XAFS scanning at the tens of millisecond timescale. In addition, these experiments often deal with small absorber concentrations (roughly <1% wt) and can then benefit from the fluorescence detection mode on the scanning spectrometers. Therefore, the interest of using EDXAS spectrometers on these timescales has decreased.

6.2.3 Pump-probe approach

Dynamical processes on timescales faster than 1 millisecond can be investigated by the scanning spectrometers using the pump-and-probe approach where the time resolution is defined by the delay between the pump (start) and the probe (stop), see Figure 6.1c. The time resolution is then limited either by the duration of the pump or by the duration of the probe. In practice, the time resolution is very often limited by the duration of the X-ray bunch, *i.e.*, around 100 ps. The time evolution of the absorption is recorded at a single energy. But an EDXAS spectrometer can track the process in "film" mode, *i.e.*, as the dynamical process evolves in time, complete EXAFS spectra are continuously recorded one after the other. By combining a high flux source, an EDXAS spectrometer allows to obtain a sufficient signal-to-noise ratio in a single "film". This is particularly important for experiments that cannot be cycled indefinitely [26] or where the specimen is destroyed after the run.[27] An EDXAS spectrometer can also operate in pump-and-probe mode. Again, the technique is particularly suited to single shot experiments where the specimen is lost at the end of the run or/and when the dynamical process differs from shot to shot.[28]

6.2.4 Space–resolved XAFS (Principle of micro/nano-XAFS)

In the optical setup of micro-XAFS or nano-XAFS (Figure 6.1d), an X-ray beam from synchrotron storage ring is focused down by a set of Kirkpatrick–Baez (KB) mirrors or Fresnel zone plates, which is composed of two elliptically bended mirrors for vertical and horizontal focusing. A specimen is mounted on a piezo stage and set at the focusing point of the X-ray beam. Fluorescent X-rays from the specimen is detected on an appropriate detector (such as a Ge detector) inclined to the specimen. Focusing size of X-ray beams depends on the accuracy of the total reflection KB mirrors and beamline at a synchrotron facility. Recently, hard X-ray beams for XAFS measurements can be focused down to micron even nanometer size, and XAFS measurements using the focused X-ray beams are called as micro-XAFS or nano-XAFS, depending on X-ray beam size. The space resolution of the XAFS measurements is determined by both X-ray beam size and the scanning distance of a piezo stage that a specimen is mounted. Many synchrotron facilities offer the space-resolved XAFS capabilities, for example, SPring-8 (Hyogo, Japan) has several 100 nm-order X-ray beams installed in the hard X-ray region for nano-XAFS analysis.[29]

Fluorescent X-rays from a specimen irradiated by the focused X-rays are often so weak that micro/nano-XAFS measurements often require much longer acquisition time, particularly, in the case of EXAFS data collection. The drifts of the optics and specimen stage are serious to deteriorate the space resolution of nano-XAFS measurements with long acquisition time. The temperature drift often affects the drifts of the optical system, so a set of KB mirrors is often placed in a closed chamber filled with inert helium gas.

The visualization of the heterogeneous structure of solid materials is one of the important goals of microscopic analysis. Microscopic techniques such as STM and AFM provide atomic-level information on solid surfaces, which is very important for understanding heterogeneous catalysis. Electron microscopes (such as TEM) show the morphology and lattice parameters of nano-scale solid materials. However, chemical information (oxidation state, symmetry, local coordination, etc.) cannot be obtained from these microscopic techniques. Micro/nano-XAFS shows the powerful ability to monitor the chemical information of heterogeneous solid materials on the micron/nano scale. For powder samples (which is one of the typical forms of solid materials), micro/nano-XAFS can separate structural information at the particle level. When the size of the focused X-ray beam is similar to the size of a single solid particle, micro/nano-XAFS can provide the structural information in the single particle, which corresponds to the average value of all atoms of the target element in the particle. By focusing an X-ray beam to a size smaller than the sample particles, one can obtain the spatial information inside a single particle, and a two-dimensional image of the elemental distribution and chemical state of target elements.

6.2.5 Time- and lateral-resolved XAFS in a single shot

A newly developed set-up (at the BAM line at BESSY-II, Germany, Figure 6.1e) combines both time- and simultaneous spatial-resolution while acquiring XAFS spectra in a single shot.[30] The synchrotron radiation light from 1.7GeV storage ring is filtered by a double-layer monochromator to produce a primary broadband beam. The primary broadband beam passes through the sample. Then the transmitted beam is diffracted by the convexly curved Si (111) crystal to produce a divergent beam. The divergent /spread beam is collected by an energy sensitive area detector (so-called color X-ray camera) or by an area-sensitive detector which is based on a CCD camera, in θ–2θ geometry. The most attractive features of this configuration are: (1) this static setting even avoids moving the sample for spatial resolution; (2) since the beam need not to be focused, a simple and cheap Si (111) wafer can be used as the crystal analyzer. This system can easily achieve a temporal resolution of less than 1 s and a one-dimensional spatial resolution of at least 50 μm.

The incident beam interacts with the sample, and the transmitted part is diffracted by the convexly bent Si (111) crystal. For the single-shot XANES method, the divergent beam is collected by an energy-sensitive area detector (a pnCCD chip of a color X-ray camera (CXC)). For single-shot EXAFS, an area-sensitive detector based on a CCD camera is used. The bent Si (111) crystal and the detector are placed under conventional θ–2θ diffraction geometry. The achievable experimental energy resolution is limited by its inherent resolution ($\Delta E/E \simeq 1.4\times10^{-4}$) and the curvature radius of bent Si(111) crystal. However, stronger bending results in a larger reflected energy range and worse energy resolution. The detectable energy range is limited by the size of the detector. Therefore, for higher or lower expected energy, the radius of curvature and geometry must be adjusted. The calculation of the radius of curvature must be performed before the measurement. Then, the Bragg equation is used to calculate the angle of incidence. Although X-rays diverge due to the convexly shaped crystals, the radius of curvature (r) can be easily calculated by using virtually convergent X-rays beam. The intersection point corresponds to the center of the circumference, and its distance to the crystal surface is the radius. Taking the geometric relationship, the radius, r, is as $\boldsymbol{r} = \frac{\boldsymbol{BW}}{\sin\boldsymbol{\beta} - \sin\boldsymbol{\alpha}}$, where BW is the beam width, α and β are the adjacent angles to the incoming angles θ_1 and θ_2, respectively.

6.3 Applications

Time-resolved XAFS

Currently, the time-resolution of *in situ* XAFS ranges from seconds to picoseconds. Picosecond time-resolved XAFS can directly probe the spin state dynamics coupled with electronic and structural modifications. For example, Aqueous $[FeII(phen)_3]^{2+}$ can be excited from the low-spin (LS) ground state to high-spin (HS) by using a Ti:Sapphire femtosecond laser, and picosecond X-ray pulse can be utilized as a probe for the transient spin-state dynamics.[31] The changes in the electric state and molecular structure are reflected in a Fe *K*-edge XAFS spectrum of the low-spin and high-spin analogy ground states. The spin crossover transition from the $^1A_1$ low-spin state to $^5T_2$ high-spin state was inferred from the changes of the XANES features of the Fe(II) 1s-3d transition before and after pulse excitations. The quantitative analysis of EXAFS data revealed that the Fe-N distance of the low-spin state is 1.98 Å and expanded to 2.15 Å after 50 ps of photo-excitation, consistent with the value in the high-spin analogue (2.15 Å). The Fe-C shell is also expanded from 2.88 Å to 3.01 Å in the photoexcited high-spin state, corresponding to the photo-induced movement of the phenanthroline ligands. This study exemplifies that the picosecond time-resolved XAFS technique is a powerful tool for capturing the spin crossover transitions in ultrafast photo-induced reactions as well as in the magnetic photocatalysts, nanomagnets, and many other disordered magnetic systems.

By taking advantage of the time structure of storage ring of synchrotron radiation sources, pump/probe scheme has been successfully utilized in time-resolved XAFS technique in detecting rapid processes.[32] In 2001, the first nanosecond-scale time-resolved laser pump/X-ray probe XAFS studies were conducted by Chen and coworkers using X-ray pulses from a third-generation synchrotron source.[33] The experiments were conducted at a wiggler beamline at APS with 100 ps at full-width-at half-maximum of the duration of the X-ray pulses. This time-scale is suitable for capturing intermediate structures with sub-nanosecond resolution, and allows to take "snapshots" of molecules when the population of the laser-generated transient species reaches a maximum. The transient molecular structure of a reaction intermediate produced by photoexcitation of $NiTPPL_2$ (NiTPP, nickeltetraphenylporphyrin; L, piperidine) in solution can be determined by pump/probe scheme. The Ni *K*-edge XANES spectra for the laser pumped $NiTPPL_2$, along with those of the ground states $NiTPPL_2$ and NiTPP show a distinct shoulder at 8.340 keV, which is attributed to the 1s to $4p_z$ transition in a square planar transition metal complex. It appears for the ground state NiTPP and the laser-pumped specimen, however, but with a lower amplitude. Also, it disappears as the pump laser beam is blocked. The laser-irradiation creates a transient square planar NiTPP-like structure, whose XANES spectrum could be reconstructed by mixing the ground state $NiTPPL_2$ and NiTPP-like spectra with a ratio of ~7:3. Fourier-transformed EXAFS spectra clearly show that the nearest Ni-N distances could be best described by two distances of 2.10 Å and 1.92 Å, with a relative ratio of 3.2:1. In contrast, the spectra of ground state NiTPP and $NiTPPL_2$ could be best fit with a single Ni-N distance (1.91 and 2.09 Å for NiTPP and $NiTPPL_2$, respectively). The XAFS measurement results confirmed that light excitation can lead to the rapid removal of the two axial ligands, resulting in a transient square planar intermediate NiTPP with a lifetime of 28 nanoseconds. Therefore, the transient structure of the photolysis intermediate adopts the same structure as the ground state in a non-coordinating solvent, and then recombines with two ligands to form a more stable octahedral coordinated NiTPPL2.

Starting with this groundbreaking attempt, X-ray transient absorption (XTA) spectroscopy has been rapidly developed. It uses a laser pulse as a pump to initiate a chemical reaction synchronously among different molecules and an X-ray pulse as a probe to detect transient structures associated with the reaction dynamics as a function of the delay time between the pump and probe. After more than a decade of development, X-ray transient absorption spectroscopy has become an increasingly important tool in tracking molecular structures in chemical reactions and to map out the structural factors that may affect the reaction pathways.[34,35,36]

Space-resolved XAFS

Besides the time-resolved XAFS technique, space-resolved XAFS has also been developed and applied to studies of functional materials. This valuable technique is capable to quantify and map the specimens of mixed oxidation states at high spatial resolution, elemental selectivity, and chemical speciation. By using the scanning μ-X-ray fluorescence XAFS technique, Martınez-Criado *et al.* did micro-x-ray absorption near-edge structure imaging for detecting metallic Mn distribution in GaN.[37] The GaN specimens with Mn concentrations ranging from 0.05% up to 11% were prepared by low temperature plasma-assisted MBE method. The scanning μ-X-ray fluorescence (XRF) and XANES measurements were performed at the ID21 and ID22 X-ray microprobe beamlines of the European Synchrotron Radiation Facility (ESRF). The monochromatic beam of ID21 beamline was focused on the specimen to a spot of 0.6 × 0.9 μm^2 size with a focus depth of ~0.4mm by means of Au Fresnel zone plates as focusing lenses. At the ID22 beamline, the hard X-ray monochromatic beam was focused on the specimen to a spot of 1.5 × 3.5 μm^2 size with a focus depth of 300 μm by Kirkpatrick-Baez mirrors. Two-dimensional X-ray fluorescence maps were taken at preferential monochromatic energies, which excited different Mn oxidation states. The XRF images were obtained by measuring the Mn $K\alpha$ line intensities at different excitation energies. It is shown that for the specimen with lowest Mn content (0.05%), Mn element is distributed homogeneously on the micrometer scale for all the chosen energies. On the contrary, for the highest Mn concentration (11%), different Mn patterns appear as a function of the incident energy. For GaN specimens with lower Mn content, a preferential homogeneous distribution of mixed Mn^{2+} and Mn^{3+} states was observed in the Mn oxidation state maps. For the highest Mn-doped GaN specimen, the XANES image features the supplementary Mn^{2+} and Mn^0 dominant centers and lacks signal from Mn^{3+}. The presence of specific cluster-like features in Mn^0 pattern suggests the surface segregation of metallic Mn centers.

On the same beamline of ID22 of the ESRF, Segura-Ruiz and coworkers used a hard X-ray nanoprobe to investigate the local structure of single Co-implanted ZnO nanowires with a Co concentration of 0.6 at.% relative to Zn.[38] The monochromatic X-ray beam was focused to 100 × 100 nm^2 spot size using a pair of Kirkpatrick-Baez Si mirrors. The uniform Zn and Co distributions along the wires are confirmed by X-ray fluorescence maps. The EXAFS analysis reveals the substitution of Co^{2+} ions in Zn sites and the high structural order in the host lattice. The polarization-dependent Zn *K*-edge XANES spectra at different points along the nanowires (Figure 6.2) reveal the high structural order of the host lattice along both radial and axial directions. Specifically, the XANES data with the electric field of the absorbed photons perpendicular to the c-axis (E⊥c) probe the symmetry of the ab-plane, namely p_{xy} conduction band states. The spectral features in Figure 6.2a reflect the typical wurtzite structure of the ZnO lattice, and signatures of Zn metallic precipitates after the ion implantation process are observed. No structural disorder in the radial direction along the implanted nanowires was found according to the line shapes and peaks energies. For the axial direction (E//c), the XANES data probe the symmetry of the c-plane, namely p_z conduction band states. Again, the spectra exhibit the peaks associated to the hexagonal structure, without any evidence of lattice damage along the nanowires (Figure 6.2b).

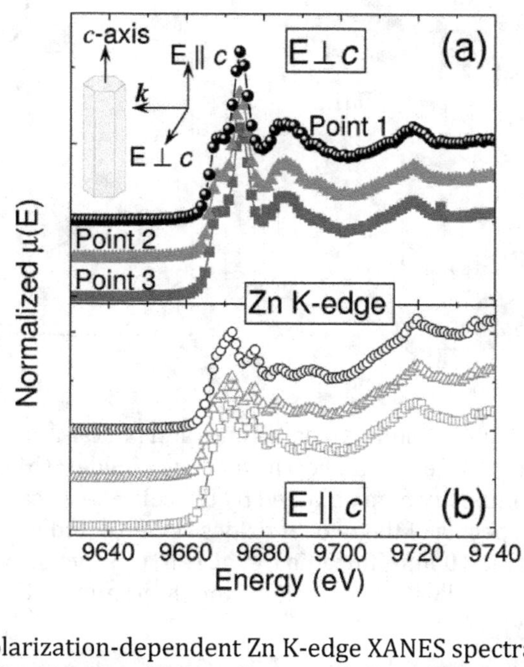

Figure 6.2 Polarization-dependent Zn K-edge XANES spectra of Co-implanted ZnO nanowires [Segura-Ruiz, et al. (2011) Nano Letters 11(12), 5322–5326, reproduced with permission from American Chemical Society].

Measurements on Fe_xO_y compounds: time- and lateral-resolution

By combining an energy dispersive device with an energy sensitive area detector, it is feasible to capture time- and lateral-resolved XAFS spectrum in a single shot. As color X-ray camera is an energy-sensitive area detector and offers the possibility of simultaneously providing time- and spatial-resolution in one dimension, this means that one can distinguish different oxidation states within the analyzed specimen area. The scheme of this principle together with the specimen information and obtained spectra are displayed in Figure 6.3. Two different Fe-oxide compounds [39] γ-Fe_2O_3 and FeO in powder were prepared and placed in a 1 mm-thick specimen holder to obtain

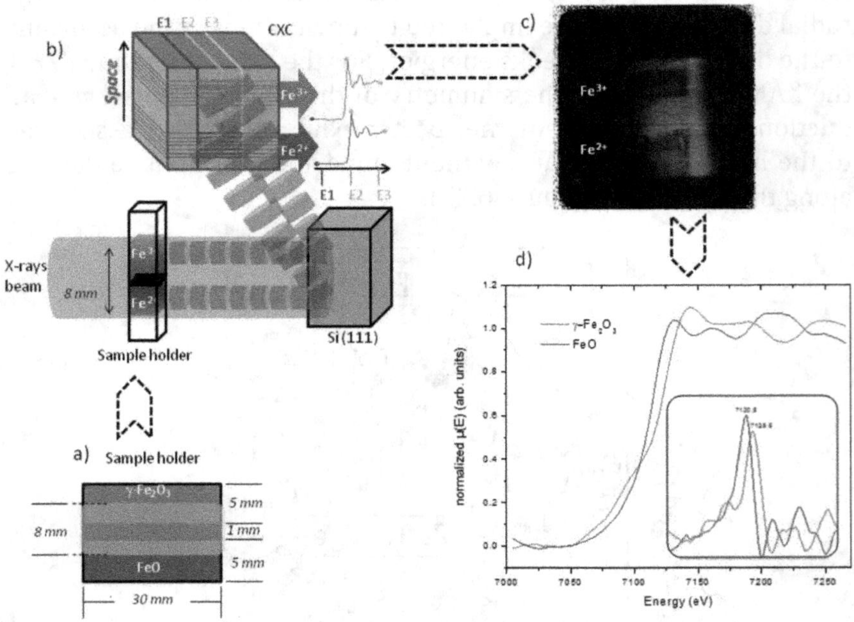

Figure 6.3 Principle for simultaneous time- and lateral-resolved XANES measurements: (a) γ-Fe_2O_3 and FeO in specimen holder; (b) sketch of the used setup; (c) intensity map obtained by the color X-ray camera; (d) correspondent spectra of the two Fe-oxides (γ-Fe_2O_3 and FeO); measurement time 10 min [Buzanich, et al. (2016) J. Synchrotron Rad. 23, 769–776, reproduced with permission from International Union of Crystallography].

an edge jump of about 10. For this purpose, both powders were diluted with boron nitride in order to fill the whole 1 mm-thick holder.

In this case the different oxides were separated by 1 mm spacing (Figure 6.3a). A beam of 8 mm height was used to cover both oxidation states. The transmitted signal carries information about both oxidation states. After being diffracted by the bent crystal, the signal is recorded by the pnCCD (Figure 6.3b). As the measurements were carried out in 'deposited energy mode', it results in the intensity map as displayed in Figure 6.3c. Thus, the two oxidation states could be distinguished in one shot (Figure 6.3d), exhibiting the expected edge shift of about 5 eV, which is better visible by the first-derivative plot within the same figure. The achievable lateral resolution in this geometry is limited by the pixel size of the pnCCD detector.

6.4 Conclusion and Perspective

Time and space-resolved XAFS techniques provide unique tool for the study of dynamics and heterogeneity of various systems, which are critical issues for property, functionality, and catalysis of nanomaterials and devices such as catalysts, functional materials, fuel cells, batteries, biomaterials, and marine resources. Instead of using step-scan of monochromator in conventional XAFS, QXAFS mode continuously sweeps the energy range through the absorption spectrum, so that time-resolved XAFS measurement can be achieved. The minimum time resolution of QXAFS is governed by the mechanical motion of monochromator scan speed and the response time of the ionization chamber, and typically in a range from about one millisecond to a few seconds when the concentration of sample is high enough. The performance of the present monochromator and ionization chamber seems to be close to the limit. So new technology for the monochromator and detector may be required to achieve higher time resolution for QXAFS. Also, the measurement of dilute samples or thin films needs a faster and more efficient fluorescence X-ray detector.

The method combined with QXAFS is expected to further spread to synchrotron radiation X-ray based techniques such as X-ray computed tomography, resonant inelastic X-ray scattering, and X-ray emission spectroscopy. Time-resolved imaging techniques will be key for the investigation of the dynamic aspect of the actual heterogeneous chemical reaction. 2D spatiotemporally resolved QXAFS studies have just begun, and applications are expected to increase in various research fields. Recently, a 3D spatial-resolved XAFS method with a resolution of several tens of nm has been applied under ex situ conditions. 3D spatiotemporally resolved XAFS is expected to be realized with such technology developments.

Unlike the conventional XAFS and QXAFS methods in which monochromatic X-ray beam is obtained by scanning the double-crystal monochromator, in the energy-dispersive XAFS technique, a polychromator (typically a bended crystal) is utilized to disperse the incident white X-ray beam, and all the data in the whole energy range of the absorption spectrum is simultaneously collected by a position-sensitive multichannel detector. Instead of placing a monochromator ahead of the sample, alternatively, by placing an energy-dispersive polychromator after the sample, it is possible to perform the time- and simultaneous lateral-resolved XAFS measurements in a single shot. This setting provides a larger analysis area (8–10 mm) in one shot. It can analyze layered samples and can even study differences in oxidation state and even local coordination environment in carbon monoxide. Tests conducted on a metal reference foil have shown that this set-up has a good signal-to-noise ratio and a time resolution of a few seconds is feasible because the sample can be fully absorbed (edge jump is at least 4).

The spatially resolved micro-XAFS and nano-XAFS technologies using focused X-ray beams can provide new insights into 2D space visualization and 2D mapping of morphology, composition, domain structure, coordination structure, oxidation state, fractire, corrosion, aggregation, etc. In-situ and ex-situ XAFS technology can also image the depth profile of the film. These technologies can be combined with other technologies, such as TEM / STEM-EDS and environmental pressure HAXPES, from a combined point of view, they can solve complementary analysis results. Element-selective nano-XAFS is a promising technology that can visualize the uneven presence and distribution of chemical states in single micro- / nano- particles and multiple nanomaterials. In the near future, time-resolved and space-resolved XAFS experiments will play an increasingly important role in in-situ / operational studies of heterogeneous catalysis and other systems.

[1] G. Bunker, (2010) *XAFS – A Practical Guide to X-ray Absorption Fine Structure Spectroscopy.* Cambridge University Press.

[2] S. Bordiga, E. Groppo, G. Agostini, J. A. van Bokhoven, & C. Lamberti, (2013) Reactivity of surface species in heterogeneous catalysts probed by in situ X-ray absorption techniques. *Chem. Rev.* 113, 1736–1850.

[3] S. R. Bare, S. D. Kelly, B. Ravel, N. Greenlay, L. King, & G. E. Mickelson, (2010) Characterizing industrial catalysts using in situ XAFS under identical conditions. *Phys. Chem. Chem. Phys.* 12, 7702–7711.

[4] D. Koziej, M. Huebner, N. Barsan, U. Weimar, M. Sikora, & J. D. Grunwaldt, (2009) Operando X-ray absorption spectroscopy studies on Pd-SnO_2 based sensors. *Phys. Chem. Chem. Phys.* 11, 8620–8625.

[5] E. Kleymenov, J. Sa, J. Abu-Dahrieh, D. Rooney, J. A. van Bokhoven, E. Troussard, J. Szlachetko, O. V. Safonova, & M. Nachtegaal, (2012) Structure of the methanol synthesis catalyst determined by in situ HERFD XAS and EXAFS. *Catal. Sci. Technol.* 2, 373–378.

[6] W. Gawelda, (2006) Thesis, *Faculte des Sciences de Base*, Lausanne, Switzerland.

[7] R. Frahm, J. Stoetzel, & D. Luetzenkirchen-Hecht, (2009) *Workshop on Energy Dispersive X-ray Absorption Spectroscopy Scientific Opportunities and Technical Challenges*, p. 23. ESRF, Grenoble, France.

[8] J. Stoetzel, D. Luetzenkirchen-Hecht, & R. Frahm, (2011) A new standalone QEXAFS data acquisition system for in situ studies. *J. Synchrotron Rad.* 18, 165–175.

[9] A. M. Flank, A. Fontaine, A. Jucha, M. Lemonnier, D. Raoux, & C. Williams, (1983) EXAFS in dispersive mode. *Nucl. Instrum. Methods Phys. Res.* 208, 651–654.

[10] F. D'Acapito, F. Boscherini, A. Marcelli, & S. Mobilio, (1992) Dispersive EXAFS apparatus at Frascati. *Rev. Sci. Instrum.* 63, 899–901.

[11] S. Pascarelli, T. Neisius, & S. De Panfilis, (1999) Turbo-XAS: dispersive XAS using sequential acquisition. *J. Synchrotron Rad.* 6, 1044–1050.

[12] J. C. Labiche, O. Mathon, S. Pascarelli, M.A. Newton, G.G. Ferre, C. Curfs, G. Vaughan, A. Homs, & D.F. Carreiras, (2007) The fast readout low noise camera as a versatile X-ray detector for time resolved dispersive extended X-ray absorption fine structure and diffraction studies of dynamic problems in materials science, chemistry, and catalysis. *Rev. Sci. Instrum.* 78, 091301.

[13] S. Pascarelli, G. Aquilanti, L. Dubrovinsky, G. Guilera, O. Mathon, M. Munoz, M. A. Newton, M. Pasquale, & A. Trapananti, (2006) *Proceedings of the 13th International Conference on X-ray Absorption Fine Structure* (XAFS13), 9–14 July 2006, Stanford, California, USA.

[14] M. Hagelstein, S. Cunis, R. Frahm, W. Niemann, & P. Rabe, (1989) The energy dispersive X-ray absorption spectrometer DEXAFS at HASYLAB. *Physica B*, 158, 324–325.

[15] D. Bhattacharyya, A. K. Poswal, S. N. Jha, Sangeeta & S. C. Sabharwal, (2009) First results from a dispersive EXAFS beamline developed at INDUS-2 synchrotron source at RRCAT, Indore, India. *Nucl. Instrum. Methods Phys. Res. A*, 609, 286–293.

[16] R. F. Pettifer, O. Mathon, S. Pascarelli, M. D. Cooke & M. R. J. Gibbs, (2005) The measurement of tiny displacements using X-ray absorption spectroscopy, *Nature* 435, 78–81.

[17] S. Pascarelli, & O. Mathon, (2010) Advances in high brilliance energy dispersive X-ray absorption spectroscopy. *Phys. Chem. Chem. Phys.* 12, 5535–5546.

[18] G. Aquilanti, O. Mathon, & S. Pascarelli, (2009) Achievements in high-pressure science at the highbrilliance energy-dispersive X-ray absorption spectrometer of ESRF, ID24. *J. Synchrotron Radiat.* 16, 699–706.

[19] M. Muñoz, V. De Andrade, O. Vidal, E. Lewin, S. Pascarelli, & J. Susini, (2006) Redox and speciation micromapping using dispersive X-ray absorption spectroscopy: application to iron chlorite mineral of a metamorphic rock thin section. *Geochem. Geophys. Geosyst.* 7, Q11020-1–Q11020-10.

[20] F. Baudelet, C. Brouder, E. Dartyge, A. Fontaine, J. P. Kappler, & G. Krill, (1990) Magnetic properties of $ReFe_2$, $CeFe_2$, $GdFe_2$, $LuFe_2$ compounds studied by magnetic X-ray dichroism. *Europhys Lett.* 13, 51–757.

[21] B. Levason, & J. Evans, (2014) Following chemical structures using synchrotron radiation. *Coord. Chem. Rev.* 277–278, 1–300.

[22] S. Nagai, K. Dohmae, Y. Ikeda, *et al.*, (2008) In situ redispersion of platinum autoexhaust catalysts: an on-line approach to increasing catalyst lifetimes? *Angew Chem. Int. Ed.* 47, 9303–9306.

[23] S. Pascarelli, T. Neisius, & S. De Panfilis, (1999) Turbo-XAS: dispersive XAS using sequential acquisition. *J Synchrot Radiat* 6, 1044–1050.

[24] M.A. Newton, A.J. Dent, In: J.A. Rodriguez, J.C. Hanson, P.J. Chupas (eds), (2013) *Energy-dispersive EXAFS: principles and application in heterogeneous catalysis, in-situ characterization of heterogeneous catalysts.* John Wiley & Sons, Ltd, Hoboken.

[25] J. Stoetzel, D. Lützenkirchen-Hecht, R. Frahm, (2010) A new flexible monochromator setup for quick scanning X-ray absorption spectroscopy. *Rev. Sci. Instrum.* 81, 073109-1–073109-4.

[26] J.L. Her, Y.H. Matsuda, M. Nakano, Y. Niwa, & Y. Inada, (2012) Magnetic field-induced spin-crossover transition in [MnIII(taa)] studied by X-ray absorption spectroscopy. *J. Appl. Phys.* 111, 053921-1–053921-4.

[27] C. Marini, F. Occelli, O. Mathon, R. Torchio, V. Recoules, S. Pascarelli, & P. Loubeyre, (2014) A microsecond time resolved X-ray absorption near edge structure synchrotron study of phase transitions in Fe undergoing ramp heating at high pressure. *J. Appl. Phys.* 115, 093513.

[28] R. Torchio, F. Occelli, O. Mathon, *et al.*, (2016) Microscopic properties of shocked Fe studied by single pulse synchrotron XAS measurements to 500 GPa. *Sci. Rep.* 6, 26402.

[29] Y. Terada, H. Tanida, T. Uruga, A. Takeuchi, Y. Suzuki, & S. Goto, (2011) High-resolution X-ray microprobe using a spatial filter and its application to micro-XAFS measurements. *AIP Conf. Ser.* 1365, 172–175.

[30] A. G. Buzanich, M. Radtke, U. Reinholz, H. Riesemeier, & F. Emmerling, (2016) Time- and spatial-resolved XAFS spectroscopy in a single shot: new analytical possibilities for in situ material characterization, *J. Synchrotron Rad.* 23, 769-776.

[31] S. Nozawa, T. Sato, M. Chollet, K. Ichiyanagi, A. Tomita, H. Fujii, S. Adachi, & S. Koshihara, (2010) Direct Probing of Spin State Dynamics Coupled with Electronic and Structural Modifications by Picosecond Time-Resolved XAFS. *J. Am. Chem. Soc.* 132, 61.

[32] C. Bressler, & M. Chergui, (2004) Ultrafast X-ray absorption spectroscopy. *Chem. Rev.* 104, 1781.

[33] L.X. Chen, W.J.H. Jager, G. Jennings, D.J. Gosztola, A. Munkholm, & J.P. Hessler, (2001) Capturing a Photoexcited Molecular Structure Through Time-Domain X-ray Absorption Fine Structure. *Science* 292, 262.

[34] M. Newcomb, J.A. Halgrimson, J.H. Horner, E.C. Wasinger, L.X. Chen, & S.G. Sligar, (2008) X-ray absorption spectroscopic characterization of a cytochrome P450 compound II derivative. *Proc. Natl. Acad. Sci. U. S. A.* 105, 8179.

[35] M.R. Harpham, S.C. Nguyen, Z.R. Hou, J.C. Grossman, C.B. Harris, M.W. Mara, A.B. Stickrath, Y. Kanai, A. M. Kolpak, D. Lee, D.J. Liu, J. P. Lomont, K. Moth-Poulsen, N. Vinokurov, L.X. Chen, & K.P.C. Vollhardt, (2012) X-ray Transient Absorption and Picosecond IR Spectroscopy of Fulvalene (tetracarbonyl) diruthenium on Photoexcitation. *Angew. Chem. Int. Ed.* 51, 7692.

[36] L.X. Chen, & X.Y. Zhang, (2013) Photochemical processes revealed by X-ray transient absorption spectroscopy. *Journal of Physical Chemistry Letters* 4, 4000.

[37] G. Martinez-Criado, A. Somogyi, A. Homs, R. Tucoulou, & J. Susini, (2005) Micro-X-ray absorption near-edge structure imaging for detecting metallic Mn in GaN. *Appl. Phys. Lett.* 87, 061913.

[38] J. Segura-Ruiz, G. Martinez-Criado, M.H. Chu, S. Geburt, & C. Ronning, (2011) Nano-X-ray absorption spectroscopy of single Co-implanted ZnO nanowires. *Nano Lett.* 11, 5322.

[39] A. G. Buzanich, M. Radtke, U. Reinholz, H. Riesemeier, & F. Emmerling, (2016) Time-and spatial-resolved XAFS spectroscopy in a single shot: new analytical possibilities for in situ material characterization. *J. Synchrotron Rad.* 23, 769–776.

Chapter 7

Quantitative EXAFS Analysis

7.1 Introduction

Extended X-ray-absorption fine-structure (EXAFS) refers to the rippled oscillation structure in the X-ray absorption coefficient spectrum above the edge. Analysis of the EXAFS provides fine details of the partial pair distribution functions of atoms surrounding the absorber, including distances, mean square deviations in distance, coordination numbers, and coordinating species.[1] Because EXAFS is caused by the interference between the outgoing wave of the photoelectron excited from the core level of absorbing atom and the backscattered electron wave by the adjacent atoms, EXAFS is equally applicable to all forms of condensed matter, even to amorphous systems. Because the basic physical process of X-ray absorption spectroscopy (*i.e.*, the photo-excitation of a deep core electron) is element specific. Consequently, EXAFS offers a valuable probe to the local structure of absorbing element in a wide variety of scientific disciplines, with applications in life science, earth and environmental science, chemistry, materials science, cultural heritage, and more. EXAFS has become one of the core competencies of any synchrotron radiation facility.

The analysis of EXAFS begins with a good quality experimental data, and at least some understanding of the physical and chemical environment of the specimen that gave that spectrum. Because EXAFS can be measured in a number of modes, and over a wide range of specimen conditions and concentration of the absorbing element, care must be taken to ensure that the spectrum being analyzed has been corrected for systematic measurement errors such as over-absorption effects [2,3] and detector saturation.[4] Here, a good measurement of $\mu(\omega)$ is simply assumed. The basic data processing steps used today was outlined by Lytle, Sayers and Stern.[5] The fine structure $\chi(\omega)$ is isolated by subtracting the slowly varying background from the measured absorption data. The energy axis of the fine structure spectrum is converted to photoelectron wavenumber, k, and then $\chi\ (k)$ is Fourier transformed in distance R-space (Figure 7.1). The complex $\tilde{\chi}(R)$ data, which resemble a pair distribution function, are interpreted to yield quantitative information about the local configuration environment around the absorbing atom by using EXAFS formula. Many works give more complete and detailed discussions of EXAFS analysis. [6]

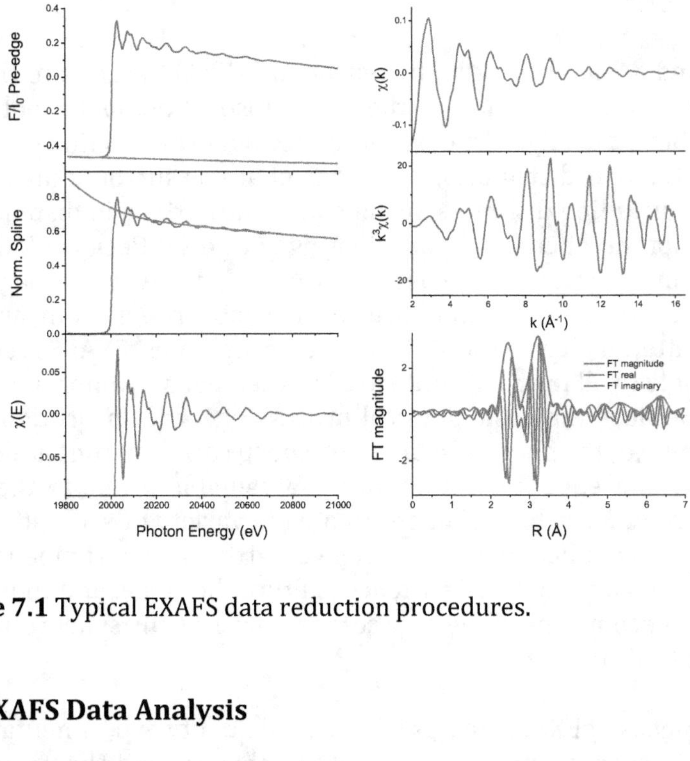

Figure 7.1 Typical EXAFS data reduction procedures.

7.2 EXAFS Data Analysis

7.2.1 Data reduction

For all EXAFS data, whether measured in transmission or (fluorescence/electron) emission mode, the data reduction and analysis are essentially the same. First, the measured intensity signals are converted to $\mu(\omega)$, normalized to unity edge jump, and then reduced to $\chi(k)$. Normalized $\mu(\omega)$ is suitable for XANES analysis, while EXAFS modeling using the EXAFS equation requires background-subtracted $\chi(k)$. In this section, these steps of data reduction will be given in some details.

For transmission measurements, the Beer–Lambert law, $I = I_0 e^{-\mu t}$, can be used to convert measured intensities to a quantity proportional to the absorption coefficient, $\mu(\omega)$. Here I_0 is the X-ray intensity incident on a specimen, I is the X-ray intensity transmitted through the specimen, and t is the specimen thickness. As the specimen thickness t is a constant and can be ignored, then this can be rearranged to give $\mu(\omega) = ln(\frac{I_0}{I})$.

Usually neither I_0 nor I is an absolute intensity by itself, but signals from ion chambers, photodiodes, or other detectors that are proportional to these intensities. Since the $\mu(\omega)$ will be re-scaled early in the data reduction, this not a problem. For measurements made by detecting emission of the excited atom (including X-ray fluorescence, Auger, and optical luminescence), $\mu(\omega)$ is given as $\mu(\omega) = \frac{I_f}{I}$, where I_f is the emitted signal of interest.

No matter whether $\mu(\omega)$ is measured in transmission or fluorescence (or electron emission), the data reduction and analysis are essentially the same. First the raw data is reduced and then can be analyzed using the EXAFS equation. The basic steps of data reduction are addressed below:

Self-absorption correction

For EXAFS measured in fluorescence $\mu(\omega) \propto I_f/I_0$ is a slight oversimplification. The probability of fluorescence is proportional to the absorption probability, but the measured fluorescence (photon) has to travel from the absorbing atoms in the specimen to the detector. Since all matter attenuates X-rays, the fluorescence intensity, and therefore the EXAFS oscillations, can be damped due to this self-absorption effect. Assuming that the specimen is at 45° to both the incident beam and the fluorescence detector, the measured fluorescence intensity can be expressed as

$$I_f(\omega) = I_0\left(\frac{\Omega}{4\pi}\right)\gamma\frac{\mu_\chi(\omega)}{\mu_{tot}(\omega)+\mu(\omega_f)}\left\{1 - e^{-\left(\mu_{tot}(\omega)+\mu(\omega_f)\right)t}\right\}, \qquad (7.1)$$

where γ is the fluorescence efficiency, Ω is the solid angle of the detector, ω_f is the energy of the fluorescent X-ray, $\mu_\chi(\omega)$ is the absorption from the element of interest, and $\mu_{tot}(\omega)$ is the total absorption in the specimen, $\mu_{tot}(\omega) = \mu_\chi(\omega) + \mu_{other}(\omega)$. For concentrated specimens, $\mu_\chi(\omega)$ and μ_{other} are comparable, and the EXAFS will be severely damped by self-absorption. For very concentrated specimens, $\mu_{tot}(\omega) \approx \mu_{\chi}(\omega)$, and the XAFS can be completely wiped out. The self-absorption can be corrected using the above equation if it is not too severe. Then again, transmission XAFS measurements are almost always preferred at high concentrations.

In some cases, monitoring the Auger electron current may be a useful method for measuring XAFS. The escape depth of electrons is usually a few nanometers, which makes these Auger measurements more sensitive to the surface than X-ray fluorescence measurements. In addition, these Auger measurements are usually best suited for metal samples. After the conversion of measured intensities into $\mu(\omega)$, possibly systematic measurement errors (such as self-absorption effects and detector dead-time) need to be corrected, and energy calibration should be performed with the reference standard.

Deglitching

The raw EXAFS spectra sometimes contain abnormally sharp spike points called "glitches". Their origin might be diffractions from the crystal monochromator or from the single crystal specimens (including crystal substrates), or multi-electron excitations. There are also other experimental faults causing glitches, like bad linearity of the detector and electronics, pinholes of the specimens, or higher harmonics of the beam. These faults make the observed EXAFS unreliable and must be carefully eliminated in *experiments*. The glitches, intrinsic to the specimen, might be many times larger than the real EXAFS signal can cause "spectral leakage" in the Fourier transform curves, and hence they must be removed before further processing. Data can be de-glitched in several different ways. The simplest perhaps is to edit the data just after the pre-edge subtraction stage using the editor which is available upon plotting. Points can simply be selected and moved or deleted, according to the user's preference and the I_0 intensity. Another way is to create new expected data through fitting the data points near the glitch region, i.e., smooth interpolation of the data on both sides of the glitch. By selecting four points, two on either side of the glitch to be removed, the glitches can be removed by fitting the pre-edge or EXAFS region with a cubic or polynomial function which will be interpolated through the middle two points to replace the glitch.

Energy calibration

Sometimes the monochromator drifting or other reason causes the energy readings unreal. The energy of the XAFS spectrum can be calibrated by placing a standard with known threshold energy (for metals, commonly their foils are utilized) between the transmission ion chamber and a reference chamber, see the experiment setup in the previous chapter. The threshold energy E_0, of the reference is typically selected at the energy of the maximum derivative of $\mu(\omega)$. If it shifts by a certain amount of energy related to the standard (table) value, then the energy in the spectrum of the specimen should be shifted in the same way.

7.2.2 Pre-edge Subtraction

The most interesting region in the EXAFS data is the part above the absorption edge, and the background absorption due to any instrumental background and absorption from other edges should be removed. This generally involves a fitting of the pre-edge absorption data to some functional form and extrapolating this function into the post-edge data. A traditional function used for the fitting is the Victoreen empirical formula $[\mu_{bk}(\omega) = A + C\omega^{-3} + D\omega^{-4}]$ (the constant A is utilized to compensate for this distortion caused by only a fraction (~20%) of I_0 measured by the ion chamber), McMaster relation or a linear line. The so-called Victoreen pre-edge function is especially helpful for data measured in fluorescence with a solid-state detector for dilute species, where elastic and Compton-scattered X-ray intensity can leak into the energy window of the fluorescence peak of interest, and where this leakage decays as the incident energy increases. In some cases, a baseline file is used, for example, in the case of a fluorescence spectrum of a dilute aqueous specimen measured with a fluorescent ion chamber detector. In this case a very significant background would be present in the data which could be mimicked by running a specimen of the buffer in which the specimen was dissolved, and subsequently subtracted as a baseline file. The first operation is to subtract the pre-edge function. For instance, a curve is fit through the region of the data that lies before the initial rise of the absorption edge, the goal being removal of the background signal to isolate an X-ray absorption spectrum for the absorbing atom alone (*e.g.* a Mo only spectrum for a Mo EXAFS experiment). The pre-edge subtraction selected is a polynomial. The Gaussian type should be used for solid state detector arrays where there is significant scatter is admitted to the single-channel-analyzer window at the low energy end of the scan. This gives a Gaussian tail to the data, and the Gaussian pre-edge function can be used to remove this. The polynomial pre-edge subtraction works by fitting a polynomial of order -1 through the range 0.0 (*i.e.* the start of the data) to 19980eV, just below the Mo K edge. The program then fits the selected region with the polynomial, extrapolates through the whole data set, and subtracts it.

7.2.3 Normalization

In order to represent the absorption of one X-ray by the element and absorption edge of interest, the experimental $\mu(\omega)$ is required to be normalized for both XANES and EXAFS analysis. Quantitative analysis of near-edge features also requires accurate normalization of X-ray absorption data. The resulting normalized $\mu(\omega)$ will be close to 0 below the edge and approximately 1 above the edge. The first step in this process is to determine the edge or threshold energy E_0. Because the chemical difference can easily move the threshold by several eV, and because the calibration between the monochromator and the beamline differs by a similar amount, the most important feature of E_0 identified at this time is consistency. Although the rough approximation has no theoretical basis, the most common method is to take the maximum value of the first derivative of $\mu(\omega)$. This is easy to replicate and easy to apply to any spectrum, however, may need improvement during fitting EXAFS. A common method to determine this step size is to fit the data to a few hundred eV above and below the absorption edge using a low-order polynomial, and then extrapolate it to point E_0. The difference between the two curves at E_0 is the step size $\Delta\mu(E_0)$, and then the resulting EXAFS spectrum is normalized to edge jump 1.

Alternatively, X-ray absorption data can also be normalized using the tabulated mass absorption coefficients.[7] As the atomic background absorption $\mu_0(\omega)$ is related to energy, the normalization by the edge-step introduces an energy-dependent attenuation to the amplitude of $\chi(\omega)$. For heavy elements, this attenuation is negligible, but for light elements, this attenuation may be important. This energy dependence of edge normalization can be further corrected by using X-ray cross-sections of free atoms tabulated by McMaster *et al.*[8] The background function and a single scale factor have to be adjusted so that the normalized data can fit the tabulated X-ray absorption cross sections. The function to be minimized is given by

$$\frac{1}{n_1}\sum_{i=1}^{n_1}\left[\mu_{tab}(E_i) + \mu_{bg}(E_i) - s\mu_{raw}(E_i)\right]^2 + \frac{1}{N-n_2+1}\sum_{i=n_2}^{N}\left[\mu_{tab}(E_i) + \mu_{bg}(E_i) - s\mu_{raw}(E_i)\right]^2,$$

$$\text{with } \mu_{bg}(E_i) = \left[\sum_{i=0}^{m} C_i\left(E - E_{edge}\right)^i\right] + Aerfc\left(\frac{E-E_{em}}{\zeta}\right),$$

where the index i refers to the point number, and n_1 and n_2 are the point numbers corresponding to energies below and above the edge. μ_{tab} is the tabulated absorption coefficient, μ_{bg} is the calculated background function, μ_{raw} is the measured absorption coefficient and s is a scale factor. The two halves of equation are calculated separately, summing the mean-square deviation below the edge and the meansquare deviation above the edge. The background composes two part: a series of Legendre polynomials centered at the edge E_{edge} the mth order (usually m = 2–3), and an error function. The complementary error function ($erfc$) describes the rapid drop in the shape of the pre-edge due to residual elastic scattering in fluorescence mode. It is centered at the energy of the X-ray emission line of the absorbing element (E_{em}), with the spectral width (ζ) and the amplitude (A) treated as variable parameters. The new normalization method demonstrates its advantages over conventional normalization methods in terms of better reliability and consistency, smaller errors and non-sensitivity to the shape of the background function, thus it allows to accurately compare the data collected in transmission mode with the data collected using fluorescence ion chambers or solid-state fluorescence detectors.

7.2.4 Post-edge background subtraction

Till now, the normalized $\mu(\omega)$ data still contain the atomic absorption $\mu_0(\omega)$. In order to model the EXAFS, the EXAFS $\chi(k)$ must be separated from the absorption coefficient, using

$$\chi(\omega) = \frac{\mu(\omega)-\mu_0(\omega)}{\Delta\mu_0}. \tag{7.2}$$

Here, $\mu_0(\omega)$ represents the absorption coefficient of the absorbing atom without photoelectron scattering from the neighboring atoms, and $\Delta\mu_0$ is the edge step. Generally, $\mu_0(\omega)$ is neither measurable nor can be calculated theoretically. Commonly a piece-wise polynomial or *spline* is used for $\mu_0(\omega)$. In doing so, the data are divided into several segments (the divided points are called nodes), and separate cubic polynomials are fit to the data over each segment. The cubic polynomials are constrained so that the curves in each region are connected continuously, *i.e.*, the values, the first and second derivatives match at the intersections. Although this *ad hoc* approach can be abused, it requires a small amount of care to make it robust. The main challenge is to determine the flexibility of the spline curve to ensure that it is not close enough to $\mu(\omega)$ to remove EXAFS. In other words, $\mu_0(\omega)$ should follow the slowly changing part of $\mu(\omega)$, rather than $\chi(\omega)$, which changes faster with the photoelectron energy ω. Then, the EXAFS oscillatory curve $\chi(E)$ can be revealed after the removal of the atomic background absorption. In the background removal process, it is important not to remove the EXAFS oscillations, nor to leave low frequency components in the extracted data. For example, in EXAFSPAK,[9] typically 3~5 spline ranges are selected, a k^4-*weighting* for the spline fit ensures an accurate fit at the crucial high energy end, and the Victoreen polynomial is utilized to normalize the data. The Victoreen polynomial is an expression for the absorption cross-section above a particular edge. The absorption coefficient, μ_{vic}, is given by $\mu_{vic} = C_{vic}\lambda^3+D_{vic}\lambda^4$ where λ, the X-ray wavelength in Ångströms is $\lambda=(hc/E)=12398.4/E(\text{eV})$, and C_{vic} and D_{vic} are the tabulated Victoreen coefficients for a particular edge and element. When the Victoreen is used, the oscillatory part of the X-ray absorption, the EXAFS $\chi(k)$, is calculated by $\chi(k) = (\mu_{exp}-\mu_{spline})/\mu_{vic}$ where μ_{exp} is the experimental absorption after pre-edge background removal and μ_{spline} is the spline fit. The use of the Victoreen polynomial ensures that any background problems with the data, or inadequacies with the pre-edge subtraction (the former are especially common for fluorescence data) do not distort the amplitude of the EXAFS. When the Victoreen is not used the EXAFS is given by $\chi(k) = (\mu_{exp}-\mu_{spline})/\mu_{spline}$.

Another method that is effective in most cases is to rely on the Fourier transform to express this idea mathematically, $\mu_0(\omega)$ should partially match the slow change of $\mu(\omega)$, but not the quickly varying parts of $\mu(\omega)$. The Fourier transform provides a weight for each frequency that makes up the EXAFS waveform. The Fourier transform converts EXAFS χ from wavenumber k to distance R. To determine the background $\mu_0(\omega)$, the spline function should remove the low R component of χ while retaining the high R component. Conveniently, the approximate distance to the nearest neighbor atom can be utilized to make the distinction between "low-R" and "high-R" physically meaningful. According to practical rules of thumb, the distance between atoms is rarely much smaller than 1.5Å-especially for the heavier elements to which EXAFS is usually applied. Therefore, a spline can be used for $\mu_0(E)$ that makes the resulting χ have the smallest possible weight below some distance R_{bkg} (typically, 1 Å), while ignoring the higher R components of χ. While this method [10] is not always perfect, it can be easily applied to any spectrum and has a certain physical meaning.

7.2.5 EXAFS isolation in k-space

In order to perform Fourier transform, EXAFS data has to be converted into k-space. After the pre-edge, post-edge background removal and moralization, the isolated XAFS, $\chi(\omega) = [\mu(\omega) - \mu_0(\omega) - \mu_b(\omega)]/\mu_0(\omega)$, can be converted into the k-space $\chi(k)$. The EXAFS can now be viewed in k-space, and an additional option to simultaneously view I_0 is often useful to identify crystal glitches in the data. k, the photo-electron wave number is a measure of the kinetic energy of the photo-electron, with wavelength λ_e , and is given by $k = \frac{2\pi}{\lambda_e} = \sqrt{\frac{2m}{\hbar^2}(\omega - E_0)}$.

Theoretically, the absorption threshold E_0 is defined as the energy at which the photoelectron is excited as a free electron and the open continuum channel for photo-electric absorption becomes available, producing a continuum photo-electron. This has an exact value from theory, subject to convergence issues. This E_0 is close to, but somewhat different from the absorption edge of the element under investigation, as it is affected by the chemical state or valence. In a solid, complications arise because the threshold from emission does not arise at zero kinetic energy, but at the Fermi level, and it is necessary to correct for the location of the Fermi level relative to the mean potential in the solid. There is no general way to determine this threshold energy, and in practice the ionization or threshold energy E_0 for the electron in theory, practically it is often selected using the deflection point (where the $d\mu(\omega)/d\omega$ reaches the maximum) of the absorption curve at the near edge or the middle point of the edge jump.

The errors in the determination of E_0 could further be compensated in the curve-fitting procedure, where a parameter ΔE will be optimized.

E_0 can be refined by comparing the experimental and FEFF simulated EXAFS signals, as implemented in the EXAFS Data Analysis software-package (called EDA).[11] The initial position of E_0 can be selected at the maximum of the first derivative of the X-ray absorption coefficient, and the EXAFS signal $\chi(k)$ can be extracted with this initial E_0. Besides, the theoretical total EXAFS signal $\chi_{FEFF}(k)$ can be computed for the reference compound by using the FEFF code. Then, the theoretical EXAFS signal is compared with the experimental one. If the phases of two EXAFS signals deviate significantly, especially in the low-k range, the E_0 position for the experimental EXAFS signal needs to be re-adjusted. Therefore, the whole extraction procedure should be repeated till the alignment accuracy of the two EXAFS signals is better than 0.5 eV. After finding a good location for the E_0 of the EXAFS spectrum of the reference compound, the same E_0 value should be used for other EXAFS signals in the analysis.

Additionally, all fast Fourier transform algorithms require equal spacing in *k-space*, and the data must generally be interpolated to accommodate this, typically Δk =0.05 Å^{-1} and an array size of $N = 2048$ k values are used in EXAFS analysis. This gives a spacing of R points in the discrete $\chi(R)$ of $\delta R \approx 0.0307$ Å. This is generally the case even when data collection specifies equal k-spacing, as energy calibration, which is generally performed subsequent to the collection, will change the k-spacing slightly. The smoothing makes the interpolation slightly better behaved in the case of noisy data.

7.2.6 EXAFS Fourier transforms

To find the start point of EXAFS, one needs to distinguish the X-ray-absorption near-edge structure (XANES) from the EXAFS. From a real-space, multiple-scattering point of view, there is no obvious distinction. At all energies the photoexcitation in XAFS follows Fermi's golden rule, *i.e.*, through dipole interaction, the incident photon promotes an electron from its initial, deep-core state into an unoccupied valence band of the final state. That is true at all energies there is a formal solution to Fermi's golden rule involving multiple-scattering theory. The theory of the real-space, multiple-scattering formalism

[12,13] involves a substantial amount of linear algebra. In theory, a mathematical distinction between XANES and EXAFS can be defined on the basis of the mathematical properties[14] of one particular matrix used in the real-space, multiple-scattering approach. As an example, in copper metal, this strict mathematical criterion defining the XANES region extends to about 16 eV above the edge or about 2 $Å^{-1}$ in photoelectron wavenumber. Although this may be a strict definition of the boundary between XANES and EXAFS, it is not a very practical definition. In practice, the EXAFS region is chosen to begin 35 eV above the edge, or about 3 $Å^{-1}$. It is an energy where the formal solution is convergent as a path expansion. Above the energy the photoelectron with enough speed is relatively insensitive to the computational details of the potential model used in the theory. Finally, it is far enough away from the edge that common background removal algorithms [15,16] are stable and reliable. Generally, the EXAFS starting from higher energy has less multiple-scattering effects, but a wider EXAFS range is required for a higher *R*-space resolution in their Fourier transform.

The *k*-weighting compensates for the reduction of EXAFS oscillations in the high-k region and reduces the effects of edge jumps and E_0 ambiguity. The advantage of Fourier transform is the separation of different shells by the interatomic distances. The choice of the *k*-weighing power is arbitrary, but a proper *k*-weight is usually chosen to make the size of the oscillation approximately constant within the data range. Then the Fourier transformed spectrum, $\chi(R)$ is dominated by the oscillatory structure of $\chi(k)$. The Fourier transform of an unweighted spectrum having a low-*R* peak suggests a quick attenuation of the $\chi(k)$ spectrum. The *k*-weighting can alter the emphasis of different contributions to the $\chi(k)$ spectrum. The scattering amplitude of light elements (such as O and C) dominates at low-*k*, and become quite small at high-*k*. Heavier elements (like the transition metals) contribute scattering mainly at very high values of *k*. For very heavy elements (such as Pb or Sn), their scattering amplitudes show minima around 5 to 7 $Å^{-1}$.

After the *k*-weights, the $k^n\chi(k)$ function is Fourier transformed into the R-space. This is a very important step in EXAFS analysis, since the EXAFS oscillations in *k*-space are the sum of different sinusoidal functions of many shells, the Fourier transform technique could separate their contributions in *R*-space. Note that in the FT curve, the peak position is not identical to the interatomic distance as the real pair radial distribution function does but shifts to the lower R-side by the phase-shift in the EXAFS equation. This shift is typically by 0.3–0.5 Å for the first shell atoms. It is worth mentioning that a real EXAFS data set has a limited k-range, while the resolution of the Fourier transform function increases with longer k-range, thus the k_{max} should be as large as possible. Anyway, the k_{max} is not infinite but truncated at a finite value, which inevitably produces some wiggles (called "ghost peaks") in the Fourier transform curves. To reduce such extra peaks, window functions are necessary in the Fourier transform. The most popular window function is the Hanning window:

$$W(k) = \begin{cases} \sin^2\left[\frac{\pi(k-k_{min})}{2d_1}\right], & k_{min} < k < d_1 + k_{min} \\ 1, & d_1 + k_{min} < k < k_{max} - d_2, \\ \sin^2\left[\frac{\pi(k_{min}-k)}{2d_2}\right], & k_{max} - d_2 < k < k_{min} \end{cases} \quad (7.3)$$

Note that for heavy coordination atoms in the fifth or sixth row of the Periodic Table of the elements, a side lobe peak also appears at the left side of the main peak, due to the nonlinear behavior of the phase-shift functions of the heavy atoms. The EXAFS data can also be Gaussian smoothed, which is useful if fast Fourier transforms are to be used later on in the data analysis session.

The EXAFS Fourier transform is generally given by:

$$\rho(R) = \frac{1}{4\sqrt{\pi}} \int_{k_{min}}^{k_{max}} \chi(k) k^3 e^{i2kR} W(k) dk\,, \quad (7.4)$$

The Fourier transform is inherently complex and gives a complex function for ρ(R) even though $\chi(k)$ is a strictly real function. Note that although |ρ(R)| almost looks like a radial distribution function, $g(R)$, and EXAFS does depend on the partial pair distribution—the probability of finding an atom at a distance R from an atom of the absorbing element, |ρ(R)| is not a pair distribution function $|\chi(R)|$. This is due to the fact that the EXAFS equation includes the non-smooth k-dependence of the scattering factor $f_{eff}(k)$ and phase-shift $\Phi(k)$. Additionally, the positions of the ρ(R) peaks are shifted to lower R from the interatomic distances of $g(R)$.

Phase corrected Fourier transformation

It is common practice to phase-correct Fourier transforms. This is done using an EXAFS phase function, typically a theoretical one, for the dominant absorber-backscatterer interaction in the EXAFS. The effect of phase-correcting the transform is to move the peaks to values close to the actual absorber-backscatterer distances (R), if the absence of any phase correction the peaks will be at positions R+Δ, where Δ is approximately half the average phase shift slope for a given interaction. The Fourier transform equation including phase-correction is:

$$\rho(R) = \frac{1}{4\sqrt{\pi}} \int_{k_{min}}^{k_{max}} \chi(k) k^3 e^{i2kR+i\phi(k)} W(k) dk, \tag{7.5}$$

where $\phi(k)$ is the appropriate EXAFS phase function combining the central atom phase and the backscatterer atom phase. In general, only the magnitude spectrum, $|\rho(R)|$, of the Fourier transform is displayed, and it is important to realize that all phase information is missing in these plots. The magnitude spectrum is given by:

$$|\rho(R)| = \sqrt{\Re[\rho(R)]^2 + \Im[\rho(R)]^2}\,. \tag{7.6}$$

If the FT is made with correction of the phase shift and amplitude functions, inter-atomic distance can be read off immediately from the transform peak position. Moreover, the phase corrected FT gives more sharp and symmetric FT peaks because nonlinearities in the phase shift and amplitude functions are removed. If E_0 is properly chosen, the peak position of imaginary part (sine wave) should coincide with the peak position of the absolute value and the real part should be zero at that position. Thus, E_0 should be adjusted until the imaginary part and absolute part coincide.

Radial distribution function

The Fourier transform of the EXAFS signal is often incorrectly treated as the pair radial distribution function (RDF) of atoms. The Fourier transform fundamentally differs from the pair radial distribution function because (1) the Fourier transform of the EXAFS signal carriers information on both the pair and multiatomic distribution function,[17] (2) the peak position does not correspond to actual interatomic distances owing to the phase shift $\Phi(k)$ related to the backward scattering of a photoelectron on atoms,[18] and (3) the peak shape is distorted because of the contribution from the nonlinear backscattering amplitude $F(k)$.[19] Moreover, (4) each peak does not necessarily correspond to the specific group of atoms, EXAFS is element-selective, providing information on the local environment of the selected element only while the pair radial distribution function contains signals coming from all the pairs of elements within the specimen, so that the different contributions have to be disentangled to obtain the local environment of an individual element. In particular, the complex form of the amplitude and phase shift of the backscattering on heavy elements results in two-hump peaks in Fourier transform, where the hump heights depend strongly on the Fourier transform interval.[20] Thus, the Fourier transform can provide only a rough measurement of the distribution of atoms surrounding the absorbing center, and this distribution can be comprehensively interpreted only by simulation.[21]

Currently, the following approaches are extensively used to analyze EXAFS signals: (1) the model dependent technique based on the parameterization of the pair radial distribution function in an analytical form, *e.g.*, Gaussian curve, or in the form of cumulative expansion;[22,23] (2) model independent methods for *ab initio* reconstruction of the pair radial distribution function such as the regularization method,[24] freestyle Monte Carlo method,[25] inverse Monte Carlo method,[26] and nonlinear conditional fitting;[27] and (3) the splice method.[28]

In the splice approach,[29] the pair radial distribution function G(R) is determined by inverting the reduced EXAFS function $\chi'(k)$ by taking the Fourier transform:

$$G(R_i) = \int \chi'(k) e^{-2ikR_i} dk \quad , \tag{7.7}$$

where $\chi'(k) = \dfrac{\chi(k) k R_i^2 e^{i\Phi(k,R_i)}}{S_0^2 F(k,r) e^{-2R_i/\lambda(k)}}$ and integration is performed from zero to infinity. Because the experimental EXAFS signal is usually determined from $k_{min} > 0$ to $k_{max} \approx 10–20$ Å^{-1}, and its amplitude decreases with increasing the wave number k, we set $\chi(k > k_{max}) = 0$ in the integral. The $\chi(k < k_{min})$ values are extrapolated by using the cumulative model, whose parameters are found from the best fit to experimental data from k_{min} to k_{max}. The accuracy of determining the pair radial distribution function is restricted by (i) the applicability of the cumulative approximation in the range $0 < k < k_{min}$ and (ii) the magnitude of the experimental EXAFS signal at $k = k_{max}$. Indeed, if $|\chi(k_{max})| >> 0$, the cutoff of the function $\chi(k)$ at $k = k_{max}$ results in distortion of the desired pair radial distribution function in integration.

7.2.7 Inverse Fourier transforms (Fourier filtering)

In many analyses, the EXAFS signal for a single shell of physical atoms around the absorbing atom needs to be isolated for further quantitative analysis or to extract the amplitude and phase functions of a specific scattering path from a reference. A particular R range (peak) is selected and inversely Fourier transform back to k space, so that most of the spectrum is *filtered* out and only a narrow band of R values is left in the resulting filtered $\chi(k)$. This was the basis of most early EXAFS analyses for all the simplest systems where there exists only one type of element in each Fourier transform peak. However, it can be surprisingly difficult to effectively isolate the EXAFS contribution from an individual scattering atom if the FT peaks of two coordination shells have some overlap. For this reason, many modern analyses of EXAFS are performed in R-space, *i.e.*, $\chi(k)$ is Fourier transformed to $\chi(R)$ for data modeling, without trying to isolate shells of atoms.

The phase and amplitude functions can be obtained either from standard reference or from theoretical model calculation. To proceed Fourier filtering, the Fourier transform must first be re-calculated without phase-correction, since the back-transform will be distorted at low k if phase-correction is used. This Fourier transform proceeds exactly as before, but without the phase correction. The data are also decomposed into total effective phase and total effective amplitude, and a brief description of the method follows. For any EXAFS signal, $\chi(k) = A(k)\sin\Phi(k)$, where the total effective phase $\Phi(k) = 2kR + \phi(k)$, and $A(k)$ is the total effective amplitude, it now becomes:

$$\chi(k) = \frac{1}{2i}A(k)e^{i\Phi(k)} - \frac{1}{2i}A(k)e^{-i\Phi(k)}. \tag{7.8}$$

After the Fourier transform of $\chi(k)$ into $\chi(R)$ (see above), the first term in the above equation corresponds to positive R values, while the second corresponds to negative R values. The latter are set by the program to zero, and the backtransform into k-space is then calculated. Then a function $z(k)$ is obtained, corresponding to only the first part of the equation, which can be utilized to obtain the EXAFS total phase and total amplitude functions:

$$\Phi(k) = \tan^{-1}\left\{\frac{\Im[z(k)]}{\Re[z(k)]}\right\}$$
$$\mathrm{A}(k) = \sqrt{\Re[z(k)]^2 + \Im[z(k)]^2}\,. \tag{7.9}$$

The total phase as calculated above will generally possess discontinuities every time the total phase function changes by more than 2π; these are detected and automatically removed. If structural information about the compound is known, *i.e.* it is to be used as a model compound, then the total phase and amplitude functions can be used to calculate EXAFS phase and amplitudes for use in curve fitting analysis of unknown specimens.

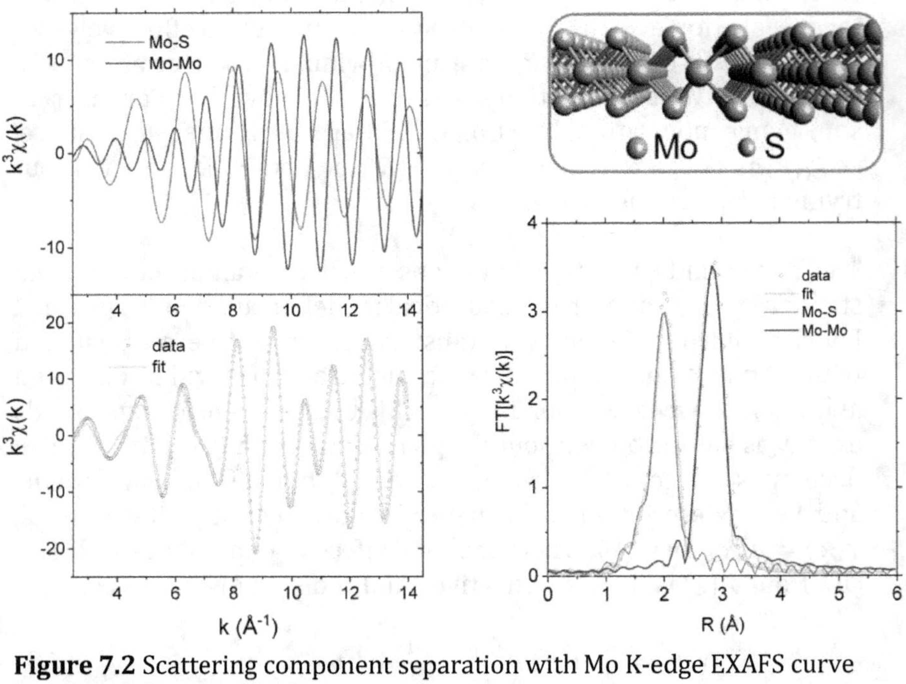

Figure 7.2 Scattering component separation with Mo K-edge EXAFS curve fitting for a $[MoS_2]$ cluster.

7.3 Fitting Procedure

Once the single-shell oscillation has been isolated, the quantitative structural parameters can be obtained through a curve-fitting procedure according to the EXAFS standard formula (Figure 7.2). Besides the structural parameters N, R, σ^2 that appear explicitly in the EXAFS equation, there is another parameter ΔE_0 to be optimized. This parameter at least plays dual roles of correcting the errors in determining the threshold of absorption energy and compensating the inaccuracy of phase-shift. Before curve-fitting the amplitude and phase-shift functions should be known. As mentioned before, these functions can be determined experimentally from model compound through either fitting or ratio method, from tabulated values or from theoretical calculations. Thanks to the development of the accurate ab initio calculations of FEFF code, nowadays the theoretical amplitude and phase-shift are more routinely used than the experimental ones.

There are various algorithms for fitting the filtered experimental EXAFS function, among which the least-squares algorithm is most commonly used. However, the least-squares algorithm suffers from some drawbacks, for example, many local minima in the sum of squares function may exist, and the obtained structural parameters are often physically impossible. To overcome this problem, some new algorithms such as simulated annealing method have been developed. In practice, one problem in the multi-shell EXAFS curve-fitting is the correlation among various parameters. For instance, both the coordination number and the Debye-Waller factor are related to the amplitude of the oscillation and hence are strongly correlated; if both are treated as free parameters, the accuracy would be decreased. It often occurs that when too many parameters are to be fitted, the correlations may force the fitting results to converge on values with unphysical meaning. Therefore, reasonable constraints between parameters are necessary, and the international XAFS working group has extensively discussed this issue and published standards and criteria in EXAFS data analysis. [30]

7.3.1 Ratio-difference method

For EXAFS data with a well-separated coordination shell, structural parameters can be extracted using the ratio method. In the analysis of a temperature series, the variation in σ^2 as a function of temperature can be obtained through the log-ratio analysis, while the change in bond length can be measured through the phase difference method. The signal from the first coordination shell should be first isolated from the data of both the unknown and reference by Fourier filtering. Then the polynomials are fitted to the logarithm of the amplitude ratio and the phase difference of the $\chi(k)$ functions. The amplitude and the phase functions are straightforwardly extracted as $A(k)=2|\chi(k)|$, $\Phi(k)=\arg|\chi(k)|+\pi/2$. Since the phase shift function is the same between the reference and the unknown specimen and if the edge position E_0 is the same, the phase difference between the specimen and the reference can be written as

$$\Phi(k)-\Phi_{ref}(k)=2kr-2kr_{ref}. \tag{7.10a}$$

The log ratio of amplitude is described as

$$\ln\left(\frac{A(k)}{A_{ref}(k)}\right)=\ln\left(\frac{Nr_{ref}^2}{N_{ref}r^2}\right)+2k^2\left(\phi_{ref}^2-\phi^2\right)+2\left(\frac{r_{ref}}{\lambda_{ref}}-\frac{r}{\lambda}\right). \tag{7.10b}$$

7.3.2 Theoretical curve fitting analysis

In order to obtain the local structural parameters for the absorbing element, a theoretical curve is created from the EXAFS equation in its most complete form:

$$\chi(k)\cong S_0^2\sum_{i=1}^{n}\frac{N_iS_i(k,R_i)f_i(k,R_i)}{kR_i^2}e^{-\frac{2R_i}{\lambda(k,R_i)}}e^{-2\sigma_i^2k_i^2}\sin[2kR_i+\phi_i(k,R_i)+\phi_c(k)].$$

The variables mean

k	Photo-electron wave number, a measure of the photoelectron wavelength, $k = 2\pi/\lambda_e$, and calculated from the X-ray energy above the threshold by $k = [0.262(E-E_0)]^{1/2}$.
S_0^2	Amplitude reduction factor due to shake up/off losses (otherwise known as the scale factor).
N_i	Coordination number for atom type *i*.
$S_i(k,R_i)$	Total central atom loss factor. ***FEFF*** calculates a value for this, in other theories it is assumed to be unity.
$f_i(k,R_i)$	Effective EXAFS scattering amplitude function.
R_i	Absorber-backscatterer distance for atom *i*.

$\lambda(k, R_i)$	Photoelectron mean free path or inelastic scattering correction. This is calculated explicitly by ***FEFF.***
σ_i	Debye-Waller factor, the mean square deviation in R_i.
$\phi_i(k, R_i)$	Backscatterer EXAFS phase function.
$\phi_c(k, R_i)$	Central atom EXAFS phase function includes ϕ_c and ϕ_i for each backscatterer. ϕ_c depends only on the edge and the type of atom that is absorbing, ϕ_r, on the other hand, depends on R_i and upon the type of backscatterer atom.

The empirical phase shift and amplitude functions can be calculated with FEFF code.[31] A single adjustable parameter in XAFS analysis, the passive electron amplitude reduction factor (S_0^2) can be obtained by fitting of experimental EXAFS data of corresponding foil with theoretical ones, and then held fixed for all the other data, since it is a property of only the absorbing atom. Correlations between (E_0, δr_j) and (N_j, σ_j^2) fitting parameters could be reduced by weighting the XAFS data by k^n (n=1, 2, 3). In relation to the correlation problem one should pay attention to the maximum freedom N_{idp} of the fitting parameter determined by $N_{idp} = \frac{2\Delta r \Delta k}{\pi}$, where Δr and Δk are the range of Fourier transformation in *r* and *k* space used in the data analysis. During the least square curve fitting, the sum of the residuals Δ is minimized, here $\Delta=(N_{idp}/v)(1/N)\Sigma(\chi_{obs}-\chi_{cal})^2/\varepsilon^2)$ where $v=N_{idp}-p>0$, *p* is the number of parameters used for fitting, the estimated error ε^2. The steps required for data analysis are discussed reasonably thoroughly by Sayers and Bunker.[32]

7.3.3 Asymmetric Distribution and disorder effects

The disorder can be classified into three categories: 1) thermal disorder, 2) structural disorder, and 3) site disorder, like Fe ions in Fe_3O_4. So, the EXAFS equation derived from Gaussian distribution is often not suitable for the large disorder systems.

Model independent method (Cumulant)

Even in perfect crystals atoms at given time-moment are rarely at their equilibrium positions: there is a lot of paths with slightly different lengths. It makes sense to group them in coordination shells and replace summation over paths belonging to the same coordination shell with an integral: $\chi_p(k) = \int_0^\infty \frac{g(r)}{kr^2} F(k,r) e^{-\frac{2r}{\lambda(k)}} \sin(2kr +$

$\phi(k)\big)dr$, where $g(r)$ is radial distribution function. By using the effective distribution $P(r,\lambda) = \frac{g(r)}{r^2} e^{-\frac{2r}{\lambda(k)}}$, and the cumulant expansion into a MacLaurin series: $ln \int_0^\infty P(r,\lambda) e^{2ikr} dr = \sum_{n=0}^\infty \frac{(2ik)^n}{n!} C_n$, the EXAFS function of a system with an asymmetric distribution is expressed in the form of a cumulant expansion:

$\chi_P(k) = S_0^2 \frac{F(k,r)}{k} exp\left(C_0 - 2C_2k^2 + \frac{2}{3}C_4k^4 - \frac{4}{45}C_6k^6\right) \sin\left(2kC_1 - \frac{4}{3}C_3k^3 + \frac{4}{15}C_5k^5 + \phi(k)\right)$, (7.11) with $exp(C_0) \cong \frac{N}{C_1^2} exp\left(-\frac{2C_1}{\lambda(k)}\right)$.

$C_1 = \langle r \rangle$ is the mean value of the distribution (average distance);
$C_2 = \langle (r - C_1)^2 \rangle = \sigma^2$, is the variance of the distribution (MSRD factor);
$C_3 = \langle (r - C_1)^3 \rangle$, is the asymmetry of the distribution (skewness);
$C_4 = \langle (r - C_1)^4 \rangle - 3C_2^2$, describes symmetric deviatios from the Gaussian shape.[33]

The log ratio of amplitude and phase difference between the specimen and reference can be writen,

$$\ln\left(\frac{A(k)}{A_{ref}(k)}\right) = \ln\left(\frac{Nr_{ref}^2}{N_{ref}r^2}\right) + 2k^2\left(\sigma_{ref}^2 - \sigma^2\right) + 2\left(\frac{r_{ref}}{\lambda_{ref}} - \frac{r}{\lambda}\right) + \frac{2}{3}\left(C_4 - C_{4ref}\right) \tag{7.12a}$$

$$\Phi(k) - \Phi_{ref}(k) = 2kr - 2kr_{ref} - \frac{4}{3}\left(C_3 - C_{3ref}\right). \tag{7.12b}$$

Effective pair potential for the absorber-backscatterer pair with an anharmonic potential of the form

$$V(r) = \frac{\alpha}{2}(r - r_0)^2 - \beta(r - r_0)^3, \tag{7.13}$$

where r_0 is the position of the minimum of $V(r)$. The Einstein temperature θ_E is connected to the potential parameter by

$$\theta_E = \frac{\hbar\omega}{k_B} = \frac{\hbar}{k_B}\sqrt{\frac{\alpha}{\mu}}, \tag{7.14}$$

where μ is the reduced mass of the absorbate-scatterer pair. By means of a quantum mechanical calculation using second order perturbation theory,[34] the temperature dependence of the moments can be expressed by

$$R(T) = r_0 + \frac{3\hbar\beta}{2\alpha\sqrt{\alpha\mu}}\coth\left(\frac{\theta_E}{2T}\right) = r_0 + 3\frac{\beta}{\alpha}\sigma^2(T)$$

$$\sigma^2(T) = \frac{\hbar^2}{2\mu k_B\theta_E}\coth\left(\frac{\theta_E}{2T}\right)$$

$$C_3(T) = \frac{\hbar^6\beta}{\mu^3 k_B^4\theta_E^4}\left(\frac{3}{2}\coth^2\left(\frac{\theta_E}{2T}\right) - 1\right)$$

Further details concerning temperature dependence of the moments can be found in references.[35,36]

Model dependent analysis

For a largely disordered system, it is better to use a model pair distribution function. The simplest method of analysis is to use the double or multi-Gaussian functions assuming that the pair distribution function is a convolution of Gaussian function. The other way is to construct a model distribution function such as

$$G(r) = \frac{1}{\sqrt{2\pi\sigma_T^2}} e^{-\frac{(r-R)^2}{2\sigma_T^2}} \tag{7.15}$$

$$P(r) = \begin{cases} \frac{1}{m!\sigma_s^{m+1}}(r-R)^m e^{-\frac{r-R}{\sigma_s}}; & for \quad r \geq R \\ 0; & for \quad r < R \end{cases} \tag{7.16}$$

where m=0, 1, 2. σ_T and σ_S are thermal and structural disorders, respectively.

$$\rho(r) = G(r) \otimes P(r) = \int_0^\infty G(x)P(r-x)dx = \int_0^\infty P(x)G(r-x)dx \tag{7.18}$$

$$\chi(k) = \frac{NF(k)}{k}\int_0^\infty \frac{\rho(r)}{r^2} e^{-\frac{2r}{\lambda(k)}} \sin(2kr + \delta(k))dr \tag{7.19}$$

However, the determination of these model distributions the fitting procedure becomes complicated and the correlation between the parameters grows stronger.

7.3.4 The fitting metric

Many EXAFS analysis programs employ a conventional, Levenberg-Marquardt, non-linear minimization algorithm in fitting with theoretical scattering factors from FEFF.[37] A ξ^2 fitting metric is often utilized to describe the difference between the experimental data and simulated data. The curve fitting purpose is to minimize a ξ^2 fitting metric by adjusting the structure parameters in the simulated EXAFS spectrum. A difference function (g) between the data and simulation is computed at each data point included in the evaluation of the fit. Fitting can be evaluated either in k-space with real-valued functions, or in the R-space with the difference function including the real and imaginary parts of the Fourier transformed data and calculation, as in Eq. 7.20. The difference function for Fourier filtered data is also complex. The fitting metric is then evaluated as the sum of squares over all data points N included in the fit. ϵ_i is the measurement uncertainty at data point i. The information content N_{idp} is the number of independent points of the data. Within the context of a Levenberg-Marquardt minimization, the the size of N_{idp} is commonly estimated by taking the Nyquist criterion,[38] *i.e.* the bandwidth of a signal processing problem. In the context of EXAFS analysis $N_{idp} \approx 2\Delta k \Delta R/\pi$, where Δk is the k range over which the Fourier transform window is non-zero, and ΔR is the range over which the fit is evaluated. Typically, the EXAFS data are quite over-sampled compared to the information content. N, the number of data points, is typically a few hundred while N_{idp} is around 10 or 20. Then, the prefactor $\frac{N_{idp}}{N}$ scales ξ^2 to the number of independent measurements in the data. For details of statistics in EXCURVE and GNXAS fitting please refer to Ref. [39] and [40].

$$g(k_i) = \chi^{exp}(k_i) - \chi^{the}(k_i)$$

$$g(R_i) = Re[\tilde{\chi}^{exp}(R_i) - \tilde{\chi}^{the}(R_i)] + Im[\tilde{\chi}^{exp}(R_i) - \tilde{\chi}^{the}(R_i)]$$

$$\xi^2 = \frac{N_{idp}}{N} \sum_{i=1}^{N} \left(\frac{g_i}{\epsilon_i}\right)^2. \tag{7.20}$$

7.3.5 Constraints on parameters of the fit

Most implementations of EXAFS analysis using FEFF consider the terms of EXAFS equation – N, S_0^2, σ^2, E_0, and ΔR – to be the parameters of the fit. While this may be an adequate strategy when analysis is limited to a single element coordination shell, it quickly becomes untenable for larger fitting models. As more scattering paths are considered, the number of floating parameters quickly exceeds the information content of the fit. The constraints between parameters become necessary in the use of theoretical scattering factors beyond a single element shell and the application of complex models to the EXAFS analysis. With constraints, the parameters of the EXAFS equation are written in terms of the parameters of the fit, rather than the individual floating parameters of the fit.

The first simplest constraints are those terms in EXAFS equation that relate to the central atom, such as S_0^2 and E_0. The attenuation factor S_0^2 is due to the incomplete overlap of wave functions of the passive electrons – all the electrons other than one excited by the incident photon – in the presence of the core-hole left behind by the photoelectron with their unperturbed wave functions. This is not of a specific path but a property of the central atom. Therefore, S_0^2 can be reasonably set the same for each path included in the fit, *i.e.*, a single parameter S_0^2 is floated in the fit as EXAFS equation is evaluated for each path. Similarly, only one such variable parameter E_0 is required in a fit and its value is used for the evaluation of each path, since the parameter E_0 aligns the energy grid of the EXAFS simulation with the energy grid of the experimental data.

For a cubic material, the lattice constant is the only structural parameter. That constant and the symmetries of the cubic space group determine the locations of all atoms. So, the constraints can be typically applied to the evaluation of ΔR and σ^2. Consider the analysis of EXAFS data measured on a simple cubic structure, say copper metal or a rock-salt structure such as MnS. The lengths of all paths can be expressed using a volumetric lattice expansion coefficient α such that, for any path, $\Delta R_\Gamma = \alpha \cdot R_{0,\Gamma}$, where $R_{0,\Gamma}$ is the half-path length from the input configuration. Here, α is the parameter of the fit and the ΔR for every path are written in terms of α. While this particular formulation is only valid for cubic crystals, conceptually similar constraints using more variable parameters can be applied to systems of lower symmetry.

One common strategy to constraint σ^2 parameters is to treat a pair of atoms as masses bound by a quantum oscillator, thus expressing σ^2 for a path in terms of an Einstein temperature.[41] This is particularly effective as it allows evaluation of all σ^2 parameters using a single variable representing the Debye temperature of the material. This is attractive for a model described in that same reference relates the σ^2 values for all paths in a single element crystal to its acoustic phonon spectrum. Unfortunately, few EXAFS analysis problems involve single element materials, limiting the utility of that approach. Another model considers the temperature dependence of σ^2 for a path. The σ^2 parameters of collinear multiple-scattering paths are related to the σ^2 values of geometrically related single-scattering paths with two reasonable assumptions:[42] the thermal motion in the perpendicular direction is small compared to the separation between the atoms, and the thermal motions of the atoms are uncorrelated. The relations between the σ^2 parameters are obtained with those assumptions, allowing parameterization of all of these paths using only two variable σ^2 parameters in the fit. In certain situations, the σ^2 of non-collinear multiple-scattering paths is likely related to the σ^2 of the single-scattering paths made up of legs in the multiple-scattering path.

The constraints of ΔR and σ^2 represent methods of encoding prior knowledge into a fit. In the case all the ΔR parameters using a single lattice expansion parameter, the prerequisites are that the material has a cubic crystal structure and the lengths of the paths must be constrained by cubic symmetry. The main purpose of applying the constraints in the fit is to reduce the number of fitting parameters that must be determined by the fit, and to make the fit more statistically robust.

The bimetallic structure problem

In the quantitative determination of structural parameters of supported metal clusters, X-ray absorption edges of the metallic components are generally well separated in energy. For supported bimetallic cluster composed of metals *A* and *B*, the absorption spectrum for each element can be measured separately without interference from another. In cases where this cannot be done due to elements being adjacent in the periodic table, alternative approaches based on data analysis techniques[43] or energy dispersive detection[44] schemes have been devised to deal with the problem. Bulk miscibility of bimetallic cluster systems can be classified into 4 catalogs: immiscible (Cu-Re, Cu-Ru, Cu-Os); low miscibility (Cu-Ir, Cu-Ag), intermediate miscibility (Cu-Rh, Re-Pd, Re-Pt) and complete miscible (Cu-Pt, Cu-Au, Ir-Rh, Pt-Ir, Pt-Sn). The coordination parameters usually have the following relationship:

$$\begin{aligned} &2\delta(AB)+2\delta(BA)=2\delta(AA)+2\delta(BB)\\ &R_{AB}=R_{BA}\\ &\sigma_{AB}\approx\sigma_{BA}\\ &x_A N_{AB}=x_B N_{BA} \end{aligned} \tag{7.21}$$

The coordination number $N_{A\text{-}B}$ is defined as the number of the nearest-neighbor B atoms around A. N_A is the sum for all neighbor metal atoms around A. N_{total} is an average metal coordination number defined for the metal cluster as $N_A W_A + N_B W_B$, where W_A is the atom percent of A. The coordination numbers satisfy the condition for equal coordination of *A-B* and *B-A*, $N_{A\text{-}B} W_A = N_{B\text{-}A} W_B$, within the error limit. The A and B atoms were also significantly coordinated to each other. The total coordination numbers for the *A-B* close-packed cluster, N_{total}, should be equal to $\left(\sum_{j=1}^{j=12} jN_j\right)/N_T$, where N_j is the number of atoms with j^{th} neighbors and N_T the total number of atoms in the cluster. In the case of cuboctahedral (a polyhedron with 8 triangular faces and 6 square faces) clusters with *fcc* structure, the cluster size with N_T atoms follows the equation $d_o = 1.105(T)^{1/3}$,[45] where d_o is the diameter of the equivalent sphere. Combining these two equations, one can obtain a graphical relationship between N_{total} and N_T, and apply it to find particle diameters.

7.4 XAFS Data Modeling

By the mid-1990s, there were already three classic theoretical software packages that allowed high-quality theories to be applied to EXAFS data analysis: EXCURVE,[46] GNXAS,[47] and FEFF.[48] All these packages share certain similarities: the EXAFS signal is represented by a cluster expansion as a sum of geometric contributions, and a muffin-tin formalism is employed to compute scattering potentials.

The muffin-tin potential was first [49] introduced to solve the problems of electron motion in periodic potentials.[50] Each atom in a crystal is represented by a sphere which just touches neighboring atoms or is slightly overlapped.[51] The space between the spheres (interstice) is flat and of uniform charge density. From this potential surface, the scattering amplitude and phase shift are calculated. It was unclear whether the potential of muffin tin may become the basis for the success of EXAFS analysis. Chemical bonding means both breaking the spherical symmetry and the variable density of electrons in the interstice. Recent advances in the calculation of near-edge structures clearly indicate that non-muffin tin potential is critical to accurately replicate the XANES spectrum. [52,53] However, at energies far above the absorption edge, the photoelectron has enough kinetic energy that it is almost insensitive to the details of the potential surface. The potential of muffin tin is still very common and useful for EXAFS applications.

The earliest attempts to provide practical theory for EXAFS analysis made some important approximations.[54] Among these was the treatment of photoelectrons as plane waves rather than spherical waves propagating outward. This is a reasonable approximation when the kinetic energy of the photoelectron is large enough to neglect the curvature of the spherical wave. However, it becomes a poor approximation for an EXAFS signal dominated by signal at low photoelectron wavenumber, either due to large σ^2 in EXAFS equation or to the nature of F(k) of the scattering atom. Each of the three-theory program implements a curved-wave approach.

Other significant improvements to the calculation of theoretical scattering factors included an effective treatment of intrinsic losses (typically included in EXAFS equation by the S_0^2 term),[55] and of the quantum mechanical, many-body, exchange correlations. The latter yields an energy dependent self-energy, which has an impact on the evaluation of $\phi(k)$ in EXAFS equation $\chi(k)$ and therefore of the evaluation of R in the analysis of the EXAFS.

Given the muffin tin description of the potential surface and the curved-wave theory and the method of dealing with intrinsic losses and many-body effects, the next step is to perform some kind of cluster expansion theory in the context of multiple scattering in real space.

Of the three user-oriented theory packages, FEFF has seen the widest adoption. GNXAS continues to be popular, with dozens of citations each year, but has never been as widely used as FEFF. The use of EXCURVE seems to have diminished in recent years. Given that all three packages were successful in bringing real-space, multiple-scattering theory to bear on the EXAFS analysis problem in a user-oriented manner, what accounts for their differences in popularity and deployment?

7.4.1 The exact curved wave theory in EXCURVE

EXCURVE is an exact curved-wave approach to the user-oriented EXAFS theory developed in 1975.[56] The early version did not scale well in terms of computation time, so its solution to the Dyson equation $G = \frac{G^0}{(1-G^0T)}$ was restricted to single, double, and triple scattering paths in terms of the free electron propagator G_0 and atomic scattering matrix T. Later this exact theory was extended to low order multiple scattering. [57] Although now less populous than GNXAS and FEFF packages, EXCURVE pioneered many important developments in EXAFS. EXCURVE was the first professionally supported commercial software for EXAFS analysis with a sophisticated graphical interface. It was also the first one providing the constrained EXAFS fitting.[58] It was the earliest software utilized to co-refine [59] EXAFS and diffraction data measured on the same material, which remains a thriving research area nowadays. The continued development on a less restricted branch – DL_EXCURV [60] becomes the basis of automate EXAFS analysis – ABRA.[61] The newly christened DL_EXCURV simulates EXAFS spectra using rapid curved wave theory and Rehr-Albers theory from the parameters of the radial shells of atoms surrounding the absorbing atom. Now it can run on multiple platforms including Windows, Linux and Mac systems. By interfacing with the Daresbury Laboratory Visualize software package, the DL_EXCURV program also provides a graphical user interface and a powerful environment for the visualization and processing of scientific data.

In single scattering mode, the package reports the radius and occupancy of a series of concentric shells centered on absorbing atoms, while in multiple scattering mode, a complete three-dimensional structure determination will be performed. Traditionally, the phase shift is caused by the superposition of neutral atomic solutions, but for systems involving large amounts of charge transfer between constituent atoms, a completely self-consistent potential is required. This potential will consider both the charge transfer and the Ewald potential, which will be significantly different at various ion sites within the crystal. The use of such potentials is essential for the analysis of EXAFS data of ion systems, such as transition metal oxides, where strong charge transfer occurs between the constituent atoms. Therefore, current efforts are to improve the methods of generating scattering potentials. The code was developed in a modular manner to ensure that it will be easier to expand in the future.

7.4.2 The n-body decomposition in GNXAS

GNXAS provides a flexible scientific tool for the EXAFS data analysis, which is based on multiple scattering calculations and strict original experimental data fitting procedures. The atomic phase shift is calculated using the muffin-tin potential obtained from atomic self-consistent relativity calculations with the account for the neighbors. Inelastic losses are included in the computation through the complex Hedin-Lundqvist potential. The multi-scattering signals associated with multiple atom configurations are calculated by using advanced algorithms. It allows to simultaneously fit multiple spectra including many edges. The original data can be fitted directly without any pre-analysis. The advanced fitting program also considers complex background multi-electron excitation features. Various model peaks, including non-Gaussian models and extremal cases, can be utilized for the pair, triplet and quadruplet distribution functions. Reliable g(r) functions in the short range can be extracted from liquid phase or disordered systems. It also provides a rigorous statistical error analysis and two-dimensional correlation maps. Its flexibility and advanced features are essential for performing modern EXAFS data analysis and accurately extracting all valuable information contained in experimental data. Calculation of theoretical scattering factors and analysis are tightly integrated in GNXAS, coming in a single package from a single group of developers.

GNXAS performs an *n*-body decomposition of the Green's function in the one-electron propagator function $G = G^0 + G^0TG^0$. Given the cluster of atoms from which the muffin-tin potentials are computed, the program considers the sub-clusters of atoms with 1, 2, or more atoms along with the absorbing atom. For each n-atom configuration,

$G = G^0 + G^0TG^0$, *i.e.*, scattering to all orders within that sub-cluster, is calculated. For a two atom sub-cluster, this is called the γ^2 signal. The γ^2 signal from an absorber/scatterer pair includes the single scattering contribution (χ^2), the triple scattering contribution (χ^4), and all successive odd orders of scattering involving the rattling of the photoelectron between the two atoms. For n larger than 2, γ^n is the irreducible signal from an n-atom configuration, meaning that the lower order γ^m (with m < n) signals are removed.[62]

To utilize n-body decomposition, the input cluster must be analyzed to extract all n-body configurations. Even in a limited cluster, there are countless n-body configurations. In practice, it is sufficient to limit the decomposition to a low-order (n≤4) configuration and within the first few coordination shells.[63] In some case, n-body configurations produce identical signal, as from the atoms of the same species in a well ordered coordination shell, this kind of degeneracy is accounted for. Non-Gaussian disorder is considered by integrating over appropriate n-body distribution functions. The EXAFS data is then fit using these γ^n signals as theoretical fitting standards modified by parameters of the coordination environment.

7.4.3 The path expansion in FEFF

FEFF program can be utilized for ab initio multiple scattering calculations of XAFS for clusters of atoms. The calculated scattering amplitudes and phases can be used with many modern XAFS analysis programs. While FEFF6-lite and FEFF8-lite versions suitable for EXAFS analysis are free available, later versions of FEFF always come with a licensing fee.

One of the innovations that FEFF brought to the EXAFS analysis problem is a fast convergent separable representation of the photoelectron propagator.[64] This is a substantial improvement both in terms of speed of computation and by enabling consideration of multiple scattering to any order. The second was an efficient exhaustive algorithm for the enumeration of all single and multiple scattering paths in an input cluster.[65] Even in a small cluster of atoms, the number of possible scattering paths is infinite if all orders of scattering paths are considered. FEFF introduced the concept of an importance factor for each path computed as a sparse, trapezoid integration of the magnitude of the calculated complex $\chi(k)$. In this way, the paths could be assessed for spectral weight and those contributing very little to the total $\chi(k)$ could be discarded. It is useful to understand how the path expansion relates to the n-body decomposition of GNXAS. FEFF calculates each path up to some order on the right side of the equation, whereas the function calculated by GNXAS is their sum on the left side. In one way, FEFF's path expansion is less convenient as the sum of paths has far more terms than the sum of n-body configurations. However, consideration of individual paths allows for a more careful treatment of σ^2 for multiple scattering paths. A full treatment of all aspects of FEFF's approach[66] to the theory of EXAFS was implemented to one specific analysis package, iFEFFit.[67]

In order to calculate the $f_{eff}(k)$ and $\Phi(k)$ needed for the EXAFS analysis, the FEFF program starts from a cluster of atoms, builds atomic potentials from this, and simulates a photoelectron with a particular energy being emitted by a particular absorbing atom and propagating along a set of scattering paths. Then, the scattering amplitudes $f_{eff}(k)$ and phase-shifts $\Phi(k)$ can be calculated theoretically using the FEFF formalism described above. These theoretical scattering factors along with a calculation of the mean free path λ_k, can be used in the EXAFS equation to refine structural parameters R, N, and σ^2 to match a set of experimental data. Typically, E_0 (the energy for which k=0) is also adjusted to best match data, to compensate for the crude estimate made in the initial processing. While one can do the refinement with the measured $\chi(k)$, working with the complex $\chi(R)$ is generally preferred, as higher coordination shells can be selectively ignored. Some aspects of the analysis shown here may depend on details of these particular programs, but similar results can be obtained with any of several EXAFS analysis programs. As mentioned above, FEFF includes many subtle but quantitatively important effects. Because one starts with a cluster that represents the atomic structure, the FEFF fitting process is a *refinement* of the path lengths and coordination numbers for those paths. Conveniently, FEFF breaks up the results in a way that can be put into the standard form of the EXAFS equation, even for multiple-scattering paths. This allows analysis procedures to easily refine distances, apply multiplicative factors for coordination numbers and S_0^2 , and apply disorder terms. Because the outputs have a uniform format, results from different runs of the programs can readily be mixed, which is important for modeling complex structures with multiple coordination environments for the absorbing atom.

7.4.4 Theoretical modeling of EXAFS

The great success of XAFS stems from the insights they provide into the atomic and electronic structure of complex materials. This success was aided in part by the development of robust theoretical methods to simulate a broad range of X-ray spectroscopies. Although these methods have often been highly successful, a limitation is that the accuracy of the predicted spectra depends on the quality of the structural models used in the simulations. In the past, this drawback has been sidestepped by either the use of known crystallographic information, or by a combination of EXAFS modeling and chemical know-how of the nature of the material in order to build more complete structural models.[68] In recent decades, however, materials research has begun to focus on complex materials under *operando* or extreme conditions. Reliable simulations under those conditions require more sophisticated models that capture the heterogeneous nature of the materials. Thus, the field of XAFS modeling has recently progressed towards the interfacing of spectroscopy codes with existing software for calculations of structural and electronic properties.[69]

Theoretical calculation of EXAFS scattering factors

Theoretical calculations of EXAFS scattering factors address most of the limitations presented by the purely empirical analysis techniques. With this theoretical tool, EXAFS can be done on metastable, *in situ*, or in operando materials or on any material otherwise unavailable as a measurement standard. Complex systems – doped, mixed phase, surface sorbed or anything else with scatterer species or distances that are not represented in readily available standards – can be analyzed. Finally, these theoretical tools provide a rigorous way of including multiple scattering contributions and higher coordination shells.[70]

Construction of the scattering potentials and enumeration of scattering paths are two essential components of the theoretical scattering factors used for EXAFS analysis. These calculations begin by preparing a list of Cartesian coordinates of atoms as input for the theory program. This list can come from a variety of sources, including crystallographic structure data, a large molecule structure, or the output of a first principles or molecular dynamics calculation. In any case, a list of atomic species and their Cartesian coordinates is the requisite input data. If the input structure comes from a reliable source – that is, the distances between atoms are physically reasonable – then the muffin-tin potentials will be calculated well enough to yield a defensible EXAFS analysis.[71] Similarly, if input atoms are at nonphysical separations, the resulting calculation will be unacceptable for analysis.

A simplified method of generating theoretical scattering factors for use in first shell analysis [72] involves prompting the user for the absorber and scatterers and their approximate separation. Using this, a rock-salt (or another similarly simple structure) configuration is constructed and used to compute the muffin-tin potentials. The effect of a non-physically large muffin-tin radius is shown on the computed $\chi(k)$ for S neighbors at the Fe K-edge in FeS_2. For the first coordination shell, the simplified method works well, yielding nearly identical results as a calculation made on the known crystal structure of FeS_2. The second shell Fe-S path computed from the known crystal structure along with the simplified calculation using the inter-atomic distance of the second coordination shell. In that case, the simplified calculation uses a large inter-atomic distance to make the notional rock-salt structure. This yields a nonphysical muffin-tin radius and results in substantial error. This demonstrates two important aspects of the application of theoretical scattering factors to EXAFS analysis, (1) the importance of good muffin-tin potentials with sensible radii and (2) the need for a reasonable input structure for EXAFS analysis beyond the first coordination shell.

Use of modeling to enhance EXAFS analysis

While the fitting of EXAFS spectra using theoretical scattering amplitudes and phase-shifts is now a well-established methodology, the calculation of other parameters of the EXAFS equation, Eq. (2.2), such as path lengths, number of near-neighbors N_R and mean-square relative displacements (MSRD) $\sigma_R{}^2$ have only recently become feasible on a routine basis. While many theoretical methods lack the accuracy needed to rival fitted EXAFS path lengths, they may be well suited to distinguish between alternative conformations, and provide good approximations to both the N_R and $\sigma_R{}^2$, thus, making the fitting of the path lengths much more reliable in many complicated materials. In this section we discuss two ways in which theoretical modeling can enhance EXAFS fits: First, we focus on the calculation of ab initio Debye–Waller factors for systems where $\sigma_R{}^2$ can be described well using a combination of static and vibrational disorder. Second, we highlight the simulation of EXAFS spectra for systems with large dynamic disorder.

Ab initio Debye-Waller factors

The theory behind mean-square relative displacements or XAFS Debye–Waller factors and how to compute them ab initio is now fairly well established.[73] Briefly, the MSRD or $\sigma_R{}^2$ for a path R is defined by the thermal average: $\sigma_R^2 = \langle (R - \bar{R})^2 \rangle$, where R and $\bar{R}$ are the instantaneous and mean path lengths, respectively. In the quasi-harmonic approximation, the average can be computed as a sum over vibrational or phonon eigenmodes [74] or, alternatively, in terms of the projected vibrational density of states $\rho_R(\omega)$ or VDOS, and the Debye integral $\sigma_R^2(T) = \frac{\hbar}{2\mu_R}\int_0^\infty \frac{1}{\omega}\cosh(\frac{\beta\hbar\omega}{2})\,\rho_R(\omega)d\omega$, where μ_R is the reduced mass for path R. Poiarkova and Rehr [75, 76] have shown that the vibrational density of states can be calculated efficiently from the imaginary part of the lattice dynamical Green's function matrix element:

$$\rho_R(\omega) = -\frac{2\omega}{\pi} Im \left\langle 0 \left| \frac{1}{\omega^2 - D + i\varepsilon} \right| 0 \right\rangle \cong w_v \delta(\omega - \omega_v), \qquad (7.23)$$

where the vector $|0\rangle$ represents the normalized mass-weighted displacement of the atom along the multiple scattering path R. Efficient calculations of the lattice dynamical Green's function can be accomplished using a continued fraction representation obtained with the iterative Lanczos algorithm that yields an N-pole representation for the vibrational density of states where ω_v and w_v are, respectively, the pole frequencies and weights of pole v. **D**, the dynamical matrix of force constants, or Hessian, is the main ingredient required for the calculation of theoretical MSRDs and is defined as:

$$D_{jl\alpha,j'l'\beta} = (M_j M_{j'})^{-1/2} \frac{\partial^2 E}{\partial u_{jl\alpha} \partial u_{j'l'\beta}}, \qquad (7.24)$$

where $u_{jl\alpha}$ and M_j are the $\alpha = \{x, y, z\}$ Cartesian displacement and mass of atom j in unit cell l, respectively, and E is the internal energy of the system. For molecular systems, **D** is routinely computed with quantum chemistry packages like Gaussian,[77] NWChem,[78] and ORCA,[79] and the accuracy of different methods is well-established.[80,81] For solids, the dynamical matrix can be computed, *e.g.*, using ABINIT,[82, 83] VASP,[84] and Quantum Espresso.[85] The mean-square relative displacements can be extracted from the output of these and other programs by using interface programs included in FEFF and used directly to replace the values provided by phenomenological Einstein or Debye models.

In practice, however, a better approach may be to set up a fitting model where the near-neighbor MSRDs are scaled by a single parameter. This approach is quite useful for systems with mixed coordination shells with different mean-square relative displacements where the number of EXAFS parameters becomes large and highly correlated.

Simulating EXAFS of dynamic systems

The standard approach to vibrational disorder in EXAFS assumes that the local environment surrounding the absorbing atom is relatively rigid, with small distortions that are harmonic, thus resulting in an approximately Gaussian pair distribution function.[86] Although broadly applicable, this approach can give poor results for systems like ions in solution or liquid materials, where the bonds between the absorber and its near-neighbors are continuously breaking and reforming. In such cases using an optimized structure plus vibrations is not sufficient, and an explicit sampling over the dynamical fluctuations is needed. A method of choice to obtain this sampling is molecular dynamics in conditions of temperature, pressure, and concentration that closely resemble those used in the experiments. Molecular dynamics simulations rely on the calculation of atomic forces that can be computed based on either model or ab initio potential energy surfaces.[87, 88] The former results in very efficient calculations, but have the drawback of requiring the parameterization of a model potential that might or might not be transferable to other systems and conditions. The latter, on the other hand, can be applied to a very broad range of conditions although at a very high computational cost. Both approaches has been applied successfully to the simulation of ions in solution.[89, 90]

7.5 Other Analytical Methods

Most of the theory of EXAFS was in place by the later 1990s. The statistical formalism described here was developed decades earlier. EXAFS researchers have explored many new ideas in the last 3 decades. Two prominent areas that merit attention are Bayesian statistical analysis and the use of the wavelet transformation.

7.5.1 Bayesian statistics

There are a number of problematic aspects of the statistical discussion in the preceding pages. The difficulty determining an appropriate estimate of ϵ leads to several ambiguities in the analysis, including the absolute determination of goodness of fit and the scaling of error bars. In multiple data set or multiple k-weight fits, no explanation was given for how to weight the relative importance of the different components of the evaluation of ξ^2. Even the determination of the number of independent points, given after ξ^2, results from an analogy with the theory of signal processing that is unlikely to hold true for a typical EXAFS analysis problem. In short, the conventional formulation of the EXAFS ξ^2 minimization problem is usually ill-posed. Despite this problem, the application of the simple statistical model presented in this chapter has proven successful, albeit only when approached conservatively and skeptically.

This approach can be improved upon. Bayesian inference provides a rational system for evaluating assumptions made about a body of data and updating those assumptions on the basis of the measured data. Refs. [91] demonstrate an approach to assessing the value of prior knowledge and the reliability of measured data. By this algorithm, the ill-posed problem is reduced to a series of well-posed, linear equations. Then a series of non-linear equations are evaluated iteratively to determine the relationship of the prior knowledge and the data to the confidence in the result.

This approach contributes a significant solution to each of the shortcomings of the conventional approach. The problems of evaluating measurement uncertainty, confidence in prior knowledge, and confidence in different portions of the data ensemble are treated formally. This Bayesian approach even provides a rigorous way to evaluate the true information content of the data. The treatment presented in Ref.

[92] is the most thorough discussion of Bayesian methods as applied to EXAFS, but its demonstration was limited to rather simple examples of measured and synthetic data. Other authors have applied Bayesian methods to more realistic research problems.[93, 94] As a practical matter, prior knowledge is always used in some sense in the formulation of fitting models for use in EXAFS analysis.

7.5.2 Wavelet transforms

The basis used in a Fourier transform – an integration of $\chi(k)$ over the exponential e^{ikR} – is delocalized in k-space and completely localized in r-space. In EXAFS measurement, however, $\chi_\Gamma(k)$ is not infinite. Since the measured $\chi(k)$ is typically defined beginning around 3 Å^{-1} and extending for only about 5 Å^{-1} to 15 Å^{-1}, the $\tilde{\chi}(R)$ signal will be quite broad even before consideration of the σ^2 term in the EXAFS equation. The wavelet transform offers a tool for distinguishing overlapping scatterers by use of basic functions that are partially localized in both k and r. [95,96] In a situation where two overlapping scatterers are considerably different in mass, their scattering functions will be peaked in amplitude in different regions of k-space. As a general rule, light elements are peaked at low k, while heavier elements have signals that extend to much higher k. Very heavy elements show a Ramsauer-Townsend [97] effect in their scattering amplitudes, adding additional structure to $\chi_\Gamma(k)$ in some k range. If a basis can be found with components localized in k-space in a way that emphasizes one kind of scatterer over another, then the wavelet transform might provide a much-needed sensitivity in the case of scatterers at similar distances. In most applications to EXAFS analysis, Morlet wavelets [98] are used. One particularly interesting implementation of the wavelet transform [99] uses scattering functions from FEFF to define wavelets which are specialized for application to EXAFS analysis.

The use of multiple k-weights bears a certain similarity to the use of wavelet transforms. By including terms in the evaluation of ξ^2 with different sensitivities to different regions of k-space, the fitting model is often able to resolve ambiguities introduced by different scattering species at similar distances. Well-chosen wavelets, particularly those based on scattering functions from EXAFS theory, might be a more refined tool for resolving these ambiguities in a data analysis.

A typical EXAFS analysis begins with discussion of Fourier transform of the *k*-weighted EXAFS data. This first estimation can give a guess upon, how many co-ordination shells at (approximately) which distance are entered by the EXAFS spectrum. This can be analyzed afterwards by *e.g.* several fitting methods. The wavelet transform has been proposed to complement the Fourier transform.[100] While the wavelet transform analyzes the distances of the back scattering atoms contained in the EXAFS signal, the wavelet transform shows additionally at which energies the back scattering takes place. Therefore, wavelet analysis can distinguish between heavier and lighter backscattering atoms, even if they have the same distance from the central atom. The expression of the wavelet transform of the k^3-weighted EXAFS data is

$$W_\chi^\Psi(k,R) = \sqrt{2R}\int \chi(k')k'^3\Psi(2R(k'-k))dk', \quad (7.25)$$

where $\chi(k)$ is the EXAFS signal, $\psi(x)$ is the wavelet (Morlet), k is the

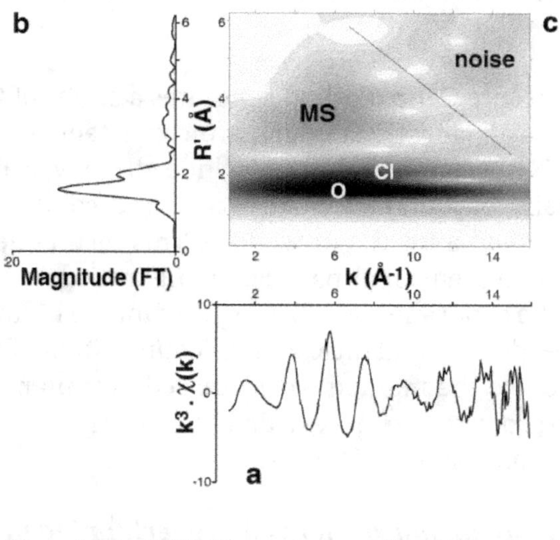

Figure 7.3 (a) k^3-weighted EXAFS spectrum collected at the Au L_{III} edge in an Au (0.01 M)-aqueous chlorine solution (pH = 9.2); (b) FT analysis of the EXAFS spectrum; (c) CCWT analysis of the EXAFS spectrum. "O" and "Cl" represent O atom and chlorine neighboring shells, respectively. "MS" corresponds to the spot related to the multiple-scattering effects arising from the first-neighbors. The solid line separates the MS features and the single-scattering contributions from the noise [M. Munoz, et al., American Mineralogist, (2003) 88, 694–70, reproduced with permission from Mineralogical Society of America].

wave number, and R corresponds to the $R+\Delta$ value in the Fourier transform. Like the Fourier transform, the wavelet transform is a mathematical one-to-one transformation of a signal, here the EXAFS spectrum. Hence, its inverse transformation recovers again the primary EXAFS signal without any loss of information.

The wavelet transform method allows the EXAFS spectrum to be visualized in three dimensions: wave vector (k), interatomic distance (R') without phase shift correction, and wavelet transform modulus (corresponding to the continuous decomposition of the EXAFS amplitude term). Therefore, even if there are spectral artifacts (such as multiple scattering features, multiple electron excitations or noise), a more direct qualitative interpretation of the EXAFS spectrum can be performed, as shown in Figure 7.3. More specifically, this method provides important information about the k range of each EXAFS contribution, such as the next nearest neighbor identification.[101]

7.5.3 Reverse Monte Carlo simulation for the analysis of EXAFS data

The Reverse Monte Carlo method generates a series of three-dimensional atomic configurations which are consistent with the experimental EXAFS spectrum.[102] It can include the physical parameters such as density of material, particle size and compositions which should be satisfied after the procedure. This technique generates a series of three-dimensional particle configurations which are consistent with the experimental EXAFS spectrum. No input information other than the density and the chemical composition of the specimen is required. Since the method works directly from an experimental spectrum it promises to be extremely powerful in modeling the structures of amorphous materials.

7.5.4 Mathematical approach to edge overlapping problems

Analysis of EXAFS data on catalysts containing Pt-Ir represents a particular challenge since these elements are adjacent in the periodic table; the L_3-edges are separated by only 348 eV. The directly extracted EXAFS data are described by an overlap of the L_3-edge with L_2-edge contribution. No quantum mechanical interference of these contributions is found and therefore the EXAFS signals of the combined edges are explained by the sum of the L_3-and the L_2-edge contributions.

$$\chi_{L_3+L_2}(E)=\frac{2}{3}\left(\chi_{L_3}(E)+\alpha\chi_{L_3}(E+\Delta E)\right), \quad (7.26)$$

with $\chi_{L2}(E)=\alpha\,\chi_{L3}(E+\Delta E)$ and $\alpha=1/2$ for EXAFS (quantum degeneracy for L_3- and L_2-edges jumps 2:1) and α= -1 for MEXAFS (quantum degeneracy 2:1 and spin polarization 1:-2→1:-1). ΔE is the energy separation of the L_3- from the L_2-edge. The pre-factor 2/3 takes into account the fact that the edge-jump of the combined $L_{3,2}$-edges is 1.5 times larger than the L_3-edge jump.

An alternative approach in the XAFS study of systems such as Pt-Ir cluster which involve overlapping X-ray absorption edges is to use energy dispersive detectors to measure the experimental data. It is now widely appreciated that the use of energy dispersive detectors affords significant advantages for the measurement of dilute specimens because of their ability to isolate a signal of interest from a high background signal resulting from scattering. The energy discrimination characteristics of these detectors can also be usefully employed to separately measure the X-ray fluorescence signals arising from different elements in the same specimen.

7.6 Difficulties in EXAFS Analysis

Although EXAFS technique has achieved great success in the local environment detection with elemental specificity, for the elements with atomic number less than 20 (Z<20, Ca), it is still difficulty to quantitatively determining their number and type of ligands. Additionally, due to the similarity in their amplitude and phase functions if their atomic number difference is small (ΔZ<5), it is almost impossible to identify the ligand type without other technical assistance.

In some cases, the theoretical amplitude has not yet completely agreed with the experimental one because of unknown factors such as an inelastic loss process and the presence of multielectron effects, and also because of a deviation from a Gaussian distribution for the interatomic distances. A more accurate expression including asymmetric distribution is sometimes required especially for the systems such as catalysts and surfaces since a much larger vibrational amplitude and static disorder, and more significant anharmonicity are expected for surface atoms compared to bulk atoms. If a highly disorder system is analyzed by the usual EXAFS analysis, the coordination number is much reduced and the interatomic distance appears shorter than the real value. For a largely disordered system, it is better to use a model pair distribution function.

There is also the correlation problem in the curve fitting analysis when one has to optimize several structural parameters which have some correlations with each other. When two or more shells have to be fitted at once, the correlation between many fitting parameters forces the fitting results to converge on physically unreasonable values. The other problem is the truncation of the data in the low k-region which has much information about the asymmetric distribution of the radial distribution curve. These problems have been published by the international XAFS working group.[103]

Accordingly, the information extracted from the EXAFS data is not easily converted into unambiguous three-dimensional structures and detailed quantitative information is not routinely obtained especially for multiphasic systems. We have to pay more attention to the EXAFS results relating to the structure of the specimens which have complex, heterogeneous, and highly disordered local structures. The EXAFS data should try to be collected over a larger range in k-space, allowing a more accurate data analysis.

---

1 G. Bunker, (2010) *Introduction to XAFS: a practical guide to X-ray absorption fine structure spectroscopy*. Cambridge University Press, Cambridge. ISBN:9780511809194.

2 P. Pfalzer, J.-P. Urbach, M. Klemm, S. Horn, Marten L. denBoer, Anatoly I. Frenkel, & J. P. Kirkland, (1999) Elimination of self-absorption in fluorescence hard-X-ray absorption spectra. *Phys. Rev. B* 60, 9335.

3 C.H. Booth, F. Bridges, (2005) Improved self-absorption correction for fluorescence measurements of extended X-ray absorption fine-structure. *Phys. Scr.* T115, 202.

4 G.F. Knoll, (2010) *Radiation detection and measurement.* John Wiley & Sons, New York. ISBN: 978-0-470-13148-0

5 F.W. Lytle, D.E. Sayers, & E.A. Stern, (1975) Extended X-ray-absorption fine-structure technique. II. Experimental practice and selected results. *Phys. Rev. B.* 11, 4825–4835.

6 E.A. Stern, S.M. Heald, (1983) In: Koch EE (ed) *Handbook of synchrotron radiation*. North-Holland, New York, pp 995–1014. ISBN-13: 978-0444867094.

[7] T.-Ch. Weng, G. S. Waldo, & J. E. Penner-Hahn, (2005) A method for normalization of X-ray absorption spectra. *J. Synchrotron Rad.* 12, 506–510.
[8] W. H. McMaster, N. Kerr-Del Grande, J. H. Mallett, & J. H. Hubbell, (1969) Compilation of X-ray cross sections. (National Bureau of Standards, Springfield, VA).
[9] https://www-ssrl.slac.stanford.edu/exafspak.html
[10] Newville M, Livins P, Yacoby Y, Rehr JJ, Stern EA (1993) Near-edge X-ray-absorption fine structure of Pb: a comparison of theory and experiment. *Phys. Rev. B* 47:14126
[11] A. Kuzmin, (1995) EDA: EXAFS data analysis software package, *Physica B* 208/209, 175-176. http://www.dragon.lv/eda/
[12] J.L. Beeby, (1964) The electronic structure of disordered systems. Proc. Royal Soc. (London), A279, 82–97.
[13] A.L. Ankudinov, B. Ravel, J.J. Rehr, & S.D. Conradson, (1998) Real-space multiple-scattering calculation and interpretation of X-ray-absorption near-edge structure. *Phys. Rev. B* 58, 7565–7576.
[14] A. Filipponi, (1991) Continued fraction expansion for the X-ray absorption cross section. *J. Phys.-Condes. Matter,* 3, 6489.
[15] T.D. Hu, Y.N. Xie, Y.L. Jin, & T. Liu, (1997) A new method for extracting X-ray absorption fine structure and the atomic background from an X-ray absorption spectrum. *J. Phys.-Condes. Matter,* 9, 5507.
[16] K.V. Klementev, (2001) Extraction of the fine structure from X-ray absorption spectra. *J. Phys. D-Appl. Phys.,* 34, 209.
[17] A. Fillpponi, A. Di Cicco, & C.R. Natoli, (1995) X-ray-absorption spectroscopy and n-body distribution functions in condensed matter. I. Theory. *Phys. Rev. B,* 52, 5122.
[18] B.K. Teo, (1986) *EXAFS: Basic Principles and Data Analysis,* Berlin: Springer, ISBN 978-3-642-50031-2.
[19] D.I. Kochubei, Babanov, Yu.A., K.I. Zamaraev, *et al.,* (1988) *X-ray Spectral Method to Study the Amorphous Body Structure: EXAFS Spectroscopy,* Novosibirsk: Nauka.
[20] A. Kuzmin, J. Purans, M. Benfatto, & C. R. Natoli, (1993) X-ray-absorption study of rhenium L_3 and L_2 edges in ReO_3: Multiple-scattering approach. *Phys. Rev. B* 47, 2480.
[21] A. Kuzmin, S. Obstd, & J. Puransdag, (1997) X-ray absorption spectroscopy and molecular dynamics studies of Zn^{2+} hydration in aqueous solutions. *J. Phys.: Condens. Matter.* 9, 10065.
[22] A. Filipponi, & A. Di Cicco, (1995) Short-range order in crystalline, amorphous, liquid, and supercooled germanium probed by X-ray-absorption spectroscopy. *Phys. Rev. B,* 51, 12322.
[23] G. Dalba, P. Fornasini, & F. Rocca, (1993) Cumulant analysis of the extended X-ray-absorption fine structure of β-AgI, *Phys. Rev. B* 47, 8502.
[24] D.S. Yang, & G. Bunker, (1996) Improved R-space resolution of EXAFS spectra using combined regularization methods and nonlinear least-squares fitting. *Phys. Rev. B* 54, 3169.

[25] P. Kizler, (1991) *Proc. 6th Int. Conf. on X-ray Absorption Fine Structure*, York, Hasnain, S.S., Ed., Singapore: Ellis Horwood, p. 78.

[26] S.J. Gurman, & R.L. McGreevy, (1990) Reverse Monte Carlo simulation for the analysis of EXAFS data. *J. Phys.: Condens. Matter.* 2, 9463.

[27] A. Kuzmin, (1995) EDA: EXAFS data analysis software package. *Physica B* (Amsterdam), 208/209, 175.

[28] C.E. Bouldin, & E.A. Stern, (1984) *EXAFS Studies of Amorphous Semiconductors, EXAFS and Near Edge Structure* III, Hodgson, K.O., Hedman, B., and Penner- Hahn, J.E., Eds., Berlin: Springer, p. 273.

[29] E.A. Stern, Y. Ma, O. Hanske-Petitpierre, & C.E. Bouldin, (1992) Radial distribution function in X-ray-absorption fine structure. *Phys. Rev. B*, 46, 687.

[30] F.W. Lytle, D.E. Sayers, & E.A. Stern, (1989) Report of the international workshop on standards and criteria in X-ray absorption spectroscopy. *Physica B* 158, 701.

[31] A.L. Ankudinov, B. Ravel, J.J. Rehr, & S.D. Conradson, (1998) Real-space multiple-scattering calculation and interpretation of X-ray-absorption near-edge structure. *Phys. Rev. B* 58, 7565.

[32] D.E. Sayers, & B.A. Bunker, (1988) Data analysis, in *X-ray Absorption: Principles, Applications, Techniques of EXAFS, SEXAFS and XANES*, eds., D.C. Koningsberger and R. Prins, pp. 211-253, John Wiley & Sons, New York.

[33] J. Haug, A. Chassé, R. Schneider, H. Kruth, & M. Dubiel, (2008) Thermal expansion and interatomic potentials of silver revealed by extended X-ray absorption fine structure spectroscopy using high-order perturbation theory. *Phys. Rev. B* 77, 184115.

[34] Y. Toshihiko, (1998) Path-integral effective-potential method applied to extended X-ray-absorption fine-structure cumulants. *Phys. Rev. B* 57, 3423.

[35] T. Yokoyama, K. Kobayashi, T. Ohta, & A. Ugawa, (1996) Anharmonic interatomic potentials of diatomic and linear triatomic molecules studied by extended X-ray-absorption fine structure, *Phys. Rev. B* 53, 6111.

[36] N.V. Hung, & J.J. Rehr, (1997) Anharmonic correlated Einstein-model Debye-Waller factors, *Phys. Rev. B* 56, 43.

[37] P. Bevington, & D.K. Robinson, (2002) Data Reduction and Error Analysis for the Physical Sciences. 3rd edition. McGraw-Hill, New York. ISBN-13: 978-9339221201.

[38] L. Brillouin, (1962) *Science and Information Theory.* Academic Press, New York. ISBN-13: 978-0486497556.

[39] R.W. Joyner, K.J. Martin, & P. Meehan, (1987) Some applications of statistical tests in analysis of EXAFS and SEXAFS data. *J. Phys. C: Solid State Phys.*, 20, 4005–4012.

[40] A. Filipponi, (1995) Statistical errors in X-ray absorption fine-structure data analysis. *J. Phys.-Condes. Matter*, 7, 9343–9356.

[41] E. Sevillano, H. Meuth, & J.J. Rehr, (1979) Extended X-ray absorption fine structure Debye-Waller factors. I. Monatomic crystals. *Phys. Rev. B*, 20, 4908–4911.

[42] E.E. Hudson, P.G. Allen, L.J. Terminello, *et al.* (1996) Polarized X-ray absorption spectroscopy of the uranyl ion: comparison of experiment and theory. *Phys. Rev. B* 54, 156–165.

[43] J. H. Sinfelt, G. H. Via, & F.W. Lytle, (1982) Structure of bimetallic clusters. Extended x-ray absorption fine structure (EXAFS) studies of Pt–Ir clusters. *J. Chem. Phys.* 76, 2779.

[44] G.N. Georg, G.H. Via, & M. Sansone, (1990) *National Synchrotron Light Source Annual Report*, 233, 17.

[45] R. Van Hardeveld, & F. Hartog, (1969) The statistics of surface atoms and surface sites on metal crystals, *Surf. Sci.* 15, 189.

[46] S.J. Gurman, N. Binsted, & I. Ross, (1986) A rapid, exact, curved-wave theory for EXAFS calculations. II. The multiple-scattering contributions. *J. Phys. C: Solid State Phys.*, 19, 1845–1861.

[47] A. Filipponi, A. Di-Cicco, & C.R. Natoli, (1995) X-ray absorption spectroscopy and nbody distribution functions in condensed matter. I. Theory. *Phys. Rev. B*, 52, 15122–15135.

[48] S.I. Zabinsky, J.J. Rehr, A. Ankudinov, R.C. Albers, & M.J. Eller, (1995) Multiple-scattering calculations of X-ray-absorption spectra, *Phys. Rev. B* 52, 2995.

[49] J.C. Slater, (1937) Wave functions in a periodic potential. *Phys. Rev.* 51, 846–851.

[50] J.J. Rehr, & R.C. Albers, (2000) Theoretical approaches to X-ray absorption fine structure. *Rev. Mod. Phys.*, 72, 621.

[51] J.G. Norman, (1976) Non-empirical versus empirical choices for overlapping-sphere radii ratios in SCF-Xα-SW calculations on ClO_4 -and SiO_2. *Mol. Phys.*, 31, 1191–1198.

[52] Y. Joly, (2001) X-ray absorption near-edge structure calculations beyond the muffin-tin approximation. *Phys. Rev. B.* 63, 125120.

[53] J.J. Rehr, J.J. Kas, M.P.Prange, A.P. Sorini, Y. Takimoto, & F. Vila, (2009) Ab initio theory and calculations of X-ray spectra. *C. R. Phys.*, 10, 548–559.

[54] I. Turek, J. Kudrnovsky & V. Drchal, (2000) Disordered Alloys and Their Surfaces: The Coherent Potential Approximation. In Hugues Dreyssé (ed.). *Electronic Structure and Physical Properties of Solids*. Springer. p. 349. ISBN 978-3-540-67238-8.

[55] J.J. Rehr, E.A. Stern, R.L. Martin, & E.R. Davidson, (1978) Extended X-ray-absorption fine-structure amplitudes: wave-function relaxation and chemical effects. *Phys. Rev. B* 17, 560–565.

[56] P.A. Lee, & J.B. Pendry, (1975) Theory of the extended X-ray absorption fine structure. *Phys. Rev. B*, 11, 2795–2811.

[57] S.J. Gurman, N. Binsted, & I. Ross, (1986) A rapid, exact, curved-wave theory for EXAFS calculations. II. The multiple-scattering contributions. *J. Phys. C: Solid State Phys.*, 19, 1845–1861.

[58] N. Binsted, R.W. Strange, & S.S.Hasnain, (1992) Constrained and restrained refinement in EXAFS data analysis with curved wave theory. *Biochemistry*, 31, 12117–12125.

[59] N. Binsted, M.J. Pack, M.T. Weller, & J. Evans, (1996) Combined EXAFS and Powder Diffraction Analysis. *J. Am. Chem. Soc.*, 118, 10200–10210.
[60] S. Tomic, B.G. Searle, A. Wander, *et al.* (2004) *New Tools for the Analysis of EXAFS: The DL EXCURV Package*. Council for the Central Laboratory of the Research Councils.
[61] G. Wellenreuther, V. Parthasarathy, & W. Meyer-Klaucke, (2009) Towards a black-box for biological EXAFS data analysis. II. Automatic BioXAS Refinement and Analysis (ABRA). *J. Synchrot. Radiat.*, 17, 25–35.
[62] A. Filipponi, A. Di-Cicco, & C.R. Natoli, (1995) X-ray absorption spectroscopy and n-body distribution functions in condensed matter. I. Theory. *Phys. Rev. B* 52, 15122–15135.
[63] A. Filipponi, & A.D. Cicco, (1995) X-ray absorption spectroscopy and n-body distribution functions in condensed matter. II. Data analysis and applications. *Phys. Rev. B* 52, 15135–15149.
[64] J. Mustre de Leon, J.J. Rehr, S.I. Zabinsky, & R.C. Albers, (1991)Ab initio curved-wave X-ray absorption fine structure. *Phys. Rev. B* 44, 4146–4156.
[65] J.J. Rehr, R.C. Albers, & S.I. Zabinsky, (1992) High-order multiple-scattering calculations of X-ray absorption fine structure. *Phys. Rev. Lett.* 69, 3397–3400.
[66] J.J. Rehr, & R.C. Albers, (2000) Theoretical approaches to X-ray absorption fine structure. *Rev. Mod. Phys.* 72, 621.
[67] M. Newville, (2001) IFEFFIT: interactive XAFS analysis and FEFF fitting. *J. Synchrot. Radiat.* 8, 322–324.
[68] A.L. Ankudinov, J.J. Rehr, J.J. Low, & S.R. Bare, (2002) Theoretical interpretation of XAFS and XANES in Pt clusters. *Top. Catal.* 18, 3.
[69] J.J. Rehr, J.J.Kas, M.P. Prange, A.P. Sorini, Y. Takimoto, & F. Vila, (2009) Ab initio theory and calculations of X-ray spectra. *C. R. Phys.* 10, 548.
[70] J.J. Rehr, J.J. Kas, F.D. Vila, & M. Newville, (2017) Theory and Analysis of XAFS. In: Iwasawa Y., Asakura K., Tada M. (eds) *XAFS Techniques for Catalysts, Nanomaterials, and Surfaces*. Springer, Cham.
[71] E. Burattini, P. D'Angelo, A.D. Cicco, *et al.* (1993) Multiple scattering X-ray absorption analysis of simple brominated hydrocarbon molecules. *J. Chem. Phys.*, 97, 5486–5494.
[72] S. Webb, (2012) SIXPACK: Sam's Interface for XAS Package, https://www.sams-xrays.com/sixpack
[73] F.D. Vila, J.J. Rehr, H.H. Rossner, & H.J. Krappe, (2007) Theoretical X-ray absorption Debye-Waller factors. *Phys. Rev. B* 76, 014301.
[74] H.J. Krappe, & H.H. Rossner, (2002) Bayes-Turchin approach to X-ray absorption fine structure data analysis. *Phys. Rev. B* 66, 18430.
[75] A. Poiarkova, & J.J. Rehr, (1999) Multiple-scattering X-ray-absorption fine-structure Debye-Waller factor calculations. *Phys. Rev. B* 59, 948.
[76] A. Poiarkova,& J.J. Rehr, (1999) Recursion method for multiple-scattering XAFS Debye-Waller factors. *J. Synchrotron Radiat.* 8, 313.
[77] *Gaussian 16*. (2019) Gaussian, Inc., Wallingford, CT.

[78] M. Valiev, *et al.*, (2010) NWChem: a comprehensive and scalable open-source solution for large scale molecular simulations. *Comput. Phys. Commun.* 181, 1477.
[79] F. Neese, (2012) The ORCA program system. *Wiley Interdiscip. Rev. Comput. Mol. Sci.* 2, 73.
[80] K.K. Irikura, R.D. Johnson, & R.N. Kacker, (2005) Uncertainties in scaling factors for ab initio vibrational frequencies. J *Phys. Chem. A* 109, 8430.
[81] I.M. Alecu, J. Zheng, Y. Zhao, & D.G. Truhlar, (2010) Computational thermochemistry: scale factor databases and scale factors for vibrational frequencies obtained from electronic model chemistries. *J. Chem. Theory Comput.* 6, 2872.
[82] X. Gonze, & C. Lee, (1997) Dynamical matrices, Born effective charges, dielectric permittivity tensors, and interatomic force constants from density-functional perturbation theory. *Phys. Rev. B* 55, 10355.
[83] X. Gonze, (1997) First-principles responses of solids to atomic displacements and homogeneous electric fields: implementation of a conjugate-gradient algorithm. *Phys. Rev. B* 55, 10337.
[84] G. Kresse, & J. Furthmüller, (1996) Efficient iterative schemes for ab initio total-energy calculations using a plane-wave basis set. *Phys. Rev. B* 54, 11169.
[85] P. Giannozzi, *et al.* (2009) QUANTUM ESPRESSO: a modular and open-source software project for quantum simulations of materials. *J. Phys. Condens. Matter.* 21, 395502.
[86] E.D. Crozier, J.J. Rehr, & R. Ingalls, (1988) In: Koningsberger DC, Prins R (eds) *X-ray absorption: principles, applications, techniques of EXAFS, SEXAFS, and XANES.* Wiley, New York, pp 373–442.
[87] D. Frenkel, B. Smit, (2002) *Understanding molecular simulation.* Academic, San Diego.
[88] R. Car, & M. Parrinello, (1985) Unified approach for molecular-dynamics and density-functional theory. *Phys. Rev. Lett.* 55, 2471.
[89] L.X. Dang, (1992) The nonadditive intermolecular potential for water revised. *J. Chem. Phys.* 97, 2659.
[90] L.X. Dang, & T.M. Chang, (1997) Molecular dynamics study of water clusters, liquid, and liquid-vapor interface of water with many-body potentials. *J. Chem. Phys.* 106, 8149.
[91] H.J. Krappe, & H.H. Rossner, (2000) Error analysis of XAFS measurements. *Phys. Rev. B* 61, 6596–6610.
[92] H.J. Krappe, & H.H. Rossner, (2002) Bayes-Turchin approach to X-ray absorption fine structure data analysis. *Phys.Rev. B,* 66, 184303.
[93] E. Zschech, W. Blau, V.K. Fjodorov, & M.A. Sheromov, (1989) EXAFS investigation of $Fe_{1-x}B_x$ amorphous alloys. nuclear instruments and methods, *Nucl. Instrum. Methods Phys. Res. A,* 282, 586–589.
[94] M. Sakamaki, T. Konishi, T. Fujikawa, *et al.* (2009) Correlation of magnetism and structure for ultra-thin Au/Co/Au films: evidence for magnetoelastic effects. *J. Phys.: Conf. Series,* 190, 012113.

[95] M. Muenoz, P. Argoul, & F. Farges, (2003) Continuous Cauchy wavelet transform analyses of EXAFS spectra: a qualitative approach. *Am. Miner.*, 88, 694–700.

[96] J. Timoshenko, & A. Kuzmin, (2009) Wavelet data analysis of EXAFS spectra. *Comput. Phys. Commun.*, 180, 920–925.

[97] A.G. McKale, B.W. Veal, A.P. Paulikas, *et al.* (1988) Generalized Ramsauer Townsend effect in extended X-ray absorption fine structure. *Phys. Rev. B* 38, 10919–10921.

[98] J. Morlet, G. Arens, E. Fourgeau, & D. Glard, (1982) Wave propagation and sampling theory Part I: Complex signal and scattering in multi layered media. *Geophysics* 47, 203–221.

[99] H. Funke, M. Chukalina, & A.C. Scheinost, (2007) A new FEFF-based wavelet for EXAFS data analysis. *J. Synchrot. Radiat.*, 14, 426–432.

[100] M. Muñoz, F. Farges, & P. Argoul, (2005) Continuous Cauchy wavelet transform of XAFS spectra. *Phys. Scr. T*115, 221-222.

[101] M. Munoz, P. Argoul, & F. Farges, (2003) Continuous Cauchy wavelet transform analyses of EXAFS spectra: A qualitative approach. *American Mineralogist,* 88, 694–70.

[102] S.J. Gurman, & R.L. McGreevy, (1990) Reverse Monte Carlo simulation for the analysis of EXAFS data. *J. Phys.: Condens. Matter.* 2, 9463-9473.

[103] S.S. Hasnain, ed., (1990) XAFS Internatiuonal workshop Report in "*X-ray absorption fine structure*", Ellis Horwood, New York, p751.

Chapter 8

NEXAFS and XANES Analysis

8.1 Introduction

X-ray absorption near-edge structure (XANES) refers to the oscillations in X-ray absorption coefficient close to the absorption edge, typically from about 5 eV below the absorption edge up to 50 eV above the edge. XANES usually exhibits much larger variation in the X-ray absorption coefficient than EXAFS, and this region is usually dominated by strong, narrow resonances due to strong transitions from the core level to unoccupied states in the valence band. This feature offers XANES technology unique capability to probe the electronic structure of unoccupied energy levels.[1] XANES in the soft X-ray region with wavelength greater than 1 Angstrom, is also called near-edge X-ray absorption fine structure (NEXAFS),[2] Nowadays, the term NEXAFS is often used for soft XAFS spectroscopy of surface science, organic matter, while XANES is commonly used for hard X-ray spectroscopy in coordination chemistry, metal atoms.

In XANES spectroscopy, a core-level electron is excited into a higher unoccupied state, if the photon energy matches the energy needed for the allowed transition. These so-called absorption edges are element-specific and depend also on the chemical environment. The absorption intensity is a measure of the number of unoccupied final states, and the shape of the absorption signal in the near-edge region contains information about the density of unoccupied states.[3,4] The absorption can be measured either in transmission for the case of thin samples, by measuring the fluorescence yield or the total electron yield by detecting the sample drain current. It is even possible to obtain soft-X ray XANES signals from nanoparticles in an organic solvent.

The XANES is associated with the excitation process of a core electron to bound and quasi-bound states, the excitation process involves multi-electron and multi-scattering effects. XANES is considerably harder to fully interpret than EXAFS, as the EXAFS equation breaks down at low-k, due to the $1/k$ term and the increase in the mean-free-path at very low-k. The lack of a simple analytic or even physical description of XANES complicates the interpretation of XANES data. Even though, XANES can be described qualitatively (and nearly quantitatively) in terms of oxidation state, coordination structure, molecular orbitals, magnetic states, chemical bonding, band-structure and multiple-scattering.

8.2 Some Experimental Facts

Without a useful "XANES Equation", one should start from the concept of available state for the photoelectron, since XANES directly detects the angular momentum of unoccupied electronic states: they can be bounded or unbounded (contiuum), discrete or broad, atomic or molecular. The transition probability is controlled by dipolar selection rules: $\Delta l= \pm 1$, $\Delta j= \pm 1$, $\Delta s= 0$. For *K*-shell absorption, where the core-level is a 1s state, the photoelectron has to end up in a *p* state according to the selection rule for electric diploe transition (*i.e.*, the photo-electric effect changes the orbital quantum number l to $l\pm 1$). Thus, even if there are available states with the right energy, there might be no 1s absorption if there are no available p states. This is rarely an important concern for EXAFS as the energies are well-above the threshold energy. But for XANES, this can play a very important role. Note that the final state is usually not atomic-like and may have mixing (hybridization) with other orbitals. This is often the interesting part of the XANES. Transition metal oxides, for example, usually have many unfilled 3d orbitals near the Fermi level, and a filled 3p band. There are empty 2p electron states from the oxygen, but these are too far away to appreciably over-lap with the metal 1s band. Therefore, the metal 3d electrons do not normally participate in the absorption process unless there is a strong hybridization of the O 2p and metal 3d levels. The XANES spectra are then especially sensitive to such hybridization.

Still, the XANES region is rich of chemical information, such as coordination environment and notably formal valence which is very difficult to experimentally determine in a nondestructive way.[5,6] EXAFS spectra are strongly affected by the pair-wise correlation of the absorber and scatterer and, thus the spectra are significantly affected by thermal vibration at typical reaction temperature above 600 K. The XANES does not suffer this disadvantage. Figure 8.1a shows the XANES spectra for several iron compounds. Clearly, the edge position and shape are sensitive to the formal valence state, ligand type and coordination environment. Therefore, XANES can be used as a fingerprint to identify phases.

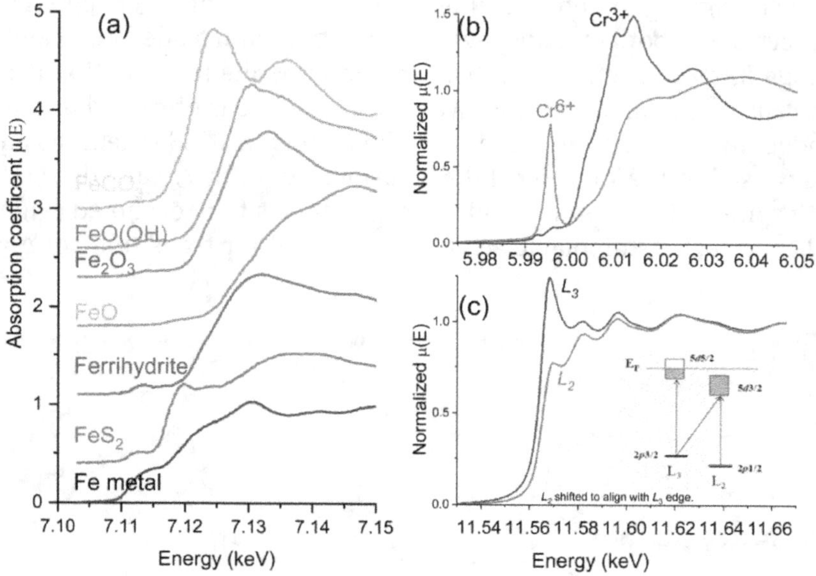

Figure 8.1 (a) Fe K-edge XANES of Fe metal and several Fe compounds. (b) Cr K-edge XANES for Cr^{3+} and Cr^{6+} oxides. Here the strong pre-edge peak in the Cr^{6+} spectrum is a consequence of the tetrahedral symmetry causing considerable overdue of the empty d-electron orbitals with the p-states that the photoelectron must fill. (c) Comparison between Pt L_3 and L_2 edge XANES spectra: 2p to 5d transitions. Pt L_2-edge is shifted to L_3 position for comparison. Same l=2 final density of states but because of selection rule, $\Delta j = \pm 1$, different total quantum number probed. Only j=3/2 probed by L_2-edge, both j = 3/2 and j =5/2 probed by L_3-edge. Pt $5d_{3/2}$ filled, so no white line.

The situation can be even more dramatic – as Figure 8.1b shows for Cr^{3+} and Cr^{6+} oxides. For ions with unfilled d-electrons bands, the p-d hybridization is dramatically altered depending on the coordination environment, which much stronger hybridization for tetrahedral coordination than for octahedral coordination. Since the photoelectron created due to a 1s core level (a *K*-shell) must have p-like symmetry, the amount of overlap with the d-electron orbitals near the Fermi level can dramatically alter the number of available states to the p-electron, causing significant changes in the XANES spectrum. For the case of Cr^{6+}, p-d hybridization results in a highly localized molecular orbital state, giving a well-defined peak below the main absorption edge, indicating a transition to a bound electronic state.

It is intriguing to note the difference between L_2 and L_3 edge XANES spectra. For both metallic tantalum and tantalum oxide, the L_2 and L_3 edge XANES spectra resemble each other, while the situation is different for metallic platinum where a white line is observed in the L_3 edge XANES spectrum but not in the L_2 edge XANES. The difference is caused by the splitting of the 5d orbitals of Pt to $d_{5/2}$ and $d_{3/2}$ states by spin-orbital coupling and the $d_{3/2}$ states is fully occupied (Figure 8.1c, inset). As dipolar transition is allowed for transitions of Δj=0,

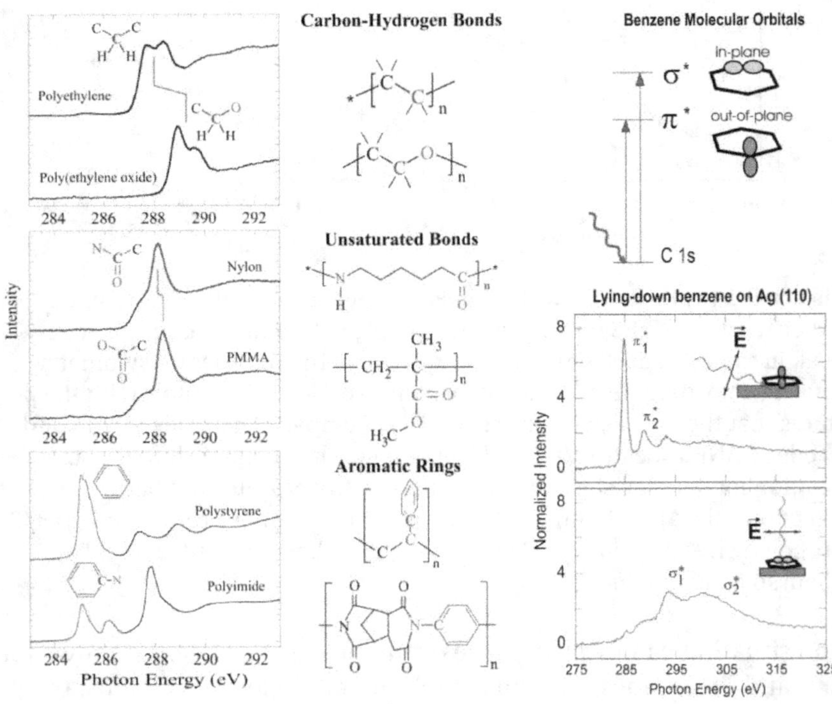

(a) (b)

Figure 8.2 (a) Carbon K-edge NEXAFS spectra of different polymers show the sensitivity to molecular functional groups. (b) Polarization dependent NEXAFS spectra of benzene chemisorbed on Ag(110), illustrating the capability to determine molecular orientations [J. Stohr, private communication].

±1, the excitation of electrons in the L_2 shell ($p_{1/2}$ state) is only allowed to the $d_{3/2}$ state while what of electrons in the L_3 shell ($p_{3/2}$) is allowed to both $d_{3/2}$ and $d_{5/2}$ states. As the latter is not fully occupied, a white line can be observed in the L_3 edge XANES spectrum. In some cases, a difference between L_2 and L_3 edge XANES spectra is observed due to crystal field splitting of d orbitals.

The *bonding environment* of the absorbing atom can strongly affect the NEXAFS line shape, and leads to considerable fine structure above absorption edge of each element.[7] This fine structure is also called resonant absorption due to strong excitations into unoccupied molecular orbitals (or valence band of a solid) from the core level. The resonances can also be reviewed as multiple-scatterings of the excited low-energy photoelectron by the molecular potential. This resonant structure at near edge is much larger than the higher energy EXAFS, which is very weak in organic materials such as polymers. The photoelectrons in EXAFS have high kinetic energies, and experience much less of the valence potential setup by the nearest neighbors, so EXAFS is mostly due to single scattering events. In contrast, the low kinetic energy of photoelectrons in XANES causes multiple scattering of adjacent atoms, so XANES (a spectral "fingerprint" technique) can be used to identify local bonding environments. This is well exemplified by the C *K*-edge NEXAFS spectra for carbon atoms with different bonding configurations, often found in polymers (Figure 8.2a). Similar to the XPS spectrum, the NEXAFS spectra exhibit chemical shift in different functional group, but more importantly, the fine structure of carbon NEXAFS spectra in disparate molecular groups is significantly different. Thus, NEXAFS and XANES can be employed to distinguish between chemical bonds and local bonds. XANES is superior to XPS in many aspects, since XPS cannot provide local structure information.

With the assistance of linearly polarize X-ray beam, NEXAFS can probe the orientation of covalent systems with directional bonds, such as low-Z molecules, macromolecules and polymers. As for dipolar transitions (operator $\epsilon \cdot r$), the X-ray absorption cross section $\sigma(\omega) = 4\pi\alpha^2\hbar\omega\sum_f\langle f|\epsilon \cdot r|i\rangle\delta(E_f - E_i - \hbar\omega)$ reaches its maximum only at the condition that the electric field vector of the linearly polarized X-ray beams is parallel to the chemical bond direction of the atom selected by its absorption edge. Benzene, C_6H_6 can serve as a good example to illustrate the polarization dependence of NEXAFS, for the unoccupied orbitals of σ and π symmetry are oriented in and perpendicular to the benzene ring plane, respectively. The XANES spectra for benzene chemisorbed on Ag(110) show a polarization dependence (see Figure 8.2b). The NEXAFS peaks due to the out-of-plane π orbitals when the electric field vector E is aligned along the surface normal, on the contrary, the in-plane σ orbital resonances become dominant when E is parallel to the surface. The slight energy difference between the π and σ anti-bond in the excited states also leads to C peak position shift. By using linearly polarized XANES, one can determine the orientation of benzene molecules on metal substrate surface. Additionally, compared with the relatively weakly chemisorption of benzene on Ag, the stronger chemisorption bonds (e.g. C_6H_6 on Mo or Pt) can significantly broaden the π resonance peaks, because the π orbitals are involved in the bond to the surface. The ***polarization dependence*** is another great asset of XANES spectroscopy.

Despite the lack of a simple analytic expression which complicates XANES interpretation, XANES can be employed to describe qualitatively (and nearly quantitatively) coordination chemistry (regular, distorted octahedral, tetrahedral coordination, as for Cr ions), molecular orbitals (p-d orbital hybridization, crystal-field theory, and so on), band-structure (the density of available electronic states), multiple-scattering (multiple bounces of the photo-electron). These chemical and physical interpretations are all related, all boil down to determining which electronic states the photoelectron can fill. The origin, content and analysis methods of near edge structure are listed in Table 8.1.

Table 8.1 The origin, content and analysis method of near edge structure

	Origin	Information	Analysis methods
Pre-edge	Inner transition from core to bounded vacancy	System symmetry, hybridized orbits	MO, crystal field theory, band structure
edge	Ionization threshold, continuum above the edge	Oxidation state of absorber	higher oxidation state upshifts the edge
XANES	Multiple scattering resonance	3d structure of neighboring atoms	Ab initio Multiple scattering

8.3 Modeling Methods and Programs

The knowledge about local atomic and electronic structure of the specific center is required for understanding the chemical and physical properties of various functional materials. As a core-shell spectroscopy, XANES can extract information about the electronic structure of the absorbing atom of interest, as well as about its local coordination environment. The interpretation of these XANES spectra is often impossible without quantum ab initio simulations. However, the present XANES computation methods still have challenge to calculate all XANES spectral features precisely and accurately, and often difficult, time-consuming, and not always reliable. At this point, although this situation is improving, it is very rare to use ab initio to quantitatively analyze XANES. Such calculations can still help explain which bonding orbitals and/or structural features cause certain spectral features.

There are typically two approaches to the XANES simulations: density functional theory (including time dependent DFT), and many-body theory.[8] The modern methods of computation for theoretical XANES can be classified into multiple-scattering, supercell DFT, Bethe–Salpeter equation method, and Multiplet calculations.

(1) Multiple-scattering (MS) calculations using model clusters
Since XANES shares the same theoretical framework with EXAFS by varying the energy of the excited electron, it is quite natural that the multiple-scattering method has been a popular approach especially for XANES simulations by solving the one-electron Schrödinger equation.[9] many systems have been satisfactorily interpreted by the multiple-scattering calculations on model clusters of up to hundreds of atoms, [10] but simulations for the system with transition elements having open d and f shells are still challenging.[11] Additionally one should carefully examine the dependence/convergence of theoretical spectra in relation to the cluster size and shape, number of scattered waves, degrees of potential approximation (full potentials or not), etc.

(2) Supercell DFT calculations with a core hole
One of conventional methods for theoretical XANES studies of wide gap materials is based on a reliable band structure calculation for a large supercell with inclusion of a core hole. Band structure methods

have been widely used for studying theoretical XANES except for transition elements having open d and f shells. Supercells exceeding 100 atoms are typically required and the effect of the core hole should be included in the self-consistent calculation. The core hole can be included for all electron methods in a straightforward manner. The theoretical transition energy can be computed as the difference in total energy between final and initial states. When core orbitals are eliminated in the ground state calculations, such as in cases of pseudopotentials or frozen core methods, special attention should be paid to maintaining the accuracy of the final state calculations.

(3) Bethe–Salpeter equation (BSE) calculations using primitive cells
In the XANES process, the Coulomb interaction between excited electrons and positively charged holes leads to the formation of bound exciton states. Sharp excitonic features can be found in experimental XANES spectra of wide gap materials when they are measured at shallow core edges with high energy resolution. Because standard density functional theory treats electrons and holes as independent, it is not possible to calculate exciton states. The two-particle Green's function can be explicitly utilized to describe the correlation between the core hole and the excited electron by solving the Bethe–Salpeter equation. [12,13] The bound exciton states can be obtained by solving the Bethe–Salpeter equation.[14] Although full BSE calculations within all electron schemes are computationally very demanding, recently some successes have been achieved in reproducing the XANES features at shallow edges.

(4) Multiplet calculations using model clusters
The strong correlation among electrons and that between the core holes in the final state lead to widely spread multiplet structures present in the $L_{2,3}$ XANES of 3d transition metal compounds having an open 3d shell. The same phenomenon was also observed in the XANES spectra for rare-earth compounds with an open 4f shell. The multiplet structures can be simulated by using either semi-empirical charge transfer multiplet calculations or *ab-initio* configuration interaction (CI) calculations using molecular orbitals. They have satisfactorily reproduced spectra of many compounds having different d or f electron numbers and coordination numbers using model clusters. However, such calculations are still limited to model clusters with a single metal ion because of the large computational demand.

Significant progress has been achieved over the past two decades in the understanding of the features of the XANES spectra.[15,16] Earlier XANES spectra were interpreted mainly in a qualitative way, considering the presence/absence of certain fingerprint features and comparing the data with the spectra of model compounds.[17,18] The appearance of progressively more sophisticated codes and the increased computing powers has changed the situation. As a structural model the local environment of the X-ray absorbing atom can quantitatively confirmed or discarded with XANES spectroscopy,[19,20] modern XANES spectroscopy is becoming a new fundamental diagnostic tool in the field of condensed matter physics and chemistry.[21,22] Among such codes (see Table 8.2) include multiple scattering FEFF-9, [23,24] XKDQ, [25] CONTINUUM, [26,27] SPR-KKR[28] and periodic basis set Wien2k, [29,30] and PARATEC.[31] The matrix inversion in full m,utiple-scattering calculations scale in compute time as the cube of system size, so XANES calculations are usually more time-consuming than EXAFS where the multiple-scattering expansion converges typically with of order 10^2 paths. As multiple core CPUs are widely available, the fast parallel, Lanczos XANES calculations already become reality.[32]

The finite difference method is attractive for XANES calculations because it avoids in a simple way the muffin-tin potential approximation.[33] In this FDMNES approach the whole space is divided in three regions: (1) an outer sphere surrounding the cluster of interest, (2) an atomic region composed by very little spheres around the atomic cores (no more than 0.65 Å of radius, much smaller than the usual muffin-tin sphere), and (3) the interatomic region where a finite difference formulation of the Schrödinger equation is performed. Thus FDMNES in principle offers much high accuracy than the conventional muffin-tin approximation–based multiple scattering theory.[34]

Several density functional theory-based codes with localized orbitals in a basis set, ADF[35] and ORCA,[36, 37] are able to directly compute the pre-edge region within the time-dependent density functional theory (TDDFT) approximation. The time-dependent density functional theory approximation can provide reasonable values for XANES cross-section up to 40 eV above absorption edge[38,39] and other techniques are being developed to access higher electronic core-excited states, *e.g.* algebraic-diagrammatic construction scheme. [40]

Table 8.2 Classification of the methods to calculate XANES spectra and available software for calculations

DFT, TDDFT	Green function	*FEFF*	*CONTINUUM*	*FDMNES*	*SPRKKR*	*MXAN*	*XKDQ*
	Finite difference	*FDMNES*					
	Basis set	*ADF*	*ORCA*	*Gaussian*	*Wien2k*	*PARATEC*	*Quantum Espresso*
Many-body	Atomic multiplets	*QUANTY*	*XTLS*	*CTM4XAS*			
	GW	*EXCITING*	*Feff*				
	BSE	*OCEAN*	*EXCITING*				
	Post HF	*MOLCAS*	*ORCA*				
	DMFT	*ONETEP*	*QUANTY*				

8.4 XANES Analysis

XANES is uniquely sensitive to oxidation state, coordination chemistry, and the partial density of unoccupied electronic states, and XANES spectra are much richer in spectral features than EXAFS, making it an excellent spectroscopic technique for almost every element.[41] Unfortunately, the EXAFS equation breaks down at low k, where λ_k diverges, so that a huge number of scattering paths contribute. With no readily parameterized XANES equation, quantitative XANES analysis has traditionally relied on fitting idealized peak-shapes to spectral features or matching spectra to linear combinations of spectra from known systems. These approaches are reasonably well justified in many specific cases, however, are not easy to generalize. Recently a reverse Monte-Carlo approach [42,43] has been employed to best-match experimental data using a suite of calculations. The advantage of a reverse Monte-Carlo approach is able to use both the XANES and EXAFS portions of a spectrum, but it is quite computationally demanding. In the soft X-ray regime, and for certain 3d metal pre-edge peaks, quantitative analysis based on multiplet analysis is also robust and very useful for assigning electronic transitions.[44] Although XANES cannot be numerically parameterized in terms of oxidation state or coordination chemistry, these can be determined using simpler spectroscopic analysis methods. For example, the pre-edge peaks in the K and L edges of 3d metal are very rich at identifying coordination chemistry and hybridization of electronic levels, even if the precise intensities of the different transitions is complicated to calculate accurately. To make analysis quantitative, fitted intensities of such peaks can be compared to calibration curves made from spectra of known materials. While such analysis is not *ab initio*, it can be robust and reliable with minimal effort.

A second common approach to XANES analysis is to treat the spectra as a signature or fingerprint of a material and determine the fractional composition of spectra from well-characterized materials. This approach uses simple linear statistical analysis, and so can build on a wide literature, including Principal Component Analysis [45] and related methods.[46, 47]

8.4.1 XANES modeling

As compared with EXAFS modelling, the theoretical modeling of XANES spectra has bigger challenges due to two main reasons: (a) XANES demands simulations are more computationally intensive and, (b) generally the information about the structure of the system under study is incomplete. These two aspects lead to the need to theoretically generate structural models through optimization or molecular dynamics simulations. As in the case of EXAFS, the former are quite useful for quasi-rigid systems, while the latter are required for systems with dynamic fluctuations and bond-breaking. As opposed to EXAFS, XANES spectra are often less sensitive to the quality of the structural model, thus permitting the direct use of theoretical structures in the simulations. The emergence of accurate and efficient DFT simulations actually eliminates the need to guess possible structures, because now people can optimize accurate models based on simple external information (such as composition).[48,49] In addition, one can use other information generated by the simulation, for example as thermodynamic properties, charge distribution and vibration properties. In light of the current trend towards multi-spectroscopy, *operando* experimental approaches,[50] this broad range of theoretical properties facilitates a more comprehensive interpretation of the experimental results.

Effects of dynamical disorder on XANES

Like in EXAFS, dynamic structural disorder plays a very important role in the accurate simulation of XANES. For systems where the experiment naturally samples a large number of conformations, XANES simulations usually must include similar sampling to reproduce the experiment. In theory, the most common method of structural sampling is molecular dynamics simulation within an appropriate time frame. These simulations can rely on the model's potential energy surface, or they can rely on the density functional theory potential (*i.e.*, density functional theory/molecular dynamics). In practice, the simulations based on model potential energy surfaces is very effective and can well understand the structural characteristics of hybrid systems with covalent and soft interactions. However, for other examples of dynamic systems with metallic bonds (such as supported metal clusters), the use of density functional theory/molecular dynamics is strongly recommended. This approach also has the added bonus of providing very useful information about the electronic structure of the material. A classic example of this type of problem is the case of Pt and PtSn nanoclusters on γ-Al_2O_3 ,[51] where modeling is needed to understand their intra-particle heterogeneity and their structural and electronic changes under realistic conditions. The density functional theory/molecular dynamics simulations reveal that the nanoscale structure and charge distribution are inhomogeneous and dynamically fluctuating over several timescales, ranging from fast (200–400 fs) bond vibrations to slow fluxional bond breaking (>10 ps).

Understanding electronic structure changes through modeling

Perhaps one of the most useful features of theoretical modeling is the possibility of directly visualizing atomistic changes in electron density distribution. This is particularly important since global probes such as EXAFS and XANES reveal the ensemble characteristics of materials while missing details of their internal structure.[52] Moreover, changes in spectral features such as edge position and white line intensity that are commonly associated with charge density fluctuations in a qualitative way can be studied quantitatively and in detail. An excellent example of the atomistic understanding that is gained through modeling arises again in the case of Pt_{10} on γ-Al_2O_3 where the cluster is roughly separated into two layers, with one layer in contact with the surface and another resting on top.[51] The atoms in the contact layer become positively charge or "oxidized," while the others are neutral or slightly negatively charged ("reduced"). The molecular dynamics simulations show that at high temperature, the number of contacts with the surface increases, thus removing more charge from the cluster which results in the expected red shift in the absorption edge position and increase in the white line intensity. Moreover, these charge fluctuations can be correlated with other observed trends such as negative thermal expansion in Pt clusters, and species segregation and differential reactivity in PtSn alloy clusters.[53] This detailed atomistic understanding can only be achieved through the close integration of modeling and experiment.

8.4.2 Principal component analysis - Edge fit

A simple approach to analysis of experimental XANES spectra is to express it in terms of linear components of model spectra. Principal component analysis is a useful method for examining the XANES spectra from samples that contain mixtures of compounds. The starting point for principal component analysis is the recognition that every spectrum can be represented as a vector. For XANES spectra, each part of the vector indicates the intensity of absorption at a particular energy. If the energy scales for the original spectra differ, interpolation can be used to place the spectra on a common scale. A data matrix, [D] is constructed in which the columns contain the individual vectors. The covariance matrix for the data, [Z], is then calculated by multiplying the transpose of the data matrix, $[D]^T$, and the data matrix together. Each member of [Z] is $[Z]=[D]^T[D]$, $[R]=[D][D]^T$, $[D]_{repro}=[R][C]$, the inner product between two vectors and represents how similar the individual spectra are to one another. The eigenvectors of [Z] are abstract components for the spectra. In themselves, there components do not have any direction physical meaning. The corresponding eigenvalues, λ, represent the reproduce the relative importance of each component. The eigenvectors are then used to reproduce the original data matrix. A column matrix [C] is formed, containing 1 or more of the most significant eigenvectors, as determined form the eigenvalues, in its rows. From the column matrix, a row matrix [R] is created. Lastly the reproduced data [D]repro which is the product from multiplication of the row and column matrices, is compared to the original data, [D]. If these data matrices agree with experimental error, then these components are sufficient to explain the data. If this procedure does not reproduce the data well, the process is repeated after adding the next most significant eigenvector to the column matrix, until the data are reproduced. One advantage of principal component analysis is that it provides a method for testing whether a particular pure material is present in the original mixtures. Target transformation is that one can test a particular spectrum without requiring any knowledge regarding the other species which may be present in the specimens. In target transformation, a test vector, Test'=$[R][\lambda]^{-1}[R]^{T}$Test. Principal component analysis is most successful for a series of spectra taken at a single beamline under identical experimental conditions. Errors in calibration and changes in monochromator crystals and/or mode of detection will all results in apparent increased experimental error. Thickness effects represent another possible complicating factor for principal component analysis.

There are many programs like EDG_FIT in *EXAFSPAK* suite allowing peak deconvolution of X-ray absorption edge spectra, this is useful to obtain accurate peak positions and areas.[54] The experimental data can be fitted to a sum of pseudo-Voigt peak shapes, plus a step function. The pseudo-Voigt is an approximation of the expected line shape function, which will be a convolution of a Gaussian line shape, defined by the monochromator resolution, with a Lorenztian lineshape from the core-hole broadening. The pseudo-Voigt peak shape is simply a sum of Gaussian and Lorenztian peak shapes:

$$I_V = mI_G + (1-m)I_L$$

$$I_G = exp\left\{\frac{-ln2(E-E_m)^2}{[(W+(E-E_m)\eta)\xi]^2}\right\}$$

$$I_L = \frac{[W+(E-E_m)\eta]^2}{[W+(E-E_m)\eta]^2+(E-E_m)^2}$$

where I_G and I_L are the Gaussian and Lorentzian lineshape components, respectively, m is the mixing factor determining the proportion of each, W is the half-linewidth, E is the abscissa (X-ray energy for edge spectra), E_m is the position of the peak, η is the peak skew, and ξ is the ratio of Gaussian to Lorentzian linewidths (lw-ratio). The step function is simply the integral of this function with η= 0 and ξ = 1 (comprised of an erf and an arctangent function for the Gaussian and Lorentzian parts, respectively). A squaric polynomial is also included to approximate any background function of the spectrum. Neglecting the latter, the spectrum, *I(E)*, is fit to the function : $I(E) = a_0 I_0(E) + \sum_i^n a_i I_{vi}(E)$ where a_i and $I_{vi}(E)$ are the peak amplitudes and peak shape functions (defined above), respectively, the summation is over all peaks included in the fit, and a_0 and $I_0(E)$ are the step function amplitude, and the step function, respectively.

8.5 Summary and Outlook

NEXAFS/XANES features are dominated by multiple-scattering resonances of the photoelectrons ejected at low kinetic energy and have large scattering cross section. As XANES measures the density of unoccupied final states, its line shape contains important information about the local chemical environment of an absorbing atom, including coordination number, symmetry, and oxidation state. For examples, many edges of many elements show significant edge shifts (binding energy shifts) with oxidation state, main edge shifts to higher energy with increased oxidation state as much as 5 eV per unit change. However, extracting structure and electron descriptors from XANES spectra is similar to solving challenging inverse problems. Empirical fingerprints still play important role in XANES decoding, the existing methods are usually qualitative or semiquantitative and not transferable. Additionally, although ab initio calculations can help explain which bonding orbitals and/or structural features cause certain spectral features, the present XANES computation methods still have challenge to calculate all XANES spectral features precisely and accurately, and not always reliable. So although XANES is a much larger signal than EXAFS, XANES is easier to crudely interpret than EXAFS, and more difficult to fully understand than EXAFS.

Machine learning might provide an alternative way to analyze XANES data in future. Machine learning methods are closely related to computational statistics and are used to solve problems such as classification, regression, clustering, distribution analysis, and dimensionality reduction.[55] Machine learning uses computer algorithms to predict the value of unknown functions based on training samples without having to explicitly program for a given task. With the rapid development of machine learning, a computing framework has recently been developed for element-specific classification of local chemical environments based on XANES spectroscopy. In addition to building structural databases, spectral databases, and structural labels, the core element of this framework is to use machine learning classifiers to systematically reveal the correlation between spectral features and the local chemical environment. The robustness and fidelity of the machine learning method is demonstrated by the average 86% accuracy in the extensive chemical space of oxides in eight 3d transition metal families.[56] Spectral features outside the pre-edge region play an important role in the local structure classification problem, especially for the late 3d transition metal elements.[57] The ability to extract key structural information in the entire spectral range makes the machine learning-based method more effective. Compared with the empirical fingerprint method based on the pre-edge region only, it is robust and transferable.

On the one hand, more and more sophisticated codes can better reconstruct the theoretical spectra, while the ever-increasing computing power can handle larger simulation models, and a higher level of theory can be used in solving the Schrödinger equation. On the other hand, complex methods are employed to deal with huge data sets generated by spatial or temporal resolution experiments. Chemometric methods (such as principal component analysis) can determine how many different species contribute to the entire collected XANES spectrum, while by using multivariate curve resolution-iterative alternating least squares algorithm approach, the XANES spectra of the pure species can be disentangled, and their contribution in each of the different collected spectra can be to determined quantitatively. The application of machine learning methods in XANES spectroscopy will have a revolutionary impact on the way we perform XANES analysis and will allow an unexpected amount of structural information to be obtained. The application of the machine learning tools is not limited to XANES spectroscopy but can, in principle, be extended to any kind of spectroscopy, such as X-ray emission.

---

[1] J.J. Rehr, & A.L. Ankudinov, (2005) Progress in the theory and interpretation of XANES. *Chem. Rev.* 249, 131-140.

[2] J. Stöhr, (1992) *NEXAFS Spectroscopy*, Springer Series in Surface Sciences, ISBN 978-3-662-02853-7

[3] C. Garino, E. Borfecchia, R. Gobetto, J.A. van Bokhoven, & C. Lamberti, (2014) Determination of the electronic and structural configuration of coordination compounds by synchrotron-radiation techniques. *Chem. Rev.* 277, 130-186.

[4] L. Mino, G. Agostini, E. Borfecchia, D. Gianolio, A. Piovano, E. Gallo, & C. Lamberti, (2013) Low-dimensional systems investigated by X-ray absorption spectroscopy: a selection of 2D, 1D and 0D cases. *J. Phys. D: Appl. Phys.* 46, 72;

[5] S. Yang, D. Wang, G. Liang, Y.M. Yiu, J. Wang, L. Liu, X. Sun, & T.-K. Sham, (2012) Soft X-ray XANES studies of various phases related to $LiFePO_4$ based cathode materials. *Energy Environ. Sci.* 5, 7007-7016.

[6] S. Bordiga, E. Groppo, G. Agostini, J. A. van Bokhoven, & C. Lamberti, (2013) Reactivity of surface species in heterogeneous catalysts probed by in situ X-ray absorption techniques. *Chem. Rev.* 113, 1736-1850.

[7] D. Schauries, V. Ney, S.K. Nayak, P. Entel, A.A. Guda, A.V. Soldatov, F. Wilhelm, A. Rogalev, K. Kummer, F. Yakhou, & A. Ney, (2013) Incorporation of nitrogen in Co: ZnO studied by X-ray absorption spectroscopy and X-ray linear dichroism. *Phys. Rev. B.* 87, 125206.

[8] A.L. Ankudinov, B. Ravel, J.J. Rehr, & S.D. Conradson, (1998) Real-space multiple-scattering calculation and interpretation of X-ray-absorption near-edge structure. *Phys. Rev. B.* 58, 7565-7576.

[9] A. Filipponi, A. Di Cocco, & C.R. Natoli, (1995) X-ray-absorption spectroscopy and n-body distribution functions in condensed matter. I. Theory. *Phys. Rev. B* 52, 15122.

[10] J .J. Rehr, & R.C. Albers, (2000) Theoretical approaches to X-ray absorption fine structure. *Rev. Mod. Phys.* 72, 621-654.

[11] A.L. Ankudinov, C.E. Bouldin, J.J. Rehr, J. Sims, & H. Hung, (2002) Parallel calculation of electron multiple scattering using Lanczos algorithms. *Phys. Rev. B.* 65, 104107.

[12] G. Onida, L. Reining , & A. Rubio, (2002) Electronic excitations: density-functional versus many-body Green's-function approaches. *Rev. Mod. Phys.* 74, 601.

[13] E.L. Shirley, (2004) Ti 1s pre-edge features in rutile: a Bethe-Salpeter calculation. *J. Electron Spectrosc. Relat. Phenom.* 136, 77.

[14] J. Vinson, J.J. Rehr, J.J. Kas, & E.L. Shirley, (2011) Bethe-Salpeter equation calculations of core excitation spectra. Phys. Rev. B. 83, 115106;

[15] Q. Wang, J.C. Hanson, & A.I. Frenkel, (2008) Solving the structure of reaction intermediates by time-resolved synchrotron X-ray absorption spectroscopy. *J. Chem. Phys.* 129, 234502.

[16] N. L. Saini, A. Lanzara, H. Oyanagi, H. Yamaguchi, K. Oka, T. Ito, & A. Bianconi, (1997) Local lattice instability and stripes in the CuO_2 plane of the $La_{1.85}Sr_{0.15}CuO_4$ system by polarized XANES and EXAFS. *Phys. Rev. B.* 55, 12759-12769.

[17] S. Bordiga, S. Coluccia, C. Lamberti, L. Marchese, A. Zecchina, F. Boscherini, F. Buffa, F. Genoni, G. Leofanti, G. Petrini, & G. Vlaic, (1994) XAFS Study of Ti-Silicalite: Structure of Framework Ti(IV) in the Presence and Absence of Reactive Molecules (H_2O, NH_3) and Comparison with Ultraviolet-Visible and IR Results. *J. Phys. Chem.* 98, 4125-4132.

[18] D. Li, G.M. Bancroft, M. Kasrai, M.E. Fleet, X.H. Feng, K.H. Tan, & B. X. Yang, (1993) High-resolution Si K-and $L_{2,3}$-edge XANES of α-quartz and stishovite. *Solid State Commun.* 87, 613-617.

[19] J.V. Lockard, A.A. Rachford, G. Smolentsev, A.B. Stickrath, X. Wang, X. Zhang, K. Atenkoffer, G. Jennings, A. Soldatov, A.L. Rheingold, F.N. Castellano, & L.X. Chen, (2010) Triplet excited state distortions in a pyrazolate bridged platinum dimer measured by X-ray transient absorption spectroscopy. *J. Phys. Chem. A.* 114, 12780-12787.

[20] E. Borfecchia, S. Maurelli, D. Gianolio, E. Groppo, M. Chiesa, F. Bonino, & C. Lamberti, Insights into Adsorption of NH_3 on HKUST-1 Metal–Organic Framework: A Multitechnique Approach. *J. Phys. Chem. C.* 2012, 116, 19839-19850.

[21] S. Bordiga, C. Lamberti, G. Ricchiardi, L. Regli, F. Bonino, A. Damin, K.P. Lillerud, M. Bjorgen, & A. Zecchina, (2004) Electronic and vibrational properties of a MOF-5 metal–organic framework: ZnO quantum dot behavior. *Chem. Commun.* 2300-2301.

[22] C.R. Natoli, M. Benfatto, S. Della Longa, & K. Hatada, (2003) X-ray absorption spectroscopy: state-of-the-art analysis. *J. Synchrotron Radiat.* 10, 26-42.

[23] J.J. Rehr, J.J. Kas, M.P. Prange, A.P. Sorini, Y. Takimoto, & F. Vila, (2009) Ab initio theory and calculations of X-ray spectra. *Comptes Rendus Physique.* 10, 548-559;

[24] J.J. Rehr, J.J. Kas, F.D. Vila, M.P. Prange, & K. Jorissen, (2010) Parameter-free calculations of X-ray spectra with FEFF9. *Phys. Chem. Chem. Phys.* 12, 5503-5513.

[25] L.A. Bugaev, I.I. Gegusin, V.N. Datsyuk, A.A. Novakovich, & R.V. Vedrinskii, (1986) Multiple scattering approach to the XANES theory of alkali halide crystals. I. Crystalline potential in the X-Ray absorption spectra problem. *Phys. Status Solidi B.* 133, 195-202.

[26] A. Bianconi, J. Garcia, M. Benfatto, A. Marcelli, C.R. Natoli, & M.F. Ruizlopez, (1991) Multielectron excitations in the K-edge X-ray-absorption near-edge spectra of V, Cr, and Mn $3d^0$ compounds with tetrahedral coordination. *Phys. Rev. B.* 43, 6885-6892.

[27] T.A. Tyson, K.O. Hodgson, C.R. Natoli, & M. Benfatto, (1992) General multiple-scattering scheme for the computation and interpretation of X-ray-absorption fine structure in atomic clusters with applications to SF_6, $GeCl_4$, and Br_2 molecules. *Phys. Rev. B.* 46, 5997-6019.

[28] H. Ebert, D. Ködderitzsch, & J. Minár, (2011) Calculating condensed matter properties using the KKR-Green's function method-recent developments and applications, *Rep. Prog. Phys.* 74, 096501.

[29] K. Schwarz, P. Blaha, & G.K.H. Madsen, (2002) Electronic structure calculations of solids using the WIEN2k package for material sciences. *Comput. Phys. Commun.* 147, 71-76;

[30] K. Schwarz, & P. Blaha, (2003) Solid state calculations using WIEN2k. *Comput. Mater. Sci.* 28, 259-273.

[31] M. Taillefumier, D. Cabaret, A.M. Flank, & F. Mauri, (2002) X-ray absorption near-edge structure calculations with the pseudopotentials: Application to the K edge in diamond and α–quartz. *Phys. Rev. B.* 66, 195107.

[32] J.J. Rehr, & A.L. Ankudinov, (2005) Progress in the theory and interpretation of XANES, *Coord. Chem. Rev.* 249, 131–140.

[33] J.D. Bourke, C.T. Chantler, & C. Witte, (2007) Finite difference method calculations of X-ray absorption fine structure for copper. *Phys. Lett. A.* 360, 702-706.

[34] Y. Joly, & S. Grenier, (2015) Theory of X-ray Absorption Near Edge Structure. In *XAS and XES: Theory and Applications*, J.A. van Bokhoven, C. Lamberti, Eds. John Wiley & Sons: Vol. 1;

[35] G. te Velde, F.M. Bickelhaupt, E.J. Baerends, C. Fonseca Guerra, S.J.A. van Gisbergen, J.G. Snijders, & T. Ziegler, (2001) *J. Comput. Chem.* 22, 931-967.

[36] F. Neese, (2009) *ORCA, an ab initio DFT and semiempirical electronic structure package*, version 2.7, University of Bonn: Bonn, Germany.

[37] F. Neese, (2012) The ORCA program system, *WIREs Comput. Mol. Sci.* 2, 73.

[38] R.G. Fernando, M.C. Balhoff, & K. Lopata, (2014) X-ray Absorption in Insulators with Non-Hermitian Real-Time Time-Dependent Density Functional Theory. *J. Chem. Theory Comput.* 11, 646-654.

[39] O. Bunău, & Y. Joly, (2012) Time-dependent density functional theory applied to X-ray absorption spectroscopy. *Phys. Rev. B.* 85, 155121.

[40] J. Wenzel, M. Wormit, & A. Dreuw, (2014) Calculating X-ray absorption spectra of open-shell molecules with the unrestricted algebraic-diagrammatic construction scheme for the polarization propagator. *J. Chem. Theory Comput.* 10, 4583-4598.

[41] G.S. Henderson, F.M.F. de Groot, B.J. Moulton, D.R. Neuville, & R.T. Downs, (2014) In: *Spectroscopic methods in mineralogy and materials sciences*, 78, 75–138.

[42] M. Benfatto, & S. Della Longa, (2001) Geometrical fitting of experimental XANES spectra by a full multiple-scattering procedure. *J. Synchrotron Radiat.* 8, 1087.

[43] G. Smolentsev, & A. Soldatov, (2006) Quantitative local structure refinement from XANES: multi-dimensional interpolation approach. *J. Synchrotron Radiat.* 13, 19.

[44] E. Stavitski, & F.M.F. de Groot, (2010) The CTM4XAS program for EELS and XAS spectral shape analysis of transition metal L edges. *Micron* 41, 687.

[45] S.R. Wasserman, P.G. Allen, D.K. Shuh, J.J. Bucher, & N.M. Edelstein, (1999) EXAFS and principal component analysis: a new shell game. *J. Synchrotron Radiat.* 6, 284.

[46] S. Beauchemin, D. Hesterberg, & M. Beauchemin, (2002) Principal component analysis approach for modeling sulfur K-XANES spectra of humic acids. *Soil Sci. Soc. Am. J.* 66, 83.

[47] M. Lerotic, C. Jacobsen, J.B. Gillow, A.J. Francis, S. Wirick, S. Vogt, & J. Maser, (2005) Cluster analysis in soft X-ray spectromicroscopy: finding the patterns in complex specimens. *J. Electron Spectrosc. Relat. Phenom.* 144, 1137.
[48] E.S. Kryachko, & E.V. Ludeña, (1990) *Energy density functional theory of many-electron systems*. Kluwer Academic Publishers, Dordrecht.
[49] F. Giustino, (2014) Materials modelling using density functional theory: properties and predictions. Oxford University Press, Oxford.
[50] A.I. Frenkel, M.W. Cason, A. Elsen, U. Jung, M.W. Small, R.G. Nuzzo, F.D. Vila, J.J. Rehr, E.A. Stach, & J.C. Yang, (2014) Critical review: Effects of complex interactions on structure and dynamics of supported metal catalysts. *J. Vac. Sci. Technol. A* 32, 020801.
[51] F.D. Vila, J.J. Rehr, S.D. Kelly, & S.R. Bare, Operando effects on the structure and dynamics of PtnSnm/γ-Al_2O_3 from ab initio molecular dynamics and X-ray absorption spectra. *J. Phys. Chem. C* 117:12446 (2013).
[52] J. García, G. Subías, & J. Blasco, (2015) XAS studies on mixed valence oxides. In *XAS and XES: Theory and Applications*, J. A. van Bokhoven, C. Lamberti, Eds. John Wiley & Sons: Vol.1.
[53] J.J. Rehr, & F.D. Vila, (2014) Dynamic structural disorder in supported nanoscale catalysts. *J. Chem. Phys.* 140, 134701.
[54] https://www-ssrl.slac.stanford.edu/~george/exafspak/exafs.htm
[55] T. Hastie, R. Tibshirani, & J. Friedman, (2009) *The Elements of Statistical Learning. Data Mining*, Inference, and Prediction, 2nd edition, Springer, New-York.
[56] M.R. Carbone, S. Yoo, M. Topsakal, & D. Lu, (2019) Classification of local chemical environments from X-ray absorption spectra using supervised machine learning. *Phys. Rev. Materials* 3, 033604.
[57] S. Bordiga, R. Buzzoni, F. Geobaldo, C. Lamberti, E. Giamello, A. Zecchina, G. Leofanti, G. Petrini, G. Tozzola, & G. Vlaic, (1996) Structure and Reactivity of Framework and Extra-framework Iron in Fe-Silicalite as Investigated by Spectroscopic and Physicochemical Methods. *J. Catal.* 158, 486-501.

Chapter 9

Atomic X-ray Absorption Fine Structure

9.1 Introduction

X-ray absorption fine structure (XAFS) spectroscopy is a local structure characterization technique based on the photoelectric effect. As ionizing electromagnetic radiation, when X-ray photons' energy are high enough, they can excite a core electron of an atom to an unoccupied state below the ionization threshold (excitonic state), or to the states above the ionization threshold (continuum). This photoelectron can be described by a spherical wave, which propagates outward and interacts with its surrounding environment. The photoelectron emitted from the absorbing atom can be backscattered by neighboring atoms, so that interference occurs between the outgoing wave and the back-scattered waves, modulating the absorption coefficient as a function of photon energy. The structure in the modulation χ enables the local geometry about the absorbing atom to be determined utilizing the well-known technique of EXAFS (extended X-ray absorption fine structure).[1] The resultant EXAFS data reveal the geometry around the absorber atom. The total X-ray absorption $\mu(\omega)$ is normally written as

$$\mu(\omega) = \mu_0(\omega)[1 + \chi_{EX}(\omega)], \tag{9.1}$$

with $\chi_{EX}(\omega)$ the EXAFS function and $\mu_0(\omega)$ the atomic X-ray absorption coefficient, or atomic background. Upon absorbing an X-ray, the absorbing atom accounts for the major part $\mu_0(\omega)$ of the absorption coefficient.

Conventionally, this atomic background was assumed to be smooth and non-oscillatory. However, Holland *et al.*[2] and later Rehr *et al.* [3, 4] and Dorssen *et al.* [5] noted that $\mu_0(\omega)$ is not a totally smooth function, since the atomic X-ray absorption coefficient can have structure due to backscattering of the photoelectron after core-hole excitation against the potential determined by the electron cloud (interstitial charge densities) of the atom from which the photoelectron is excited. Therefore, this structure is called atomic XAFS (AXAFS) and offers the possibility to obtain electronic structure information utilizing a procedure comparable to EXAFS. The atomic absorption coefficient can thus be expressed as

$$\mu_0(\omega) = \mu_{0,free}(\omega)[1 + \chi_{AX}(\omega)], \tag{9.2}$$

with $\chi_{AX}(\omega)$ the AXAFS fine structure and $\mu_{0,free}(\omega)$ the free atomic background.[6] Therefore the X-ray absorption coefficient becomes:

$$\mu(\omega) = \mu_{0,free}(\omega)[1 + \chi_{AX}(\omega)][1 + \chi_{EX}(\omega)], \tag{9.3}$$

where the oscillatory part due to scattering at the neighbors $\chi_{EX}(\omega)$ is separated from the oscillations due to scattering at interstitial charge densities $\chi_{AX}(\omega)$. In this representation of the total X-ray absorption coefficient, the scattering of the photoelectron at the interstitial charges is separated from the scattering at neighboring atoms. The absorbing embedded atom (*i.e.* the atom bonded to its environment) accounts for the major part $\mu_0(\omega)$ of the absorption coefficient. The structure in $\chi_{EX}(\omega)$ permits the determination of the local geometry of the absorbing atom. The $\chi_{AX}(\omega)$ function offers the possibility to obtain electronic structure information utilizing a procedure comparable to EXAFS.

This AXAFS effect was ignored for years, one reason was the use of special analysis routines like AUTOBK.[7] These fitting routines minimize the intensities in the Fourier transform located below the nearest neighbor distance, because it was believed that these peaks have no physical meaning. This minimization is achieved by a spline function which exhibits the low frequency oscillations. While using those routines, one has to be aware that the physical information of the scattering in the periphery of the absorbing atom is then in the oscillating spline function. If this spline function is just taken as an analytical tool and is not investigated on its own, the vital information about the AXAFS is lost.

While the regular EXAFS stems from the backscattering of the photoelectron by neighboring atoms, the AXAFS contribution is assigned to the scattering at the charge densities placed between the atoms. Therefore, the AXAFS contribution to the entire fine structure of the X-ray absorption coefficient should be highly directional for systems with strong anisotropic bonding. The scattering of the photoelectron at interstitial charge densities results in an oscillatory fine structure $\chi_{AX}(\omega)$ of the atomic X-ray absorption coefficient $\mu_0(E)$ of the absorbing atom. As suggested by Rehr and co-workers, a heuristic model of AXAFS is derived to illustrate its nature based on perturbation theory in the difference between the free and embedded atomic potentials, $\delta v = v_0(r) - v_a(r)$. The model fine structure leads to have the same form as EXAFS, $\mu_0(\omega) = \mu_{0,free}(\omega)[1 + \chi_{AX}(\omega)]$, where $\mu_{0,free}(\omega)$ is the much smoother free-atom absorption, and the AXAFS oscillatory fine structure $\chi_{AX}(\omega)$ can be described in the muffin-tin approximation by [8]

$$\chi_{AX}(\omega) = -\frac{S_0^2|f_e|}{kR_{MT}^2} e^{-2\frac{R_{MT}}{\lambda}} e^{-2\sigma^2k^2\left(\frac{R_{MT}}{R}\right)^2} \sin(2kR_{MT} + 2\delta_l^a + \phi_e), \quad (9.4)$$

where R_{MT} is the muffin-tin radius, the effective curved-wave scattering amplitude of the interstitial charge density$f_e = |f_e|\exp(i\phi_e)$ with $|f_e| \sim exp(-2\pi k/\xi)/k$, ξ is a wave function decay constant, and δ_l^a is the central-atom phase shift.

9.2 Physical Principles of AXAFS

AXAFS reflects the change in scattering of the photoelectrons by the periphery of the embedded absorbing atom itself relative to the free absorbing atom. This change stems from the bonding of the absorbing atom with its environment and contains the electronic structure information about the absorbing atom. Therefore, any changes in the environment that alter the charge or structure of the absorbing atom should be reflected in AXAFS. In order to understand the AXAFS scattering process, a description of the potential surrounding the absorber atom is necessary. The embedded atom potential can be well described using the well-known muffin-tin approximation as schematically illustrated in Figure 9.1. The embedded atom potential can be calculated by using the muffin-tin approximation. The exact potential is "cliped" by the muffin-tin approximation at the muffin-tin radius R_{MT} and it is set equal to the interstitial potential V_{int}.[9] The potentials inside and outside R_{MT} are respectively assumed to be spherical and flat and zero, that is to say, no forces are exerted on the particle in the interstitial region. V_{int} is determined by averaging the potential at R_{MT} of all the atoms in the cluster; this interstitial potential V_{int} determines the zero energy of the system [10] or the effective bottom of the conduction or itinerant band. The free atom potential

U_{free} and the embedded potential U_{emb} respectively reflect the electron distribution in the free atom, and the electron distribution after embedding the free atom into its chemical environment and allowing interaction with neighbors. Note that potentials are functions of *R*, which are denoted with U, constants with V).

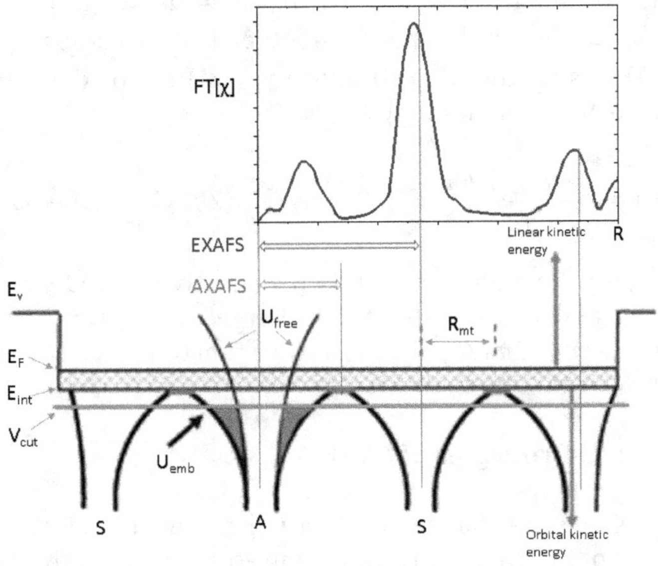

Figure 9.1 Schematic showing scattering of the outgoing photoelectron by bound electrons of the same absorber atom (AXAFS) and neighbouring atoms (EXAFS). Illustration of the muffin-tin approximation to the interatomic potentials showing locations of R_{MT}, V_{int} and E_F. The shape and intensity of the Fourier transform of the AXAFS is determined by the black shaded area between U_{emb} and U_{free} and below V_{cut} ($V_{cut} = 2V_{int} + |E_f|$) as defined in the text. Maximum scattering occurs if the linear kinetic energy of the photoelectron is equal to the orbital kinetic energy of the electron scatterer.

The AXAFS $\chi_{AX}(\omega)$ can be expressed mathematically in a first distorted wave Born approximation [11] and using the muffin-tin approximation as

$$\chi_{AX}(k) \approx -\left[\frac{2m}{k\hbar^2}\right] Im \int e^{2ikr+2i\delta} \Delta U dr, \qquad (9.5)$$

where $\Delta U = U_{emb} - U_{TFA}$, U_{emb} is the embedded atom potential and U_{TFA} is the truncated free atom potential. $V_{TFA}(r) =$

$min[U_{free}(r), V_{int}(r)]$. Here, U represents the potential as a function of the distance from the absorber, and V gives a line with a constant potential value. The free atom potential (U_{free}) reflects the distribution of electrons in free atoms. The embedded potential (U_{emb}) reflects the distribution of electrons after embedding free atoms into its chemical environment and allowing interaction with its neighbors. The AXAFS is best reflected in the Fourier transform of the $\chi_{AX}(\omega)$ producing the first peak at about half of the first shell bond length. A phase-corrected and *k*-weighted Fourier transform of $\chi_{AX}(k)$ results in

$$\left|FT\left(ke^{-2i\delta}\chi_{AX}(k)\right)\right| \approx \Delta U\Gamma, \tag{9.6}$$

where Γ is a broadening function due to the limited Fourier transform range. Equation 9.6 then reveals that the shape and intensity of |FT| can be represented by the black area between U_{TFA} and U_{emb} and below V_{cut} ($V_{cut} = 2V_{int} + |E_f|$), see Figure 9.1. The Fourier transform of AXAFS is then directly proportional to $U_{emb}(R)$ - $U_{free}(R)$ and therefore its shape and intensity directly reflect electronic and chemical bonding information. The AXAFS contribution introduces additional oscillations, although the wavelength is longer because the distance to the scattering source is shorter. In principle, this AXAFS scattering is somewhat larger than normal EXAFS due to the shorter distance and the $1/R^2$ factor out front [see Eq. 9.2]. Also, the electron density completely surrounds the absorber atom instead of being concentrated at several lattice sites around the absorber. On the other hand, the AXAFS scattering is short ranged in *k*, because it arises primarily from scattering by the deep valence electrons rather than the core electrons as in EXAFS. This makes the peak in the Fourier transform of AXAFS wider and generally smaller in magnitude than the regular EXAFS scattering.

The scattering of photoelectrons by electrons on the orbit is mainly resonant, which means that if the kinetic energy of photoelectrons is equal to the kinetic energy of the bound electrons undergoing scattering, scattering will occur. Going upwards above Fermi level E_f increases the linear kinetic energy of the photoelectron, whereas going downwards below the interstitial potential V_{int} increases the orbital kinetic energy of the valence electrons localized on a single atom. The photoelectrons at the absorption edge already have a kinetic energy equal to E_F-V_{int}. The electrons in the conduction band are not effective for scattering, because even at a threshold, their kinetic energy is much lower than that of the photoelectron. In addition, electrons in a state higher than V_{cut} have insufficient orbital kinetic energy. Therefore, only electrons in a state lower than V_{cut} (that is, with a higher binding energy) can effectively scatter photoelectrons with kinetic energy close to or far above the threshold. In short, only deeper local valence band electrons will cause AXAFS scattering.

9.3 The Field and Inductive Interactions and Charge Effects

The AXAFS can be altered by either changing U_{emb} or V_{cut}, the latter ($V_{cut} = 2V_{int} + |E_f|$) is directly determined by the position and width of the itinerant conduction band. A change in the nature or number of bonds around the absorbing atom will definitely alter these two properties of the absorber atom, and the AXAFS should change accordingly. Thus, an increase in the number of bonds about an absorbing atom will pull down U_{emb} (*i.e.* increase the "roll-over" of U_{emb}) and will enhance the AXAFS. Further, reducing its covalent character will decrease the AXAFS, because a decrease in the covalent nature will limit the "roll-over" in U_{emb}. Due to polarization of the absorbing atom by its neighbors,[12] two important effects on the electronic properties can be distinguished with AXAFS: (1) the Coulomb field effect, *i.e.* the effect induced by a Coulomb field, correlated to different charges on neighboring substituents (thereby altering the "roll over" of the interatomic potential); and (2) the bond inductive effect, induced by a difference in electronegativity of the substituent (thereby altering the value of V_{int}). For an isolated Coulomb field effect, the AXAFS intensity increases and a simultaneous shift to lower radius is observed. For the isolated bond inductive effect, the intensity goes down with a simultaneous shift to lower distance. Moreover, the difference in AXAFS shape change illustrated is typical for the two different effects ('lower R part' increased by field effect *vs.* 'higher R part' decreased by inductive effect). In fact, many more subtle long-range effects can be observed in AXAFS, which is essential for applying AXAFS to examine the interaction between metal and support. In the muffin-tin model used to describe AXAFS, the polarization of more distant atoms can modify the absorbing atoms in the same two ways as mentioned above: (1) the "through-space field effect" by the direct overlap of the nearby atom potentials, altering U_{emb}; or (2) the "through-bond inductive effect" by affecting the average V_{int} and V_{cut}.[13] This reference is appropriate because if a neighboring atom is highly electronegative, V_{int} will be more negative, which means that more electrons will be delocalized into the interstitial region, exactly the essence of the inductive effect.

Finally, the removal of an electron from the absorber atom (ionization) will lead to an increase in the AXAFS. This may not be obvious, because one may wish that the fewer electrons remaining on the absorbing atoms, the less scattering. However, the removed electrons are usually taken from the Fermi level, and the electrons are not effective scatterers anyway. But removing this electron will cause all other electrons to transfer to a higher binding energy, making them more effective scatterers. As the atomic charge increases, this shift is similar to the increase in atomic ionization potential.

9.4 Separation of AXAFS from EXAFS Data

In order to obtain accurate AXAFS signal, a very careful background subtraction procedure is required to separate the AXAFS contribution from the XAFS raw data.[14,15] The most critical aspect of an AXAFS analysis is the isolation of χ_{AX} from μ_{total}. This can be difficult, and it is clearly the major factor in preventing a broader application of this technique than that exhibited thus far. The background subtraction procedure is extensively described in the literature.[16] Detailed criteria are given how to optimize both the AXAFS χ_{AX} and EXAFS χ_{EX} contributions in the resulting XAFS data, and how to separate double electron excitations (DEE) and the Ramsauer–Townsend resonances (RTR) that are present in the free atom contribution from the oscillatory part of the spectrum. In order to perform a careful DEE/AXAFS/EXAFS separation, it is important to have the ability to use a continuous adjustable smooth parameter.

9.4.1 Separating χ from μ_{total}

Pre-edge subtraction and determination of edge

The extraction of $\chi = \chi_{AX} + \chi_{EX}$ from the experimentally obtained μ_{total} follows several consecutive steps. As defined in Eq. 9.1, once the free atomic absorption is known, χ can be calculated,

$$\chi = \frac{\mu_{total} - \mu_{free}}{\mu_{free}}.$$

First, the pre-edge background has to be subtracted before μ_{free} can be determined. In the pre-edge background removal, a modified Victoreen curve approximates the pre-edge background and extends the whole data range,[17] and then is subtracted from the raw data.[18] After the pre-edge subtraction, the edge energy (*e.g.* E_{edge}) has to be defined. Generally, it chooses the energy at the maximum in the first derivative of the absorption edge. Sometimes, the presence of resonance or electronic excitation can obscure the exact location of E_{edge}. In order to avoid this complexity, the position of the half-height (0.5~0.6) of the standardized edge is usually also selected as E_{edge}. The value of the edge energy is utilized to calculate the momentum of the photoelectron k (Å^{-1}), $k = \frac{2\pi}{\lambda_e} = \sqrt{\frac{2m}{\hbar^2}}\left(\hbar\omega + \Delta - E_{edge}\right)$. The inner potential, Δ, or the zero point of the kinetic energy of the outgoing electron is often chosen to be zero at E_{edge}.

Post-edge background removal and normalization

The post-edge background is approximated by a normal cubic spline function. The atomic background (μ_{free}) after the absorption edge has to be determined, normally using a cubic spline,

$$\sum_{j=1}^{N} \frac{\left(\mu(k_j) - \mu_{free}(k_j)\right)^2}{e^{-Wk_j^2}} < SM. \tag{9.7}$$

The cubic spline curve consists of four parameters that define the function of the spline curve: SM is the smoothing parameter, W is the weighting factor, and the start and end energies that determine the number of points (N). The more data points available, the better the definition of the spline. [19] If *SM* is set to zero, the background goes through all the experimental points and no EXAFS structure remains after background subtraction. Increasing the value of *SM* smoothes out the background and leaves more structure in the EXAFS function. In general, this structure is best characterized in the Fourier transform of the EXAFS, which presents peaks at various R values. Historically, the criteria were to decrease *SM* until all of the features below 1.5 Å in the Fourier transform were extracted from the Fourier transform (denoted as the $SM_{R<1.5}$ criterion). In most cases the $SM_{R<1.5}$ criterion led to a small decrease (10%) in the first shell EXAFS peak.

In order to isolate both the AXAFS and EXAFS signals from the smooth free atom background, a more refined procedure has to be used. New criteria have been introduced for background removal:

(a) In the Fourier transform diminish the contribution at R<0.5 Å, *i.e.* the double electron excitations and the Ramsauer–Townsend resonances, as much as possible. Check whether the double electron excitations and the Ramsauer–Townsend resonances fully remain in the background by examining the background.
(b) Make sure the EXAFS intensity at R > 1.5 Å is unchanged (unreduced).
(c) Check both k^1- and k^3-weighted spectra, for different k-ranges, including low k-values, to ensure that the EXAFS signal is not removed from the data.

These new criteria will be denoted as the $SM_{R<0.5}$ criterion. The spline function as given in Eq. 9.5 is used for the background removal, with the weight parameter W set to zero. The spline functions for the background are therefore determined primarily by the smoothing parameter SM. These criteria can be applied generally. However, they were constructed for XAFS data of specimens in which the AXAFS and the EXAFS peaks in the Fourier transform of the XAFS data have less or no overlap. For these cases, applying these criteria is easy and straightforward. Note that other background subtraction procedures using nodal splines or polynomials are not capable of straightforwardly separating the AXAFS from the atomic background in the same systematic way. This procedure is described in detail for Pt/LTL,[20] Pt foil,[16] and Pt pincer complexes.[14]

9.4.2 Separation of AXAFS from EXAFS

It is essential to isolate the AXAFS from the EXAFS signal (*i.e.* χ_{AX} from χ_{EX}), especially in the studies of the AXAFS peak changes in different systems or under different experimental conditions. The Fourier transform of the first shell EXAFS contribution often tail into the AXAFS peak, as its coordination distances often fall between 1.0 < R < 2.0 Å. Actually, this interference would reduce the amplitude of the AXAFS peak in the Fourier transform. In these cases, the difference file technique can be employed to separate AXAFS from EXAFS, and be utilized to judge the quality of the fit procedure.[21,22] The first shell contributions are fitted in R-space and subtracted from the Fourier transform of the EXAFS signal. Then the Fourier transform of the resulting difference file at low R are due to the presence of the AXAFS. In this way, the AXAFS peak can be isolated. The AXAFS oscillations can now be obtained by applying an inverse Fourier transform from $0 < R < 1.6$ Å. The AXAFS oscillations usually have a long wavelength and decrease rapidly as the values of k increase.

It should be emphasized again that, before the AXAFS and EXAFS signals are fully separated from the atomic background and double electron excitations, it is necessary to follow a very careful and systematic background subtraction procedure as mentioned above. By monitoring the disappearance of the double electron excitations from the Fourier transform of the oscillatory part of the X-ray absorption spectrum, as a function of the smooth parameter (*SM*), the described procedure allows for a more objective approach than that used for the isolation of the EXAFS from the background signal. In addition, the *R*-space fitting technique has been shown to further distinguish the AXAFS data from the first shell EXAFS contributions. Segregating AXAFS data is a prerequisite for using AXAFS to explore the influence of the support and adsorbates on the electronic structure of supported metal particles.

9.5 Applications of AXAFS

AXAFS originates from the scattering against the potential of the electron cloud of the absorbing atom itself. The embedded potential of probed atoms depends on the chemical and electronic environment of the atoms and is affected by the support characteristics. The intensity and position of the AXAFS peak directly reflects the bonding of the absorbing atom with its environment. Therefore, any change in the support oxide that alters the embedded potential of the absorbing atom will be reflected in its AXAFS spectrum. The analysis of the AXAFS contribution from the experimental XAFS data can help distinguishing the charging/potential effect, and the through-space field effect as well as the through-bond inductive effect. Therefore, AXAFS has broad applications in the fields such as catalysis, biology, surface science, etc. Recent AXAFS results demonstrate that (1) the charge on a metallic cluster can be monitored with changes as small as 0.05e (e = electronic charge),[23] (2) polarization effects from ions at a distance of two and three coordination shells from the absorber are clearly observed,[6] (3) the charge redistribution within organometallic complexes can be monitored [16] and (4) the amplitude and shape of the AXAFS (as best revealed in the Fourier transform) directly correlate with the turnover frequency for Pt-catalyzed hydrogenolysis [24] and hydrogenation reactions.[25]

9.5.1 Change of AXAFS with charging/potential

A detailed study of a PtRu electrode in an operating PEM fuel cell shows that the AXAFS of the anode in methanol tracks with the anode potential. The significant correlation between the AXAFS amplitude and the estimated anode potential did verify the interpretation of the AXAFS change. That is due to the polarization of the anode. Note that for H_2, the increase in AXAFS is much smaller, which is consistent with the smaller change in anode potential with current when using H_2 (V <30 mV), because the kinetics of the hydrogen oxidation reaction is much faster. The Fourier transform magnitude of AXAFS taken before introducing the electrode into the fuel, is very large. Before contacting the fuel, the Pt electrode is highly oxidized. The O atom strongly polarizes Pt through the field effect. The enhancement and variation of the AXAFS amplitude of Pt by nearby O atoms has been extensively discussed previously.[26,24,11]

The systematic changes in AXAFS was first observed in the study of the influence of the potential in an electrochemical cell on the electronic structure of a Pt/C electrode in 0.1 M H_2SO_4.[27] The Fourier transform (k^0, $0.5 < k < 8.5$ Å^{-1}) for the Pt L_3 XAFS spectra of the in situ Pt electrode at the indicated applied voltages relative to the reversible hydrogen electrode (RHE) shows that, the peak positions do not change with the potential, indicating a constant atomic radius having no change in the immediate chemical environment. But, the magnitudes of the peaks show a strong and systematic increase as the electrode potential is increased. To understand such a phenomenon, curved-wave cluster calculations were carried out for a Pt_{13} cluster by using the FEFF7 code with the normal muffin-tin approximation. The potential of zero charge (PZC) for polycrystalline Pt is between 0.2 and 0.3 V relative to RHE.[28] Above this potential the charge on the surface is positive and below it the charge is negative. The cyclic voltammetry data reported by Clavilier *et al.*[29] suggested a charge of ±0.05e (electrons) per surface Pt atom in the double layer region. Therefore, the FEFF calculations were performed with charges of +0.05, 0.0 and −0.05e per Pt atom to mimic the experimental data reported at the three voltages (0.0, 0.54 and 0.74 V relative to RHE). The theoretical trends are in a reasonable agreement with the experimental one with applied voltage; the FEFF simulation reasonably estimates the magnitude of the AXAFS but underestimates the average R. The muffin-tin approximation "clips off" the potential maximum, effectively increasing the AXAFS magnitude and in muffin-tin radius R_{MT} is less than atomic radius R_a. As discussed above, the amplitude of the Fourier transform of AXAFS increases with charging. Thus, the magnitude of the AXAFS tracks with the charge on an electrode in an electrochemical cell.

9.5.2 The interatomic potential model for metal–support interaction

In the catalyst preparation, the active metal particles are dispersed on the proper supports in order to stabilize the particles and promote the catalytic performance. The metal–support interaction which plays critical role in these aspects, can be described with the interatomic potential model. AXAFS is a powerful tool to monitor the interatomic potential change. For example, the intensity and shape of the Fourier transform of the AXAFS (FT[χ_{AX}]) from supported Pt particles in different zeolites (*e.g.* LTL, Y) are strongly dependent on the amounts and types of charge compensating ions (H+, K+), Si/Al ratio and extra-framework Al.[10,30,24] Further, on a flat Al_2O_3 support with different modifiers,[31,32] the Fourier transform peak of AXAFS is a function of the support properties. The essence of the interaction of the support with Pt metallic particles is a shift in the ionization potential of the Pt valence band. The amounts and types of charge-compensating ions (H+, K+), Si/Al ratio and extra-framework Al in the zeolites and the modifiers in the Al_2O_3 support all have a strong influence on the Madelung potential in the supporting oxide.[33] The Madelung potential in turn alters the ionicity of the cation–anion bonds in the oxide support. The overall trend observed in the Fourier transform of the Pt AXAFS peak is that when the electron charge of the oxygen atoms in the support becomes smaller, the peak centroid has higher the intensity and shifts to a lower R value, *i.e.*, for a support with low ionicity, such as a covalent or acidic support. The reverse is true for supports with a high ionicity, such as for ionic or basic supports. By simply changing the charge on the O atom of the supporter, the results of the ab initio calculation using the FEFF7 code performed on Pt_4/O_4 clusters with the four O atoms mimicking the support can well reproduce the changes of the AXAFS data observed in the experiment with the changes of the support properties.[24] These ab initio calculations indicate that the electronic properties of the platinum surface atoms representing the catalytically active sites are directly affected by the change in charge density on the oxygen ions in the support. This is the essence of the interaction between the metal and the support: the electronic properties of the supported metal particles directly induced by the support change.

For neutral O atoms, the field effect is very short, so cluster polarization will not occur. The difference ΔU between U_{free} and U_{emb} represents the Fourier transform of AXAFS. The position of *Vcut* will be used as a reference for the energy level. As positive charges are added to oxygen (as happens on a support with reduced ionicity: more covalent or acidic), the field effect caused by the additional positive charges now becomes longer range, so the Pt atoms in the entire cluster feel separately to a different field. Since the cluster can be partially metallic, the charge can flow to the support Os until the potential across the entire cluster is the same. Compared with the reference value, the final potential of the cluster is "pulled down". This causes the Fourier transform amplitude of AXAFS to be higher, and the center of mass of the peak shifts to a lower R value.

From the more orbital image, the lowered electric potential causes the valence d orbit to shift to a higher binding energy (larger negative potential), which causes the electronic properties of the Pt particles to change. As predicted by the interatomic potential model, one of the most important aspects of the interaction between the metal and the support is the change in the ionization potential of the Pt particles caused by the polarization caused by the oxygen ions of the carrier.

Several previous metal-support interaction models have been proposed. Solymosi [34] and Schwab [35] have proposed the idea of electronic interactions between support and metal particles in early papers. They showed that doping the semiconductor TiO_2 support will cause the activation energy of several metal-catalyzed reactions to change, which implies a change in semiconductor-metal junction. Another model proposed a method of forming metal-proton adducts [36,37] to explain the electron-deficient metal particles observed in acid zeolites. Protons are thought to delocalize on metal particles and withdraw electron density from surface atoms. However, this adduct cannot account for the increased electron density of the metal particles on the alkaline supports. Alternatively, it is suggested that the electronegativity of the support oxygen atoms increases as the zeolite alkalinity increases.[38] Charge transfer[39] between the support oxygen atoms and adjacent metal particles is believed to result in a higher electron density on the metal particles in the alkaline zeolite. The two models are similar in that, due to the interaction between the metal and the carrier, they both propose the transfer of electron density from (or to) metal particles. The only difference between these models is that in the former, the metal transfers the electron density to the support protons, while in the latter, it is believed that the transfer occurs between the metal and oxide ions. The third explanation is based on the polarization of metal clusters by nearby cations.[40,41] Calculations show that the metal atoms near the cation attract electrons, so the electron-deficient metal atoms are located on opposite sides of the cluster. In this model, there is no net change in the total electron density of the cluster; however, the polarization of the electron density causes those atoms that are involved in the electron deficient catalysis.

The actual charge transfer of the metal particles to the support will keep the center of mass of the Fourier transform peak the centroid of the Fourier transform peak of AXAFS unchanged, and the displacement is observed in the experiment. This fact and other important points made by Ponec and Bond[42] make the charge-transfer model very unrealistic. Similarly, polarization of cations is considered unlikely. The polarization of metal particles can indeed occur, the size of which depends on the morphology of the particles. However, in the current model, this polarization is not due to the presence of cations, however, is caused by support oxygen ions. All of these data point to a cluster on acidic supports acting as if it were less metallic, even though no net charge has been removed from the cluster. The current model is also consistent with other data. According to the AXAFS study, the valence state of the platinum particles on the acidic support is shifted to a higher binding energy, which is completely consistent with the infrared data of CO adsorbed on Pt / LTL, Pd / LTL, and Pt / SiO_2.[10] The displacement increases the energy difference between the interacting d orbital and the CO $2\pi^*$ orbital, thereby reducing the back donation of platinum to CO. The FTIR CO data shows that under acidic conditions, the ratio of linear / bridge bonded CO increases, consistent with the greater difficulty for the charge transfer required in the bridged CO configuration.[43] In a recent study, many different zeolite-supported Pt catalysts were studied using FTIR spectroscopy to monitor chemisorbed CO and AXAFS to monitor the platinum particles. The AXAFS intensity varies systematically as a function of the ratio of linear / bridge bonding CO.[44]

In order to investigate the electronic nature of the interaction of the support with small- and medium-sized (d < 2 nm) metallic and oxidic particles, a new *in situ* technique in XAFS has been developed that provides previously unobtainable electronic structure information. This applications of the AXAFS part of the XAFS data directly gives information regarding the embedded potential of the X-ray absorbing atom.

9.5.3 AXAFS in surface physics

The AXAFS phenomenon at the Pr, Ce and Ba K-edges of $PrBa_2Cu_3O_7$, CeO_2, and BaO was investigated by Rehr *et al.*[45] The oscillatory fine structure was found in the atomic X-ray absorption background $\mu_0(\omega)$. However, this work was questioned by Filipponi and Di Cicco [46] because they were able to show [47] that double-electron excitations corresponding to [1s4p], [1s3d] and [1s3p] double-core-hole configurations can mimic oscillations in $\mu_0(\omega)$. Recently O'Grady et al. presented peaks in the Fourier transform below the nearest neighbor distance for the Pt L_3 edge of Pt–Ru alloys [48] although possible multielectron excitations were not discussed. This problem can be avoided by investigating the *K*-edges of light atoms like N, O, C as shown in refs.[49,50] The complication of multielectron excitations discussed above does not exist for the low-Z atoms. The 2s binding energies of 37.3 eV in the case of N and 41.6 eV for O do not influence an EXAFS analysis. The oscillations of the atomic X-ray absorption coefficient $\mu_0(\omega)$ of an embedded atom for the N/Cu(110) surface system can be calculated with the ab initio FEFF code.[51] The clearly visible oscillations presented in the inset have been extracted with a smooth polynomial of third order. A spline function can be utilized to extract the low frequency oscillations up to 400 eV above the N *K*-edge. These structures cannot be related to multielectron excitations. This clear interference phenomenon has to be interpreted by backscattering at interstitial charges. The (2×3)N/Cu(110) system exhibits a pseudo-(100)-c(2×2)-reconstruction[52] and the R45°

O/Cu(100) system can be explained by a missing row reconstruction.[53] The experimental data are taken in normal X-ray incidence geometry for both systems. In this geometry the bonds of the light adsorbates to the Cu atoms in the surface plane are probed. Temperature dependent surface EXAFS investigations have shown that these four bonds for the N/Cu(110) system and the three bonds for the O/Cu(100) system are equally strong. As the surface EXAFS intensity scales with the projection of the bond vector to the electric field vector an average over these bonds is investigated. This is also the case for the backscattering at interstitial charge density. By measuring at a single azimuthal angle, it is not possible to map the localization of the interstitial charge. As shown by the temperature dependent surface EXAFS studies, the bonds in the surface plane are much stronger than the bonds to the second layer. This directionality of the bonding can explain why a strong AXAFS contribution is found for normal X-ray incidence. The backscattering of photoelectrons at interstitial charge densities was also observed in two low-Z adsorbate-metal systems, namely (2×3)N/Cu(110) and $\left(\sqrt{2} \times 2\sqrt{2}\right)R45^o$ O/Cu(100). As is a characteristic of these two reconstructed systems, directional bonding is the mainly reason for the strong AXAFS. Moreover, low-Z adsorbates turn out to be ideal candidates for the study of AXAFS phenomenon as multielectron excitations do not interfere with the AXAFS oscillations.[54]

9.6 Summary and Outlook

As XAFS can use only photons (*i.e.*, photons enter and use transmission or fluorescence to generate photons out), XAFS is the only spectral technique that provides information about the electronic and structural properties of the catalyst under reaction conditions and in the presence of reactants. X-ray diffraction (XRD) is similar in this respect, but it requires a long-distance order and only provides geometric information. X-ray photoelectron spectroscopy (XPS) and any other electron spectroscopy techniques (such as Auger spectroscopy) can provide electronic information, however, require ultra-high vacuum conditions. Until recently, XAFS has been restricted and can only provide geometric information on short-range systems through conventional extended X-ray absorption fine structure (EXAFS) technology. However, recent advances have expanded XAFS to enable it to provide electronic structures through atomic XAFS

(AXAFS), and through the use of X-ray absorption near-edge structure (XANES) differential technology (here called Delta XANES (ΔXANES)) for covering and binding sites of adsorbate. Therefore, the new development of XAS and the conventional EXAFS technology are very suitable for studying heterogeneous catalysts under actual reaction conditions and in vacuum.

EXAFS is known to be caused by electron scattering from other atoms surrounding the absorbing atom. The basic tools of EXAFS data analysis include (a) fitting in R space, (b) application of differential file technology, and (c) weighting control of the fitting process by k^1 and k^3 with the help of phase and amplitude correction. These basic tools can not only study the structure of the supported catalyst, but also the interaction between the catalyst and the support. Using calibrated theoretical reference values, EXAFS data can be analyzed to very low k values. This can more accurately determine the presence of low-Z scatterers in EXAFS data. The EXAFS structural analysis of silica-supported chromium oxide clusters will serve as an example of these analysis tools and will show various pitfalls that may complicate data analysis.

The underlying physics of AXAFS is the scattering in the periphery of the absorbing atom, so the Fourier transform peak of AXAFS generally locates at half of the distance of the first coordination shell. The contributions of AXAFS has to be carefully separated from the experimental XAFS data by following the strict rules. An intuitive explanation is given on how various interactions of the absorber atom with its neighboring atoms affect the AXAFS contribution. Two important effects are distinguished: (1) through-space field effect and (2) through-key induction effect. AXAFS data can be utilized to obtain electronic information about supported metal particles and supported metal oxides. In particular, AXAFS can provide direct insights into the interaction between metal-support and metal oxide-support.

The ΔXANES technique utilizes the XANES data of the "clean" sample (denoted here as μ) (when the substrate (M) surface does not contain adsorbate), and is subtracted from the data obtained after the chemical adsorption of the adsorbate (Ad) $\Delta\mu = \mu\,(Ad / M) - \mu\,(M)$. the total change in the XANES due to the presence of an adsorbate (Ad), $\Delta\mu = \mu(Ad/M) - \mu(M) = \Delta\mu_0 + \Delta(\mu_0\chi_{M-M}) + \mu_{0,Ad}\chi_{M-M}$, where the terms on the right include: $\Delta\mu_0$ equal to the change in the atomic

XAFS due to the adsorbate, $\Delta(\mu_0\chi_{M-M})$ equal to the change in the M–M total scattering induced by chemisorption of the adsorbate, and $\mu_{0,Ad}\chi_{M-M}$ equal to the additional scattering by the adsorbate. With the help of a complete multiple scattering *ab initio* calculation, this Δμ difference reflects the characteristics of the binding site of the adsorbate, such as the coverage and binding site geometry. [55]

Recently, the more conventional EXAFS data analysis technology has been supplemented by newer AXAFS and Δμ XANES technologies. [56] With AXAFS, the electronic effect of the carrier on the metal particles can be tracked, while with ΔμXANES, it is now possible to determine the adsorbate, specific adsorption sites, and coverage of the adsorbate. The mechanism is in the working reactor or fuel cell system. However, the advantage of XAS is in particular its ability to provide information on nanoparticle morphology (EXAFS), electron polarization (AXAFS) and adsorbate coverage (ΔμXANES) in situ or under operating conditions using only one experimental spectrum. In recent years, many complex electrocatalysts have reported studies combining ΔμXANES and EXAFS.[57] As the complexity of systems and reactions studied using Δμ technology increases, biofuel cell research and biocatalysis experiments will soon become the focus, and some experiments are already underway. We expect that the combination of ΔμXANES, AXAFS and EXAFS technologies will be more widespread in the future.

---

[1] D.C. Koningsberger, & R. Prins, (1988) in: D.C. Koningsberger, R. Prins (Eds.), *X-ray Absorption: Principles, Applications and Techniques of EXAFS, SEXAFS and XANES*, Wiley, New York, pp. 211–253.

[2] B.W. Holland, J.B. Pendry, R.F. Pettifer, & J.J. Bordas, (1978) Atomic origin of structure in EXAFS experiments. *Physica C*, 11, 633.

[3] J.J. Rehr, C.H. Booth, F. Bridges, & S.I. Zabinsky, (1994) X-ray-absorption fine structure in embedded atoms. *Phys. Rev. B*, 49, 12347.

[4] J.J. Rehr, S.I. Zabinsky, A. Ankudinov, & R.C. Albers, (1995) Atomic-XAFS and XANES. *Physica B*, 208/209, 23-26.

[5] G.E. van Dorssen, D.C. Koningsberger, & D.E. Ramaker, (2002) Separation of double-electron and atomic XAFS contributions in X-ray absorption spectra of Pt foil and $Na_2Pt(OH)_6$. *J. Phys. Condens. Matter.*, 14, 13529, and references therein.

[6] D.E. Ramaker, X. Qian, W.E, O'Grady, (1999) 'Atomic' X-ray absorption fine structure: a new tool for examining electronic and ionic polarization effects, *Chem. Phys. Lett.*, 299(2), 221-226.

[7] B. Ravel, M. Newville, J.O. Cross, & C.E. Bouldin, (1995) Analysis of DAFS fine structure and background. *Physica B* 208/209, 145-147.

[8] J.J. Rehr, C.H. Booth, F. Bridges, & S.I. Zabinsky, (1994) X-ray-absorption fine structure in embedded atoms, *Phys. Rev. B* 49, 12347(R).

[9] D.E. Ramaker, B.L. Mojet, D.C. Koningsberger, W.E. O'Grady, (1998) Understanding atomic X-ray absorption fine structure in X-ray absorption spectra. *J. Phys. Condens. Matter*, 10, 8753.

[10] B.L. Mojet, J.T. Miller, D.E. Ramaker, & D.C. Koningsberger, (1999) A New Model Describing the Metal–Support Interaction in Noble Metal Catalysts. *J. Catal.* 186, 373.

[11] D.E. Ramaker, B.L. Mojet, D.C. Koningsberger, & W.E. O'Grady, (1998) Understanding atomic X-ray absorption fine structure in X-ray absorption spectra. *J. Phys.: Condens. Matter* 10, 8753–8770.

[12] M. Tromp, M.Q. Slagt, R.J.M. Klein Gebbink, G. van Koten, D.E. Ramakerc, & D.C. Koningsberger, (2004) Atomic XAFS as a probe of electron transfer within organometallic complexes: Data analysis and theoretical calculations. *Phys. Chem. Chem. Phys.*, 6, 4397-4406.

[13] W. F. Reynolds, (1980) An approach for assessing the relative importance of field and σ inductive contributions to polar substituent effects based on the non-proportionality of field and σ inductive substituent constants. *J. Chem. Soc., Perkin Trans.* 2, 985.

[14] H. Wende, & K. Baberschke, (1999) Atomic EXAFS: evidence for photoelectron backscattering by interstitial charge densities. *J. Electron Spectrosc. Relat. Phenom.* 101–103, 821.

[15] H. Wende, P. Srivastava, R. Chauvistre, F. May, K. Baberschke, D. Arvanitis, & J. J. Rehr, (1997) Evidence for photoelectron backscattering by interstitial charge densities. *J. Phys. Condens. Matter.* 9, L427.

[16] G.E. van Dorssen, D.C. Koningsberger, & D.E. Ramaker, (2002) Separation of double-electron and atomic XAFS contributions in X-ray absorption spectra of Pt foil and $Na_2Pt(OH)_6$. *J. Phys. Condens. Matter.* 14, 13529, and references therein.

[17] B.K. Teo, (1986) in *EXAFS: Basic Principles and Data-analysis*, Springer, New York.

[18] D.C. Koningsberger, B.L. Mojet, G.E. van Dorssen, & D.E. Ramaker, (2000) XAFS spectroscopy; fundamental principles and data analysis. *Top. Catal.*, 10, 143.

[19] J.W. Cook Jr., D.E. Sayers, (1981) Criteria for automatic x-ray absorption fine structure background removal. *J. Appl. Phys.* 52, 5024.

[20] D.E. Ramaker, G.E. van Dorssen, B.L. Mojet, D.C. Koningsberger, (2000) An atomic X-ray absorption fine structure study of the influence of hydrogen chemisorption and support on the electronic structure of supported Pt particles. *Top. Catal.* 10, 157.

[21] D.C. Koningsberger, (1994) in Hercules Course, Neutron and Synchrotron Radiation for Condensed Matter Studies, Vol. II. *Applications to Solid State*

Physics and Chemistry, ed. J. Baruchel, J.L. Hodeau, M.S. Lehmann, J. R. Regnard and C. Schlenker, Springer Verlag, France, p. 213

22 J.B.A.D. van Zon, D.C. Koningsberger, H.F.J. vant Blik, & D.E. Sayers, (1985) An EXAFS study of the structure of the metal–support interface in highly dispersed Rh/Al_2O_3 catalysts. *J. Chem. Phys.*, 82, 5742.

23 W.E. O'Grady, D.E. Ramaker, (1998) 'Atomic' X-ray absorption fine structure: a new tool for examining electrochemical interfaces. *Electrochim. Acta* 44, 1283.

24 D.C. Koningsberger, J. de Graaf, B.L. Mojet, J.T. Miller, D.E. Ramaker, (2000) The metal–support interaction in Pt/Y zeolite: evidence for a shift in energy of metal d-valence orbitals by Pt–H shape resonance and atomic XAFS spectroscopy. *Appl. Catal. A* 191, 205.

25 D.C. Koningsberger, J. de Graaf, J.A.R. van Veen, D.E. Ramaker, (2001) Nature of the Metal-Support Interaction in Supported Pt Catalysts: Shift in Pt Valence Orbital Energy and Charge Rearrangement. *J. Catal.* 203, 7.

26 M. Tromp, J.A. van Bokhoven, M.Q. Slagt, G. van Koten, D.E. Ramaker, & D.C. Koningsberger, (2004) AXAFS as a probe of charge redistribution within organometallic complexes. *J. Am. Chem. Soc.*, 126, 4090.

27 W.E. O'Grady, D.E. Ramaker, & X. Qian, (1997) Systematic chemical effects observed in "Atomic" X-ray absorption fine structure. *J. Phys. Chem. B*, 101, 5624.

28 S.I. Zabinsky, J.J. Rehr, A. Ankudinov, R.C. Albers, & M.J. Eller, (1995) Multiple-scattering calculations of X-ray-absorption spectra. *Phys. Rev. B* 52, 2995.

29 J. Clavilier and D.J. Armand, (1986) Electrochemical induction of changes in the distribution of the hydrogen adsorption states on Pt (100) and Pt (111) surfaces in contact with sulphuric acid solution. *J. Electroanal. Chem.*, 199, 187.

30 D.C. Koningsberger, M.K. Oudenhuijzen, J. de Graaf, J.A. van Bokhoven, & D.E. Ramaker, (2003) In situ X-ray absorption spectroscopy as a unique tool for obtaining information on hydrogen binding sites and electronic structure of supported Pt catalysts: towards an understanding of the compensation relation in alkane hydrogenolysis. *J. Catal.* 216, 178.

31 D.C. Koningsberger, M.K. Oudenhuuijzen, D.E. Ramaker, & J.T. Miller, (2000) An atomic XAFS study of the metal-support interaction in Pt/SiO_2-Al_2O_3 and Pt/MgO-Al_2O_3 catalysts: an increase in ionisation potential of platinum with increasing electronegativity of the support oxygen ions. *Stud. Surf. Sci. Catal.*, 130, 317.

32 A. Yu. Stakheev, Y. Zhang, A. V. Ivanov, G. N. Baeva, D. E. Ramaker, & D. C. Koningsberger, (2007) Separation of Geometric and Electronic Effects of the Support on the CO and H_2 Chemisorption Properties of Supported Pt Particles: The Effect of Ionicity in Modified Alumina Supports. *J. Phys. Chem. C*, 111, 3938.

[33] D.E. Ramaker, & D.C. Koningsberger, (2010) The atomic AXAFS and Δμ XANES techniques as applied to heterogeneous catalysis and electrocatalysis, *Phys. Chem. Chem. Phys.*, 12, 5514-5534.
[34] F. Solymosi, (1967) Importance of the electric properties of supports in the carrier effect. *Catal. Rev.* 1, 233.
[35] G. M. Schwab, (1978) Electronics of supported catalysts. *Adv. Catal.* 27, 1.
[36] Z. Karpinski, S.N. Gandhi, W.M.H. Sachtler, (1993) Neopentane conversion catalyzed by Pd in L-zeolite: Effects of protons, ions, and zeolite structure. *J. Catal.* 141, 337.
[37] Z. Zhang, T.T. Wong, & W.M.H. Sachtler, (1991) The effect of Ca^{2+} and Mg^{2+} ions on the formation of electron-deficient palladium-proton adducts in zeolite Y. *J. Catal.* 128, 13.
[38] A. de Mallmann, & D. Barthomeuf, (1990) Correlation between benzene hydrogenation activity and zeolite basicity in Pt-faujasites. *J. Chem. Phys.* 87, 535.
[39] A. de Mallmann, & D. Barthomeuf, (1989) Specific Platinum Particles Properties in Basic Zeolites. *Stud. Surf. Sci. Catal.* 46, 429.
[40] A.P. Jansen, & R.A. van Santen, (1990) Hartree-Fock-Slater calculations on cation-induced changes in the adsorption of carbon monoxide on iridium tetrameric clusters. *J. Phys. Chem.* 94, 6764.
[41] E. Sanchez-Marcos, A.P.J. Jansen, & R.A. van Santen, (1990) H_2 chemisorption on Ir_4 clusters: A HFS-LCAO study. *Chem. Phys. Lett.* 16, 399.
[42] V. Ponec, & G.C. Bond, (1995) Catalysis by Metals and Alloys, *Stud. Surf. Sci. Catal.* 95, 234.
[43] R.A. van Santen, (1987) Coordination of carbon monoxide to transition-metal surfaces. *J. Chem. Soc., Faraday Trans. 1* 83, 1915.
[44] A.M.J. van der Eerden, T. Visser, Y. Ikeda, M. Lepage, D.C. Koningsberger, & B.M. Weckhuysen, (2005) Atomic XAFS as a tool to probe the electronic properties of supported noble metal nanoclusters. *J. Am. Chem. Soc.* 127, 3272.
[45] J.J. Rehr, C.H. Booth, F. Bridges, & S.I. Zabinsky, (1994) X-ray-absorption fine structure in embedded atoms. *Phys. Rev. B* 49, 12347.
[46] A. Filipponi, & A. Di Cicco, (1996) Comment on "X-ray-absorption fine structure in embedded atoms". *Phys. Rev. B* 53, 9466.
[47] A. Filipponi, & A. Di Cicco, (1995) Atomic background in X-ray absorption spectra of fifth-period elements: Evidence for double-electron excitation edges. *Phys. Rev. A* 52, 1072.
[48] W.E. O'Grady, X. Qian, & D.E. Ramaker, (1997) Systematic Chemical Effects Observed in "Atomic" X-ray Absorption Fine Structure. *J. Phys. Chem. B* 101, 29, 5624–5626.
[49] H. Wende, D. Arvanitis, M. Tischer, R. Chauvistre, H. Henneken, F. May, & K. Baberschke, (1997) Surface atomic-XAFS for nitrogen atoms on copper. *J. Phys. IV* (France) 7, C2–211.

50 H. Wende, P. Srivastava, R. Chauvistre, F. May, K. Baberschke, D. Arvanitis, & J.J. Rehr, (1997) Evidence for photoelectron backscattering by interstitial charge densities. *J. Phys. C* 9, L427.

51 S.I. Zabinsky, J.J. Rehr, A. Ankudinov, R.C. Albers, & M.J. Eller, (1995) Multiple-scattering calculations of X-ray-absorption spectra. *Phys. Rev. B* 52, 2995.

52 H. Wende, D. Arvanitis, M. Tischer, R. Chauvistre´, H. Henneken, F. May, & K. Baberschke, (1996) Bond length and adsorbate vibrations of (2× 3) N/Cu (110): A SEXAFS study. *Phys. Rev. B* 54, 5920.

53 T. Lederer, D. Arvanitis, G. Comelli, L. Troeger, & K. Baberschke, (1993) Adsorption of oxygen on Cu (100). I. local structure and dynamics for two atomic chemisorption states. *Phys. Rev. B* 48, 15390.

54 H. Wende, & K. Baberschke, (1999) Atomic EXAFS: evidence for photoelectron backscattering by interstitial charge densities. *J. Electron Spectrosc. Relat. Phenom.* 101 –103, 821–826.

55 M. K. Oudenhuijzen, J. H. Bitter, & D. C. Koningsberger, (2001) The Nature of the Pt–H Bonding for Strongly and Weakly Bonded Hydrogen on Platinum. A XAFS Spectroscopy Study of the Pt–H Antibonding Shaperesonance and Pt–H EXAFS. *J. Phys. Chem. B* 105, 4616.

56 M. K. Oudenhuijzen, J. A. van Bokhoven, J. T. Miller, D. E. Ramaker, & D. C. Koningsberger, (2005) Three-Site Model for Hydrogen Adsorption on Supported Platinum Particles: Influence of Support Ionicity and Particle Size on the Hydrogen Coverage. *J. Am. Chem. Soc.*, 127, 1530.

57 Y. Ji, A. M. J. van der Eerden, V. Koot, P. J. Kooyman, B. M. Weckhuysen, & D. C. Koningsberger, (2005) Influence of support ionicity on the hydrogen chemisorption of Pt particles dispersed in Y zeolite: consequences for Pt particle size determination using the H/M method. *J. Catal.* 234, 376.

Chapter 10

Diffraction Anomalous Fine Structure

10.1 Introduction

X-ray absorption fine structure (XAFS) is inherently sensitive to local order *and shows a strong temperature dependence*. Moreover, it is element-specific, owing to the element-specific energy of the *K*-, *L*-, and *M*-absorption edges. These facts confer advantages on the study of (partially) disordered and diluted materials, but this technique can only provide the averaged local structure information of all absorbing atoms of the same element. XAFS also has difficulty in achieving information beyond a distance of roughly 5–6 Å. The proper use of the so-called "related techniques" would unlock the full potential of EXAFS. A complete structural characterization of the investigated materials can be achieved by combining information at long (XRD), medium (PDF) and local range (XAS, DAFS). The parallel techniques provide structural information comparable to that obtained by EXAFS which are able to overcome the limitations listed above. The XAFS difficulty of no information above 5 Å can be overcome by combining EXAFS with XRD for crystalline materials and with PDF for amorphous and nanostructured materials. By combining the capabilities of X-ray diffraction (XRD), extended XAFS (EXAFS) and X-ray absorption near edge structure (XANES) into a single technique, anomalous diffraction fine structure (DAFS) can provide information similar to EXAFS and XANES for a specific subset of atoms selected by diffraction conditions, as well as site-specific XAFS-like spectroscopic and structural information for the inequivalent sites of a single atomic species within the unit cell.[1]

DAFS refers to the energy-dependent modulation of the Bragg peak intensity in the vicinity of an absorption edge. As an X-ray spectroscopic, structural, and crystallographic method, DAFS combines the long-range order and crystallographic sensitivities of X-ray diffraction with the spectroscopic and short-range order sensitivities of XAFS techniques.[2] In both DAFS and XAFS, the incoming photon promotes an electron from a compact core state to an empty bound state, or to an empty continuum state. Closely related to the XAFS, this fine structure can be utilized to obtain the same information of the local environment of the resonant atom as usually obtained from XAFS. When the core electron is excited to an empty bound state, or to the low (~30 eV) continuum states, the fine structure signals are called XANES and DANES (diffraction anomalous near edge structure). In the near-edge region, DAFS provides the same structural and spectroscopic sensitivities as XANES including the valence, empty orbital and bonding information for the resonant atoms. When the electron is promoted to the continuum states more than 30 eV over the edge, the absorption and diffraction oscillatory fine structure signals are called EXAFS and EDAFS (extended diffraction anomalous fine structure), respectively. In the extended fine structure region, DAFS provides the same short-range structural information as EXAFS like the bond lengths, coordination numbers, neighbor types, and bond disorders for the atoms surrounding the resonantly scattering atoms. Indeed, DAFS provides both the advantages of XRD and XAFS; however, it is more than the addition of absorption and diffraction, it is simultaneously a site and chemically selective probe.

The intensity of the DAFS and XAFS signals for each photon energy depends on the matrix elements between the ground state wavefunction (the core state) and the excited state wavefunction (final state). For the EDAFS and EXAFS signals, the intensities depend on how well the excited state wavefunctions fit into the potential "wells" created by the absorbing atom and its neighbors. So there are interference between the outgoing and backscattering waves, which varies as a function of $\sin(kr + \phi)$, where k is the photoelectron wavevector, r is the bond length, photoelectron wavenumber dependent phase shift ϕ due to the walls of the "wells" formed by the screened excited central atom and by the surrounding atoms. Both DAFS and XAFS signal can be precisely simulated by using the multiple-scattering theory.

At a fixed momentum transfer, the real and imaginary components of the scattering amplitude are intimately linked. For each incoming and outgoing direction, the complex scattering amplitude of the resonant (anomalous) atoms is an analytic function of the energy. Consequently, the real and imaginary components of the scattering amplitude are a Kramers-Kronig or Hilbert transform pair, according to the Cauchy's theorem. Therefore, the real and imaginary components of the scattering amplitude in any fixed direction are closely related.

Absorption and dispersion are also related. According to the optical theorem (a special case of unitarity), the sum of all the outgoing and absorbed waves must equal the incoming wave. For each photon energy, the optical theorem links the angular integral of the elastic scattering (DAFS) over all directions to the absorption (XAFS). In the special case of pure dipole scattering, the scattering amplitude has the same energy dependence in all directions. Consequently, for pure dipole scattering, the energy dependence of the imaginary component of the scattering amplitude is identical to that of the absorption, and the energy dependence of the real component is given by the Kramers-Kronig transform of the absorption. Because X-ray scattering is often predominantly dipolar, DAFS and XAFS can usually be related by angle independent Kramers-Kronig transforms. [3]

The physical effects that are employed to measure XAFS, such as X-ray fluorescence, total electron yield and direct beam attenuation, are incoherent processes. Therefore, although XAFS is very sensitive to the local environment around the absorbing atoms at resonance, the final measurement result is the incoherent sum of the absorption signals from all absorbing atoms. On the other hand, DAFS is measured in coherent elastic scattering channels. The most significant difference between DAFS and XAFS is the crystal structure factor, which is the basis of the DAFS signal and allows the selection of local structural information about certain subsets of resonant atoms in the diffractive material based on the long-range order of the diffractive material. This provides a huge impetus for obtaining XAFS-like information from resonance (anomalous) atoms in complex materials, which is difficult to obtain using XAFS.

Fine structure in the diffraction intensities[4] was first observed about 30 years after the discovery of XAFS.[5,6] With the advance in the theoretical understanding of the nexus between DAFS and XAFS,[7, ,8] quantitative analysis of DAFS became feasible.[9] Around 1990, based on the Kramers-Kronig dispersion relations, Pickering[10] and Stragier[11] developed an algorithm for refining DAFS data. Through this iterative fitting procedure, the real and imaginary parts of the complex scattering amplitude can be isolated from the measured DAFS intensity, and then can be analyzed using standard XAFS procedures.

This chapter describes theoretical and experimental considerations behind DAFS, explains how the real and imaginary components of the scattering amplitude are related and isolated from the experimental DAFS data, and illustrates how generalized multiple-scattering DAFS theory can be employed to analyze the isolated diffraction fine structure.

10.2 Theoretical Description of DAFS

In non-relativistic quantum mechanics frame, by neglecting the magnetic scattering terms, for photons with energy $E = \hbar\omega$ and with incident and scattered momenta k_1 and k_2, the total atomic scattering amplitude,$f = f_0 + \Delta f$, is the sum of the non-resonant Thomson scattering amplitude (f_0), and the anomalous scattering amplitude (Δf). In terms of the classical single electron scattering amplitude, the Thompson and anomalous scattering amplitudes can be expressed as [12]

$$f_0(\boldsymbol{k_2} - \boldsymbol{k_1}) = f_0(\boldsymbol{Q}) = -r\hat{e}_2^* \cdot \hat{e}_1 \sum_j \langle j | e^{-i(\boldsymbol{k_2}-\boldsymbol{k_1})\cdot\boldsymbol{r}} | j \rangle, \qquad (10.1)$$

$$\Delta f(k_1, k_2, E) = f'(k_1, k_2, E) + if''(k_1, k_2, E) = \frac{r_0}{m}\sum_j \sum_n \left\{ \frac{\langle j|\hat{e}_2^* \cdot \boldsymbol{p}e^{-i\boldsymbol{k_2}\cdot\boldsymbol{r}}|n\rangle\langle n|\hat{e}_1 \cdot \boldsymbol{p}e^{+i\boldsymbol{k_1}\cdot\boldsymbol{r}}|j\rangle}{E_n - E_j - \hbar\omega + \left(\Delta_n - \frac{1}{2}i\Gamma_n\right)} + \frac{\langle j|\hat{e}_1 \cdot \boldsymbol{p}e^{+i\boldsymbol{k_1}\cdot\boldsymbol{r}}|n\rangle\langle n|\hat{e}_2^* \cdot \boldsymbol{p}e^{-i\boldsymbol{k_2}\cdot\boldsymbol{r}}|j\rangle}{E_n - E_j + \hbar\omega} \right\}, \qquad (10.2)$$

where $r_0 = e^2/mc^2$, and the self-energy corrections introduce the Lamb-shift $\Delta_n(E_n + \hbar\omega)$ and the linewidth $\Gamma_n(E_n + \hbar\omega)$ of the resonant term.[13]

The scalar Thomson amplitude depends on the photon momentum transfer, $\hbar\boldsymbol{Q} = \hbar(\boldsymbol{k_2} - \boldsymbol{k_1})$, and on the photon polarization factors, $\hat{e}_2^* \cdot \hat{e}_1$, but is independent of the photon energy. The Thomson amplitude is proportional to the Fourier transform of the atom's electronic charge distribution. In contrast, the anomalous amplitude depends separately on the incident and scattered wavevectors, k_1 and k_2, as well as the photon energy, $E = \hbar\omega$. As a tensor, Δf is determined by the matrix elements between the ground state and the virtual intermediate states, thus Δf is not proportional to the Fourier transform of the total or subshell charge density.[14] However, the k_1 and k_2 dependencies of anomalous scattering are experimentally found often small, and the full photon energy- and momenta-dependent $\Delta f(k_1, k_2, E)$ is conventionally approximated by its momenta-independent forward scattering limit, denoted $\Delta f(E) = f'(E) + if''(E)$. Consequently, the total atomic scattering amplitude, *f*, depends on the photon energy, *E*, via its *f'* and *f''* terms, and on the wavevector transfer $(\boldsymbol{Q})$, via its f_0 term.

By exploiting the relationship between the real and imaginary parts of the atomic-scattering factor (*f*), the DAFS technique can also extract the information contained in XAFS spectra from resonant, energy-dependent X-ray diffraction measurements. The integrated intensity I_{hkl} of the X-ray Bragg reflection (*hkl*) from a weakly scattering crystal is proportional to the square of the crystallographic structure factor:[15]

$$I_{hkl}(\omega) \propto \left|\sum_{j=1}^{N} f_j(\omega) e^{-2\pi i(Q_{hkl} \cdot r_j)}\right|^2 A(Q_{hkl}, \omega), \qquad (10.3)$$

where the sum runs over the *N* atoms of the unit cell, the X-ray absorption correction factor for a symmetric Bragg reflection from a sample of thickness t is given by $A(Q_{hkl}, \omega) = \left[1 - exp\left(-\frac{2\mu t}{\sin\theta}\right)\right] / 2\mu$, absorption produces additional fine structure in the measured DAFS intensities due to the XAFS modulation of the X-ray absorption coefficient, $\mu(\omega) = \mu_0(\omega)[1 + \chi(\omega)]$. The crystallographic structure factor accounts for the scattering from each of the *j* atoms in the unit cell according to their positions r_j and their atomic scattering amplitudes $f_j(\omega)$ at the photon wave-vector-transfer $Q_{hkl} = k_f - k_i$. For crystalline materials, its modulus Q_{hkl} must fulfill the Bragg condition: $Q_{hkl} = 2\pi/d_{hkl}$.[16] In turns, the atomic-scattering amplitude $f_j(\omega)$ from a single atom is the sum of the Thomson scattering amplitude f_0, which is independent of the photon energy $\hbar\omega$, and an energy-dependent resonant correction Δf(E), which changes dramatically in the vicinity of a core-level excitation energy. Δf(E) is the sum of a real part *f'*(E), and an imaginary part *f''*(E).[17]

In the standard EXAFS description of the absorption coefficient, $\mu(\omega)$ is given by: [18]

$$\mu(\omega) = \mu_0(\omega)[1 + \chi(\omega)] \Rightarrow \chi(\omega) = \frac{\mu(\omega) - \mu_0(\omega)}{\mu_0(\omega)}, \qquad (10.4)$$

where $\mu_0(\omega)$ is the background absorption of bare-atom, and $\chi(\omega)$ is the oscillatory part of the X-ray absorption coefficient that arises from the local backscattering from nearby atoms. Analogously, the real and imaginary parts of $\Delta f(\omega)$ can also be separated into their atomic and oscillatory terms, and can be rewritten as:

$$\Delta f(\omega) = f_a'(\omega) + i f_a''(\omega) + f_a''(\omega)[\chi'(\omega) + i\chi(\omega)], \qquad (10.5)$$

where the subscript a denotes the smooth, atomic-like contribution to the response function, its smooth anomalous term, $f_a'(\omega) + i f_a''(\omega)$, is usually calculated using the Cromer-Liberman method.[19, 20] However the correct values of $f_a'(\omega) + i f_a''(\omega)$ are given by embedded atom anomalous amplitudes.[21] For the systems in which the embedded atom corrections are small, and consequently Cromer-Liberman values are valid.

When there are neighboring atoms, the oscillating component, $f_a''(\omega)[\chi'(\omega) + i\chi(\omega)]$, is generalized as an energy-dependent fine-structure function of the DAFS complex-valued $[\chi'(\omega) + i\chi(\omega)]$. This separation of f into smooth and oscillatory DAFS components is analogous to the usual XAFS separation of μ into smooth μ_0 and oscillatory χ terms. Because $f_a'(\omega)$ and $f_a''(\omega)$ are not independent, the latter may be determined either directly by measuring the total absorption cross-section $\sigma_{tot}(\omega)$, that is proportional to the absorption coefficient $\mu(\omega)$ measured in a standard EXAFS experiment via the optical theorem [22]:

$$f''(\omega) = \frac{\omega}{2hC}\sigma_{tot}(\omega) \text{ or } f''(\omega) = \frac{m\omega}{2\pi N e^2}\sigma_{tot}(\omega), \qquad (10.6)$$

or by measuring the photon-energy-dependent intensity of a Bragg reflection and then applying the Kramers-Kronig (principal value) dispersion relations:

$$f'(\omega) = \frac{2}{\pi} P \int_0^{+\infty} \frac{\varepsilon f(\epsilon)}{\varepsilon^2 - \omega^2} d\varepsilon, \qquad (10.7a)$$

$$f(\omega) = -\frac{2\omega}{\pi} P \int_0^{+\infty} \frac{\varepsilon f'(\epsilon)}{\varepsilon^2 - \omega^2} d\varepsilon. \qquad (10.7b)$$

In Equations (11.7a) and (11.7b), P indicates the Cauchy principal value of the integral.[23] It is possible to iteratively solve the Equations

(11.7a,b) for $f_a''(\omega)$.[24]

Once used in combination with EXAFS, DAFS can be utilized to unravel the contribution of the mixture specimen in the experimental EXAFS spectrum, in which the amorphous phase and the crystalline phase coexist. Both the amorphous phase and the crystalline phase signals exist in the EXAFS spectrum, but only the crystalline phase has a contribution to the DAFS signal. If there are different local environments in the sample, XAFS spectroscopy may not be able to provide relevant information, because all different environments are detected and averaged in XAFS due to lack of spatial selectivity. The difficulty can be overcome with the site selectivity of DAFS by selecting atomic sites, as indicated by Equation (11.3). DAFS is sensitive to both the short- and the long-range order correlations.[25] Although DAFS contains contributions of both the real and imaginary parts of the complex anomalous scattering factors, $f'(\omega)$ and $f''(\omega)$, it can be analyzed like EXAFS in the extended region, since XAFS is only proportional to the imaginary part.[26] DAFS is intimately connected to XAFS by unitarity and causality.[27] The diffraction fine structure can be calculated using the causal, unitary generalization of the multiple-scattering XAFS formalism. The original EXAFS theories were single-scattering plane-wave theories. The curved-wave and multiple-scattering effects can be quite important in the near edge region. The existing codes able to simulate the DAFS signal include, for example, FEFF, [28] ATOMS,[29] and XFIT.[30]

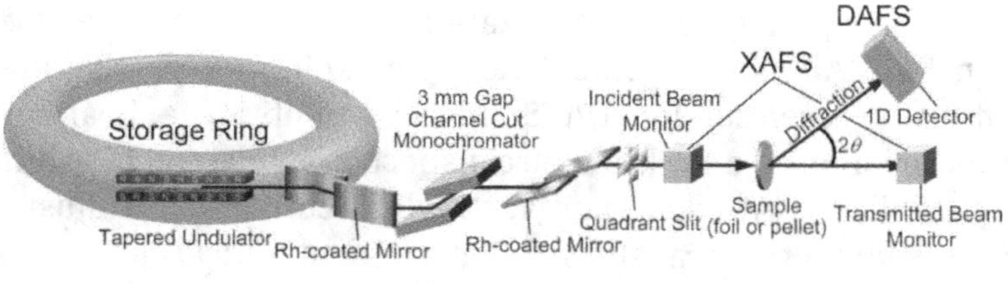

Figure 10.1 A schematic of the optical geometry for XAFS and DAFS setups [Kawaguchi, et al. (2014) J. Synchrotron Rad. 21, 1247–1251, reproduced with permission from International Union of Crystallography].

10.3 DAFS Measurements

The schematic diagram of typical experimental setup for both EXAFS and DAFS is illustrated in Figure 10.1. As a diffraction technique,

DAFS results in an XAFS-like signal for materials that have long-range order. It yields the local environment information of a selected element. The intensity of a (*hkl*) diffraction peak of a crystalline material normally varies smoothly with the wavelenght of the incident X-ray beam, λ (*i.e.*, with the photon energy *hv*) utilized to carry out the diffraction experiment. However, this statement does not hold in resonant diffraction experiments,[31,32] where the λ variation scans across an absorption edge of an element present in the material. The DAFS method combines X-ray diffraction and X-ray absorption in the same experiment.[33,34] By measuring the intensity of such a diffraction peak as a function of λ (or, alternatively *hv*), across the absorption edge of an element present in the material, an XAFS-like signal will be obtained from which the local structural parameters of the selected element can be extracted. The long-range structural information contained in diffraction peaks can be combined with the chemical and local structure selectivity of X-ray absorption spectroscopy. Thus, it provides site-selective and chemical-selective structural information.

On the experimental ground, similar to EXAFS, DAFS data of a very high signal-to-noise ratio are required to enable quantitative oscillation analysis to be performed on a diffraction yield that is only a very small fraction of the total one. So a brilliant beam is needed together with a high quality diffractometer coupled to very stable absorption-dedicated optics.[35] The synchrotron radiation source provided an intense continuous energy spectrum, strong collimation, small beam size, and a high degree of polarization. An XAFS beamline is often chosen for DAFS measurement because rapid and precise energy scanning is essential for DAFS measurements. The beamline is adapted by adding a multiple-circle spectrometer to give it diffraction capabilities. At each energy, the diffractometer is adjusted to keep the momentum transfer fixed; it is essential to accurately track the Bragg peak as a function of photon energy to obtain reliable intensity measurements.

DAFS measures the elastic Bragg diffraction intensities as a function of photon energy. To optimize a DAFS experiment, one needs to choose diffractions that balance three factors: intensity, contrast and size of the absorption correction. High intensity diffractions are preferred over low intensity diffractions for several reasons. Low intensity Bragg reflections have poor counting statistics, require long measurement times, and have large background signals due to fluorescence and diffuse scattering. However, high intensity Bragg reflections usually have poor contrast. The DAFS contrast measures the size of the fine-structure signal relative to the diffraction intensity. The intensity dependence on the smooth (A_0) and fine structure (FS) terms, $I \sim |A_0 + FS|^2 \sim |A_0|^2 + 2Re(A_0)Im(FS) + 2Im(A_0)Re(FS) + |FS|^2$. Because the $|FS|^2$ term is small and smooth, the contrast is roughly proportional to the ratio of the cross terms to $|A_0|^2$. High intensity diffractions have larger absorption corrections than weak Bragg reflections, because the absorption contributions are weighted by the $|A_0|^2$ term, whereas the fine structure contributions from the χ' and χ'' diffraction terms are weighted by $Re(FS)$ and $Im(FS)$.

In DAFS measurement, generally thin film forms have several advantages over other forms, because thin films provide more intense diffracted beams than powder specimens of the same volume; typically their broad smooth mosaic distributions allow easy tracking of the Bragg peaks as a function of photon energy; their diffraction intensities can be analyzed using the kinematic formalism; and their absorption corrections are relatively small so that optimization consists primarily of finding a compromise between high intensity and high contrast.

Additional optimization should be considered for experiments that utilize the wavevector and site selectivities of DAFS. For wavevector selective DAFS experiments, Bragg reflections from the desired components should be sufficiently isolated from the other Bragg reflections. For site selective DAFS experiments, Bragg reflections should be selected to produce a linear combination of signals from the different sites that is easy to separate. In general, many suitable Bragg reflections are available even after applying these restrictions.

10.4 DAFS Data Reduction

The goals of the DAFS analysis are to isolate the fine structure oscillations and, from them, to obtain the detailed wavevector and site selected EXAFS-like short-range order information and XANES-like structural, valence and empty orbital information. The methods to isolate the DAFS signals and to analyze the isolated signals are schematically listed below:

1) Measure the Bragg intensities as a function of photon energy and apply the standard absorption and diffraction corrections.

   (a) Measure and subtract the background.
   (b) Apply the area, Lorentz-polarization, instrument and absorption corrections.
   (c) Remove any Bragg glitches from the XAFS and determine E_0.

2) Isolate the fine structure using one of the following approaches.
   (a) Isolate the effective fine structure term χ_e directly from the diffracted intensity using spline methods analogous to conventional XAFS analysis.
   (b) Impose the Kramers-Kronig relation between the real and imaginary components of the DAFS signal and use an iterative method to extract the real and imaginary components, $f'(\omega)$ and $f''(\omega)$, from the data.

3) Analyze the isolated fine structure signals with simulated path contributions by using codes like FEFF.

Absorption corrections

For the absorption of the incident and diffracted X-ray beams as they travel through the specimen, there are three possible absorption correction methods:

(1) For small absorption corrections, intensity formula and the effective χ_e produced by the direct spline analysis method can be appropriately phase shifted.
(2) The simultaneous fluorescence measurements can be performed to reconstruct the absorption correction.
(3) The absorption correction could also be constructed self-consistently using the f'' functions obtained during the iterative Kramers-Kronig analysis of the DAFS signal.

Note that the absorption χ_μ signal is different from the DAFS χ' and χ'' signals, because XAFS weights the sites based on the number of atoms within the unit cell, whereas DAFS weights the sites crystallographically. Thus, for systems which contain multiple inequivalent sites, the measurements of the single χ_μ function are not sufficient to determine the multiple χ' and χ'' functions.

Two methods of isolating the diffraction fine structure

Two useful methods to isolate the diffraction fine structure are the direct spline method and the itinerary Kramers-Kronig methods. The isolation of the diffraction fine structure requires a careful comparison with X-ray absorption analysis, because the diffraction intensity is the square of the unit cell structure factor and involves both $f'(\omega)$ and $f''(\omega)$ amplitudes.

The direct spline method

Analogous to standard XAFS analysis, the direct spline method first splines the measured DAFS intensity, subtracts the spline from the data, and then normalizes the resulting fine structure signal to the spline. In its simplest form, the as-produced effective χ_e function contains a mixture of the χ', χ'' and χ_μ components. If the data is divided by the measured absorption correction before it is splined, the χ_e function will still be composed of a mixture of χ' and χ'' components. The measured intensity (I_M) is related to the theoretical intensity by $I_M = C(\omega)I$, where $C(\omega)$ accounts for the energy dependent instrument correction factors. The measured intensity is fit with a smooth spline $S(\omega) = C(\omega)|A_0|^2$, and the fine structure is separated from the measured intensity by subtracting the spline from the data. The fine structure is normalized by dividing by $C(\omega) = S(\omega)/|A_0|^2$ to yield the effective χ_e function,

$$\chi_e(\omega) = \frac{[I_M - S(\omega)]|A_0|^2}{S(\omega)} = 2Re(|A_0|)\sum_m (f_s'')_m [\chi_m' Re(\alpha_m) - \chi_m'' Im(\alpha_m)] + 2Im(|A_0|)\sum_m (f_s'')_m [\chi_m' Re(\alpha_m) + \chi_m'' Im(\alpha_m)]. \quad (10.8)$$

Note that if the crystallographic weights ($\alpha_m = \sum_{r_j \in \{m\ sites\}} e^{iQ\cdot r_j}$) and the real part $Re(\alpha_m)$ and imaginary part $Im(\alpha_m)$ of the smoothly varying background can be determined, the site specific χ_m' and χ_m'' signals can be determined from the effective χ_e function.

The iterative Kramers-Kronig method

As a pair of $f'(E)$ and $f''(E)$ functions are a self-consistent Kramers-Kronig transform pair, the Kramers-Kronig method iteratively finds them in agreement with DAFS data. For centrosymmetric systems with crystallographic weights, a_m, this method uses the intensity fitting function,

$$I_{fit} = A(Q,\omega)N(\omega)\left\{1 + \gamma^2 + 2\beta f' + 2\gamma\beta f'' + \beta^2\left[(f')^2 + (f'')^2\right]\right\}, \quad (10.9)$$

where $\beta = \sum_m a_m/Re(A_0)$ and $\gamma = Im(A_0)/Re(A_0)$, and the energy dependent instrument correction function $N(\omega) = m[(\omega - E_0) + b]$. To obtain starting values for β, γ, b, and m, f' and f'' are estimated using Cromer-Liberman f_s' and f_s'' values, and Eq. 11.9 is fit to the experimental DAFS data. Then the functions f' and f'' are updated by using an iterative procedure: Equating the fitting function to the data, the resulting quadratic equation is solved for f'. This f' is then Kramers-Kronig transformed to obtain a corresponding f''. With these new values for f' and f'', the data is refitted to obtain new values of β, γ, b, and m, and in turn a new value for f'. This process usually takes several iterations to converges.

The iterative Kramers-Kronig method offers several advantages. First of all, little understanding of the crystal structure is necessary to obtain the f' and f'' functions. Second, the iterative method properly considers the dependence of the diffraction intensity on the real and imaginary parts of the abnormal amplitude, and also properly considers its dependence on the square of the fine structure. Worthy mentioning that the direct spline method of isolating the diffraction fine structure introduced in the previous section ignores the square of the diffraction fine structure.

FEFF analysis of the isolated EDAFS oscillations

Once the fine structure has been isolated and normalized by one of the above methods, it can be analyzed to obtain local structural information. The analysis is performed in almost the same way as modern multi-scatter XAFS analysis, *i.e.*, a model of the structure is constructed, and FEFF is employed to calculate the amplitude, phase and degeneracy of each photoelectron scattering path. The model is fit to the DAFS data to obtain the desired XAFS-like structural and spectroscopic information via the EXAFS expression,

$$\chi_\mu(k) = S_0^2 \sum_{i=1}^{n} \frac{N_i f_i(k,R_i)}{kR_i^2} e^{-\frac{2R_i}{\lambda(k,R_i)}} e^{-2\sigma_i^2 k_i^2} \sin[2kR_i + \phi_i(k,R_i) + \phi_c(k)]. \quad (10.10)$$

Information content of DAFS compared to XAFS

As the DAFS and XAFS signals have limited photoelectron wavenumber "bandwidths", their information content is limited. The number of independent XAFS points is given by $N = 2\Delta k \Delta R/\pi$, where ΔR is the filtering window width in R-space and Δk is the window width in K-space.[36] For typical values, $\Delta R = 1.5\text{Å}$ and $\Delta k = 12\text{Å}^{-1}$, the number of independent points is approximately 10. This is the maximum number of parameters that can be determined from single shell XAFS data. Note, however, that if there are m inequivalent sites, a set of DAFS measurements for m or more inequivalent reflections can provide up to m times as much information as XAFS. If there are also p components in the sample, and each component has m inequivalent sites, then a total of (pmN) parameters can be determined per shell.

10.5 Applications of DAFS

By combining XRD and XAFS, DAFS has been extensively applied to systems of great technical significance such as semiconductor nanostructure,[37] superconductor,[38] magnetic films.[39] Structural properties at long and short distance scales, such as strain, chemical composition, and knowledge of mixing at the interface, are essential for understanding the growth mechanism and electronic and optical properties of heterostructures,[40] film,[41,42] nanostructures,[43,44] and oxides.[45] Moreover, in order to fit into the device, the nanostructures are encapsulated or embedded in the superlattice, and the capping plays a decisive role in the change of optical properties that are caused by modifying the strain and possibly inducing atomic diffusion. Strain is closely related to the composition, shape and aspect ratio of nanostructures, as well as the mutual stress applied by the nanostructures, substrate and substrate to each other. With its wavevector and site selectivities, DAFS allows to determine the local environment of atoms located in an isostrain volume selected by diffraction is a very powerful approach to disentangle strain and composition.

Wavevector selectivity

An important feature of DAFS is its enhanced wavevector selectivity. As long as different spatial regions or components of the specimen produce Bragg diffraction peaks at separable positions in reciprocal space, wave vector selectivity can be realized. Then, even if different regions contain the same atom type, one of its characteristic diffraction peaks can be utilized to determine the local atomic structure of each region. In general, XAFS measurement cannot achieve wave-vector selectivity because it is usually impossible to separate the absorption signals of different regions or components. Some applications where DAFS wave vector selectivity can provide information not available using XAFS include: (1) strain and / or composition modulation layers, (2) mixed phase powders or composite materials, (3) buried layers, and (4) buried single layer or reconstructed interface.

Materials consisting of a mixture of crystalline and amorphous phases and crystallization of amorphous materials, such as glasses, are very important subjects in catalysis,[46] materials science,[47,48] metallurgy,[49] chemistry,[50,51,52] physics[53,54] earth science,[55] environmental science,[56] and biology.[57] Since most technologies produce incomplete / entangled information, it is difficult to obtain quantitative structural characteristics of this mixed crystalline / amorphous material. Incomplete information can be obtained through technologies that are sensitive to the long-range order (such as XRD), so that only the crystalline phase is revealed. Entangled information is obtained by averaging techniques between different phases, such as EXAFS and total scattering. Obviously, in this case, it is important to combine different technologies. This problem becomes more serious when dealing with nanostructured materials embedded in a matrix, where the main scattering comes from the matrix rather than the nanostructured part of the specimen. In this case, the structure of interesting parts of the material can be determined by combining the EXAFS elemental selectivity and the DAFS wavevector selectivity.

An emblematic example is to investigate the local environment around Ge in Ge nanoparticles embedded in a SiO_2 matrix by combining different techniques.[58] The samples were prepared by co-deposition of Ge and Si oxides onto quartz substrates by radio-frequency magnetron sputtering and annealed for 1 h at 800 °C in an argon atmosphere. The presence of Ge-Ge and Ge-O bonds disclosed by Raman[59] and EXAFS[60] studies performed on the same samples for both the as-prepared and annealed material, suggests negligible Ge-

Si alloying. However, the co-presence of three different phases (Ge nanocrystals, amorphous Ge, and Ge oxide) complicated a complete structural understanding. In this case, the combined use of XRD, DAFS, and EXAFS data sets was mandatory.

XRD analysis yielded the relative fraction of amorphous and crystalline phases in both as-deposited and annealed samples. The as-deposited sample was found to be amorphous, while the annealed samples was crystalline, showing clear Ge (111), (220), and (311) diffractions. For Ge nanoparticles, the Ge(111) reflection was observed at Q_{111} = 1.9179(2) Å^{-1}, resulting in $d_{111} = 2\pi/Q_{111}$ = 3.276 Å , that is stretched with respect to the value of unstrained Ge bulk (3.2665 Å). From the full width at half maximum (FWHM) of the measured diffractions (ΔQ_{FWHM} = 0.042(1) Å^{-1}), and using the Scherrer equation, the estimated diameter $\langle D \rangle$ of the ordered Ge regions was $\langle D \rangle = \frac{0.94(2\pi)}{\Delta Q_{FWHM}} = 140$Å. Note that the Scherrer equation is usually reported as a function of the (*hkl*) Bragg angle θ, of the used X-ray wavelength λ and of the FWHM of the (*hkl*) Bragg peak measured in the scattered intensity versus 2θ plots (B_{hkl}), that corresponds to twice the peak width in the scattering angle θ: $B_{hkl} = \Delta(2\theta_{FWHM}) = 2\Delta\theta_{FWHM}$ as :[61,62]

$$\langle D \rangle = \frac{K\lambda}{B_{hkl}\cos\theta} = \frac{K\lambda}{2\Delta\theta_{FWHM}\cos\theta} = \frac{2\pi K}{\Delta Q_{FWHM}}, \tag{10.11}$$

the numerical factor K is close to unit, to be optimized in each particular nanoparticle case. In Scherrer's original article,[63] this factor was defined as $K = 2[\ln(2)/\pi]^{1/2} \sim 0.939$. The K value was then slightly modified in 1925 by Seljakow [64] and generalized one year later by von Laue to include in the model also the shape of the crystallites.[65] The fourth term of Equation (10.11) is obtained from the third term: $\Delta\theta_{FWHM} = \lambda\Delta Q_{FWHM}/[4\pi\cos(\theta)]$.

The $f'(E)$ and $f''(E)$ curves were obtained from the raw Bragg intensity using the iterative Kramers-Kronig algorithm described above, Eqs. 10.3–10.9. Then, the total absorption cross-section $\sigma_{tot}(E)$ was obtained from $f''(E)$ by inverting Eq. 10.6. In the EXAFS experiment, this quantity was obtained directly, from the ratio of the fluorescent and incident beam intensities. The k-weighted $\chi(k)$ functions extracted from the raw $\sigma_{tot}(E)$ EXAFS and DAFS data are further Fourier transformed. The phase-uncorrected Fourier transform amplitude of the $k\chi(k)$ curve obtained from the EXAFS data exhibits two clear first shell signals centered around 1.3 and 2.1 Å due to Ge-O and Ge-Ge contribution respectively, confirming the co-existence of germanium and germanium oxide phases. Conversely, only Ge-Ge contributions are present in the DAFS data. The fitting of the fluorescence EXAFS data yielded a Ge-O distance of 1.75 Å for the germanium oxide phase and an average Ge-Ge distance of 2.48 Å from both crystalline and amorphous Ge phases. In contrast, the first shell Ge-Ge distance of the crystalline Ge nanoparticles from the DAFS data was 2.44 Å, in perfect agreement with the value of 2.45 Å obtained from XRD. The Ge-Ge distance in the crystalline phase is much shorter than the average value obtained from EXAFS, suggesting that the average Ge-Ge distance distribution in the specimen is bimodal, shorter in the crystalline phase, and longer in the amorphous phase. This fact can be clearly proved by comparing the residuals obtained by Ge-Ge's contribution to EXAFS data with single-peak and double-peak distribution models. The fitting residual obtained by the single-peak model is twice that of the bimodal model, and the mismatch is near the Ge-Ge peak, concentrated on the right side of the experimental Ge-Ge peak, showing a longer amorphous Ge-Ge distance is not specified in the unimodal model.

Wrapping up, DAFS signal is selective only on the crystalline Ge phase while EXAFS signal contains a mixture of GeO, amorphous and crystalline Ge. By combining analysis of the DAFS signal and the cumulative EXAFS signal, the structural contributions of all the three phases can be separated.

Site selectivity

Another important enhancement provided by DAFS is its site selectivity. Crystallographically non-equivalent Bragg reflection site-selective DAFS measurements can be used to determine XAFS-like information about unequal sites within the unit cell, even if these sites have the same atomic species. Because the DAFS intensity contains a linear combination of contributions from various unequal sites, specific site information can be obtained by analyzing each contribution. In very favorable situations, some reflections will be dominated by one location, and separation is easy. Usually, the individual contribution is unknown and requires a crystallography procedure.

DAFS can be utilized to disentangle the contributions superimposed in a standard EXAFS experiment where a given specific element occupies multiple crystallographically independent sites in a crystalline material.[66,67] The proper selection of a set of Bragg reflections disentangles the contributions from crystallographically non-equivalent sites, providing EXAFS-like information for each site separately. In such a way, the problem of overlapping signals is circumvented, thus yielding the local structure for each non-equivalent site. Due to symmetry reasons, an atom in a given crystallographic position may not contribute to the intensity of a given (*hkl*) reflection, because of the cancellation effects in the sum giving $I_{hkl}(E)$ in Equation (10.3).

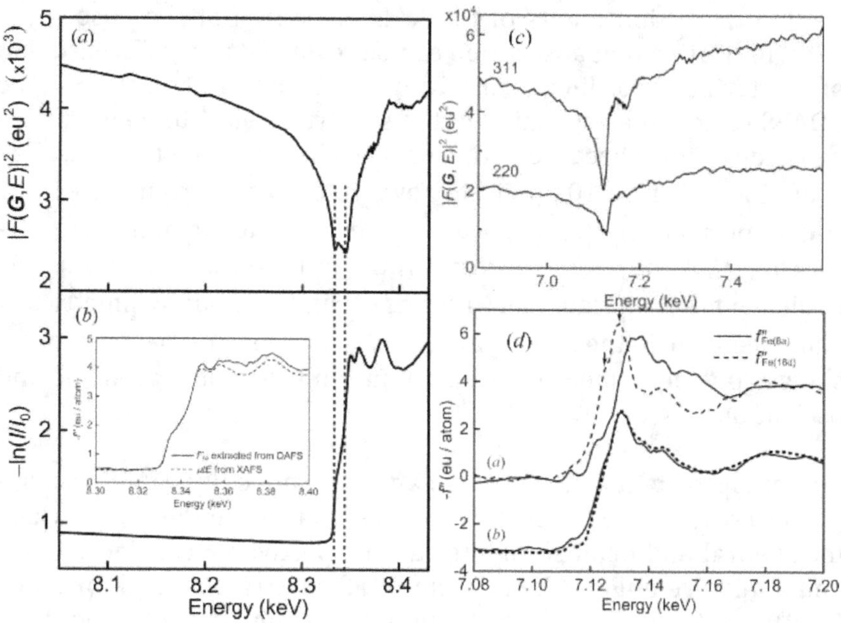

Figure 10.2 (a) DAFS spectrum obtained from Ni 111 diffraction and (b) XAFS spectrum simultaneously measured. insety: f''_{Ni} extracted from a 111 DAFS spectrum by LDR (solid line) and μt_E obtained from an XAFS measurement (broken line). The values of f''_{Ni} have some offset and μtE is scaled by theoretical values (Sasaki, 1989[Sasaki, S. (1989). KEK Report 88-14, p. 1. KEK, Japan.]) for comparison. (c)DAFS spectra of Fe3O4 from 311 and 220 diffraction.(d)(a) Fe at the 8a tetrahedral site (solid line) and 16d octahedral site (broken line) extracted from DAFS spectra. (b) Averaged spectrum at each site from the DAFS method (solid) and μtE obtained from XAFS (dotted line) with an offset of −3. [Kawaguchi, et al. (2014) J. Synchrotron Rad. 21, 1247–1251, reproduced with permission from International Union of Crystallography].

The strategy in DAFS data analysis is to reduce experimental DAFS spectra to EXAFS-like data sets for isolated crystallographically non-equivalent sites. Then the EXAFS data can analyzed by using well-established EXAFS methods.[68]

Figure 10.2ab displays the DAFS and XAFS *K*-edge spectra of Ni 111 obtained from the polycrystalline Ni foil. The DAFS spectrum monotonically decreases toward the Ni *K*-edge at approximately 8.340 keV owing to the resonant effect in f', and then rapidly increases with oscillations above the absorption edge. The two local minima around the cusp at 8.333 and 8.343 keV in the DAFS spectrum correspond to the two inflection points of the XAFS spectrum at the absorption edge, indicating that absorption can be reviewed as a differentiation of scattering according to the Kramers–Kronig relation.[1] In the EDAFS region, the oscillation of the spectrum tends to be disturbed by random noise because of tiny changes in the XRD intensities. The solid line in Figure 10.2b inset shows f''_{Ni}extracted from the Ni 111 DAFS spectrum by $|F_{111}|\sin\varphi = 4f''_{Ni}e^{-M_{Ni}}$ with temperature factor exp(-M_{Ni}).The XAFS spectrum (μtE) in the broken line must be equivalent to f''_{Ni} because all Ni atoms occupy only one symmetrically equivalent site in the face-centered cubic lattice. The spectrum of f''_{Ni} shows good agreement with that of μtE in terms of edge shape and oscillation.

To demonstrate its site-selective analysis capability, DAFS was applied to resolve the Fe local structure in magnetite (Fe_3O_4) in which tetrahedral and octahedral sites are respectively occupied by Fe^{3+} and a mixture of Fe^{2+}/Fe^{3+}.[69] The DAFS spectra obtained for Fe_3O_4 (220) and (311) are markedly different (Figure 10.2c), suggesting the different contributions from the tetrahedral and octahedral sites. The imaginary parts of the resonant dispersion terms of Fe at tetrahedral (8a) and octahedral (16d) sites in Fe_3O_4 were calculated by solving the simultaneous equations derived from Eq.10.3

$$|F_{220}|\sin\varphi = 8f''_{Fe(8a)}e^{-M_{Fe(8a)}}$$

$$|F_{311}|\sin\varphi = -4\sqrt{2}f''_{Fe(8a)}e^{-M_{Fe(8a)}} - 8f''_{Fe(16d)}e^{-M_{Fe(16d)}}$$

The spectra of f''_{Fe} at the 8a and 16d sites are compared in Figure 10.2d(a). A distinctive pre-edge peak is observed at 7.114 keV in

$f''_{Fe(8a)}$ for the tetrahedral site, due to the transition from the Fe 1s orbital to the p component of the p–d hybrid orbital.[70] In contrast, the pre-edge peak of $f''_{Fe(16d)}$ for the octahedral site is broad and weak. At 16d sites, due to the absence of the hybridization, and only the weak quadrupole transition of Fe 1s to Fe 3d occurs. In the f''_{Fe} spectrum at 16d, the two shoulders at the main edge, indicated by arrows, arise from Fe^{2+} and Fe^{3+}.[71,72] By weighting the experimental f''_{Fe} spectrum at each site [Figure 10.2d(a)] in response to the number of atoms at the site, an averaged f''_{Fe} spectrum is comparable with μtE from the XAFS spectrum (see Figure 10.2d(b)). There are subsets of (*hkl*) reflections for which only octahedral (Fe_{octa}) or only tetrahedral (Fe_{tetra}) sites contribute, in addition to another subset where both sites contribute to the crystallographic structure factor $f_{hkl} = (I_{hkl})^{1/2}$, see Eq. 10.3. The Fe *K*-edge DAFS signal across the (222) reflection is Fe_{octa} selective, the (022) and (224) ones Fe_{tetra} selective, and the (444) one unselective. Thus, the local environment of Fe in Fe_{octa} or only Fe_{tetra} sites in Fe_3O_4 were disentangled, which is not possible with a standard EXAFS experiment.

Continuously, DAFS spectroscopy is used to solve the site-specific local structure of Mn in manganese ferrite ($MnFe_2O_4$) films grown under non-equilibrium conditions.[73] This material has attracted widespread attention due to its wide application in electronics.[74] $MnFe_2O_4$ (*Fd–3m* space group, n. 227) exhibits two independent sites for the Mn cation, namely A and B. Mn K-edge DAFS signals across the (422) reflection is A-selective, the (222) one is B-selective, and the (111) one is unselective. The A and B site-selective (E) spectra were obtained, through Eq.12.28, from the measured f''(E) DAFS spectra of (422) and (222) reflections, respectively. Both XANES spectra and corresponding $k\chi(k)$ spectra significantly differed for A and B sites. This difference is even more evident in the corresponding FTs, particularly for the second shell signal. DAFS data determine that Mn atoms occupies sites A and B with a fraction of 82 and 18%, respectively. The disentanglement of the contribution of the two sites would not be possible with EXAFS. The robustness of the study was confirmed by the fact that the average of the site-specific DAFS $k\chi$(k) spectra weighted by the corresponding site occupancies of 0.82 and 0.18 for A site and B site, perfectly reproduced the standard site-unselective EXAFS spectrum.

It is of interest to apply DAFS to extend the useful *k*-region of an EXAFS spectrum in crystalline samples containing elements with close edges.[75] When two or more absorption edges in a material are sufficiently close in energy, EXAFS is of limited use as the data range above the lower-energy edge is truncated by the higher-energy edge. Energy- or wavelength-discriminating detection methods may fail to resolve the fluorescence lines which are very close in energy. A solution to this problem is using the resolution in momentum transfer of DAFS to separate the fine-structure signals from elements with closely spaced fluorescence lines, as for $BaTiO_3$ exhibiting the Ti K-edge around 4.97 keV and the Ba L_{III} edge around 5.25 keV. A different approach to solve this problem has been presented later and consists in collecting high-energy resolution fluorescence detected (HERFD) EXAFS.[76,77]

The multiple-wave DAFS method combines the X-ray multiple-wave diffraction and diffraction anomalous fine structure techniques. With multiple-wave DAFS, a superior wave vector sensitivity of the fine structure can be obtained on a GaAs single crystal. [78]

DAFS has been widely employed to characterize semiconductor thin films and heterostructures.[79] In the study epilayers or multilayers, DAFS can provide selective structural information by choosing the Bragg peaks of the strained epitaxial phase, that occurs at a different Bragg angle than the substrate.[80,81] The heterostructure is probed throughout its entire thickness,[82,83] while the glancing angle geometry is adopted [84] to be sensitive to the last few topmost monolayers of the epitaxial films. In the DAFS study at both the Ga and As *K*-edges in a buried 21 nm-thick $In_{0.215}Ga_{0.785}As$ layer grown coherently on GaAs,[85] the analysis of DAFS oscillations at the Ga *K*-edge resulted in $r_{Ga\text{-}As}$ = 2.44 Å and $r_{As\text{-}Ga}$ = 2.44 Å while $r_{As\text{-}In}$ = 2.53 Å was obtained from the As *K*-edge experiment. The first shell Ga-As bond length from DAFS revealed a strain-induced contraction of 0.013 Å relative to the $r_{Ga\text{-}As}$ in bulk $In_xGa_{1\text{-}x}As$ of the same composition, in good agreement with the uniform bond-length distortion model for strained layers which was proposed to explain the In *K*-edge EXAFS study on the same specimen.[86] The In-As first shell bond length shows a contraction of 0.015 Å relative to the $r_{In\text{-}As}$ bond length in bulk $In_xGa_{1\text{-}x}As$ with the *x* value. The effect of built-in strain on III-V epitaxial semiconductors was further investigated with DAFS experiments at the Ga and As *K*-edges of a strained-layer superlattice of $(GaP)_2(InP)_3$ and three single epilayers of $GaAs_{1-x}P_x$ partially relaxed,[87] with a different amount of residual parallel strain ($\varepsilon_{//}$) in the $0.004 \le \varepsilon_{//} \le 0.007$ range; $\varepsilon_{//} = (a - a_0)/a$, where a and a_0 are the unstrained lattice parameters of the epilayer and of the GaAs(001) substrate, respectively. The first shell $r_{Ga\text{-}P}$ bond distance in the superlattice was stretched by about 0.04 Å in agreement with the predictions of the elastic theory. The same group has successively extended the application of DAFS from 2-D to 1-D and 0-D systems by investigating InAs/InP(001) and InAs/GaAs(001) quantum wires and GaN/AlN and Ge/Si(001) quantum dots by working in glancing angle geometry.[88,89] The DAFS method can access important information about the composition and strain of such nanostructures with low coverage epitaxial layers. The general formalism utilized to analyze diffraction oscillations is valid for any type of crystallographic structure, which proves that DAFS spatial selectivity provides a unique tool for studying systems that cannot be achieved by other X-ray techniques.

Finally, as a diffractive analogue of XAFS, DAFS gives location-specific local structural information. As a diffractive analogue of X-ray magnetic circular dichroism (XMCD, *i.e.*, McXAFS), McDAFS provides site-specific local magnetic information. McXANES and McDANES detect the empty density of magnetic states near the threshold. McEXAFS and McEDAFS probe the radial distribution of the magnetization of atoms surrounding excited atoms. The primary preference of McDAFS over XMCD is to obtain site selectivity thanks to the crystallographic sensitivity of DAFS. McDAFS technique allow element-specific and site-specific determinations of the density of magnetic states near threshold (from the McDANES) and of the radial distribution of the magnetic neighbors surrounding the excited atom (from the McEDAFS). The beauty of McDAFS is that it provides the only way to directly measure the complete site-specific local magnetic environment, and the properties of specific elements at specific sites. McDAFS wavevector selectivity should also prove to be useful in magnetic multilayer studies, as it will allow the separation of signals from the same element type in different layers.

10.6 Summary and Outlook

With the stimulation by the spectacular new synchrotron radiation sources, DAFS technology is becoming one of the rapidly developing structural spectroscopy and crystallographic methods. The infusion of crystallographic methods into X-ray absorption spectroscopy clearly allows to determine the local atomic configuration and chemical state around a certain element at a particular crystallographic site of a unit cell and also in a particular crystalline phase of a sample. The DAFS theory in the extended range has been well developed and fits well with the experiment, but practical near-edge theory is still needed further development. It is also interesting and worthy developing the reverse benefits that the X-ray spectroscopic sensitivities can bring to crystallography. This widespread work on the smooth and oscillatory scalar and tensor scattering amplitudes of atoms in condensed matter has turned these previous problems into new solutions.

Truly, comprehensive techniques should make a full use of all the degrees of freedom of the photons including their momenta, energies and spins. Polarization analyzed DAFS uses all of these degrees of freedom in the special case of elastic scattering—there is momentum transfer, but no energy transfer. The three other possibilities are closely related: (1) elastic forward scattering—no energy or momentum transfer, (2) absorption and inelastic forward scattering—energy, but no momentum transfer, and (3) inelastic resonant scattering (X-ray Raman)—both energy and momentum transfer. These combined techniques have high potential to overcome the XAFS limitations, such as (a) a "quality" decrease of the information obtained under high temperatures; (b) the difficulty in identifying low-Z neighbors and in discriminating among neighbors with similar Z-elements; and (c) the inability to collect full XAFS data at the *K*-edges falling in the UV and ultra-soft X-ray region, preventing the determination of the local environment around very low Z-elements (from hydrogen to boron) and making XAFS studies on carbon,[90,91] nitrogen, and oxygen [92,93] very challenging.

As a complementary technology to XAFS, X-ray emission spectroscopy (XES) provides valuable information about the electronic structure (local charge and spin density) and the nature of the binding ligand. XES is one of the photon-in/photon-out spectroscopies. The core electrons are excited by incident X-ray photons, and then the core hole of the excited state is decayed by emitting X-ray photons. The emitted photon energy is equivalent to the energy difference between the energy levels of the involved electrons. The purpose of X-ray emission spectroscopy is to analyze the energy dependence of emitted photons. The combination of XAFS and XES can solve the low-Z neighbor problem. It is possible to obtain EXAFS-like information from low Z elements (PDF, EXELFS, XRS), local-range contributions coming from amorphous and crystalline phases present in the same sample can be disentangled (DAFS), and EXAFS-like contributions coming from the same element hosted in different crystallographic sites can be separated (DAFS). EXAFS-like data on very low Z-elements can be obtained with pair distribution function (PDF) or by exploiting inelastic scattering techniques such as extended energy-loss fine structure (EXELFS), and X-Ray Raman Scattering (XRS), and may be achieved in the near future with β-environmental fine structure (BEFS). Additionally, XAFS can also be simultaneously utilized with other spectroscopic (IR, Raman, UV-Vis) and scattering (XRD, SAXS) methods in time-resolved experiments.

The high energy resolution fluorescence detection (HERFD) technique is particularly important for the X-ray absorption spectrum measurement, as the intensity of the fluorescent line corresponding to a specific excitation state decay process can be recorded with a narrow energy resolution. Generally, a narrower band from the emission line of the specimen can be selected by using a crystal analyzer. By collecting the emission line with an energy resolution higher than the natural width of the emission line, the significantly improved intrinsic resolution of the absorption spectrum can help to resolve the structures invisible in the conventional XAFS spectrum.[94] This technique was successfully utilized to study the L_3-edge XANES of Pt and Au.[95,96] The K_β emission line of metals is sensitive to the spin state of the photo-absorbing atoms. The main and the satellite emission lines provides information about different spin states. Therefore, the XANES spectrum recorded by the emission of the main emission line or the satellite emission line is spin-dependent, and these methods can be used to record spin-resolved EXAFS.[97,98] The chemical state of the photon-absorbing atoms directly affect the position and spectral shape of the emission band. In a sample with light-absorbing atomic species in different oxidation states, spectral contributions from different positions can be distinguished by using high-resolution X-ray fluorescence.[99] Sometimes the multi-edges of the sample interfere with data collection in the relevant spectral region. For example, in the manganese ore system, the *K*-edge of rare earth elements and the *L*-edge of manganese often overlap.[100] High energy resolution fluorescence detection technology can overcome this difficult absorption by collecting appropriate emission lines to avoid overlap.

By combining diffraction and XAFS into a single experiment in which scattering and XAFS respectively provide information about spatial modulation and electronic structures, resonant elastic X-ray scattering (REXS) provides a unique element, site, and valence specific probe to study spatial modulations of charge, spin, and orbital degrees of freedom in condensed matters. The resonant X-ray scattering near the absorption edge is also reviewed as virtual transitions from the core level to the unoccupied state close to the Fermi level, which depend largely on the charge, spin, and orbital configuration. In addition, the resonance process greatly enhances the scattering cross section and provides an opportunity to study the selected atomic species in the crystal.[101,102] By combining the pump-probe approach and REXS, time-resolved femtosecond resonant XRD can probe the interesting physics in nonequilibrium states of materials. For example, in the stripe-ordered nickelate REXS, despite distinct spin and charge energy scales, the spin and charge orders are strongly coupled even during the nonequilibrium process.[103]

As a photon-in/photon-out spectroscopy, resonance inelastic X-ray scattering (RIXS) measures the energy and momentum changes of scattered photons. In the RIXS experiment, X-ray photons are scattered inelastically by matter. The energy, momentum, and polarization changes of the scattered photons are transferred to the inherent excitation of the material under study, so RIXS provides information about many basic excitations, such as charge transfer excitations, crystal field and orbital excitations, magnetic oscillators, plasmons, and phonons. The RIXS process can also be described as a resonance X-ray Raman or resonance X-ray emission process. In addition, similar to REXS, the resonance in RIXS greatly enhances the cross-section and brings the selectivity for atomic species. In this respect, RIXS provides broad prospects for the study of strongly related electronic systems.[104] Combining XAFS and XES, RIXS scans both incident energy and emitted energy. The obtained map from the scan shows XAFS and XES data as well as EXAFS. It is commonly used to determine the transferred energy by measuring the difference between the incident energy and the emitted energy.

However, to cover all resonant X-ray scattering processes, the above classification must be further expanded to include photon spin angular momentum transfer and magnetic coupling of photons to both charge and condensed matter systems. The exciting future development of resonance X-ray scattering seems very bright.

---

[1] L.B. Sorensen, J.O. Cross, M. Newville, B. Ravel, J.J. Rehr, H. Stragier, C.E. Bouldin, & J.C. Woicik, (1994) In G. Materlik, C.J. Sparks, and K. Fischer, editors, *Resonant Anomalous X-Ray Scattering: Theory and Applications*, pages 389-420, Amsterdam, Elsevier Science, North-Holland.

[2] H. Stragier, J.O. Cross, J.J. Rehr, L.B. Sorensen, C.E. Bouldin, & J.C. Woicik, (1992) Diffraction anomalous fine structure: a new X-ray structural technique. *Phys. Rev. Lett.*, **69**(21), 3064–3067.

[3] G. Wendin, (1980) Anomalous X-ray scattering. *Phys. Scri.* 21, 535-542.

[4] Y. Cauchois and C. Bonnelle, C.R. Acad. Sci. (Paris) 242 (1956) 100 and 1596.

[5] H. Fricke, (1920) The K-characteristic absorption frequencies for the chemical elements magnesium to chromium. Phys. Rev. 16, 202.

[6] G. Hertz, (1920) über die Absorptionsgrenzen in derL-Serie. *Z. Phys.* 3, 19.

[7] T. Fukamachi, S. Hosoya, T. Kawamura, & J. Hastings, (1977) Measurements of integrated intensity near the absorption edge with synchrotron radiation. *J. Appl. Crystallogr.* 10, 321.

[8] S.I. Salem, & V.L. Hall, (1980) Extended X-ray fine structure in crystalline copper. *J. Phys. F.* 10, 1627-1629.

[9] I.J. Pickering, M. Sansone, J.J. Marsch, & G.N. George, (1993) Site-specific X-ray absorption spectroscopy using DIFFRAXAFS. Jpn. J. Appl. Phys. 32 Suppl. 32-2 , 206.

[10] I. J. Pickering, M. Sansone, J. Marsch, & G. N. George, (1993) Diffraction anomalous fine structure: a new technique for probing local atomic environment. *J. Am. Chem. Soc.* 115, 6302-6311.

[11] H.J. Stragier, (1993) *DAFS: A New X-Ray Structural Technique*, Ph.D. thesis, University of Washington.

[12] D.H. Templeton, (1991) in *Handbook on Synchrotron Radiation*, edited by G. Brown and D.E. Moncton (Elsevier, Amsterdam), p. 201.

[13] L. Davidovich, & H.M. Nussenveig, (1980) in Foundations of Radiation Theory and Quantum Electrodynamics edited by A.O. Barut, Plenum, New York, p. 83. references therein.

[14] H.K. Wagenfeld, (1987) The angular dependence of X-ray dispersion corrections. *Z. Phys. B* - Condensed Matter 65, 437-441, and references therein.

[15] E. Borfecchia, D. Gianolio, G. Agostini, *et al.* (2013) Characterization of MOFs. 2. Long and local range order structural determination of MOFs by

combining EXAFS and diffraction techniques, in Metal Organic Frameworks as Heterogeneous Catalysts (eds F. X. Llabr´es i Xamena and J. Gasc´on), The Royal Society of Chemistry, Cambridge.

16 J. Als-Nielsen, & D. McMorrow, (2011) *Elements of Modern X-ray Physics*, 2nd Edition, John Wiley & Sons, Ltd.

17 C. Garino, E. Borfecchia, R. Gobetto, *et al.* (2014) Determination of the electronic and structural configuration of coordination compounds by synchrotron-radiation techniques. *Coord. Chem. Rev.*, 277–278, 130–186.

18 D. Gianolio, (2015) How to start an XAS experiment:, in *X-Ray Absorption and X-Ray Emission Spectroscopy: Theory and Applications*, vol. I (eds J.A. van Bokhoven and C. Lamberti), John Wiley & Sons, Ltd., pp. 99–124.

19 D.T. Cromer, & D. Liberman, (1970) Relativistic calculation of anomalous scattering factors for X rays. *J. Chem. Phys.* 53, 1891, and D.T. Cromer, D.A. Liberman, (1981) Anomalous dispersion calculations near to and on the long-wavelength side of an absorption edge. *Acta Cryst. A* 37, 267.

20 S. Sasaki, (1989) Numerical Tables of Anomalous Scattering Factors Calculated by the Cromer and Liberman Method. *KEK Report* 88-14.

21 J.J. Rehr, C.H. Booth, F. Bridges, & S.I. Zabinsky, (1994) X-ray-absorption fine structure in embedded atoms. *Phys. Rev. B* 49, 12347(R).

22 J.D. Jackson, (1999) *Classical Electrodynamics*, 3rd edition, John Wiley & Sons, Inc., New York.

23 R.G. Newton, (1976) Optical theorem and beyond, *Am. J. Phys.*, 44, 639–642.

24 J.O. Cross, W.T. Elam, J.C. Woicik, & L.B. Sorensen. (1999) Reliability of structural parameters determined from DAFS data using the iterative dispersion integral algorithm, *J. Synchrotron Radiat.*, 6, 335–337.

25 A.I. Frenkel, A.V. Kolobov, I.K. Robinson, J.O. Cross, Y. Maeda, & Ch.E. Bouldin, (2002) Direct separation of short range order in intermixed nanocrystalline and amorphous phases, *Phys. Rev. Lett.*, 89, 285503.

26 M.G. Proietti, H. Renevier, J.L. Hodeau, J. García, J.F. Bérar, & P. Wolfers. (1999) Diffraction-anomalous-fine-structure spectroscopy applied to the study of III-V strained semiconductors. *Phys. Rev. B*, 59, 5479–5492.

27 L.B. Sorensen, J.O. Cross, M. Newville, B. Ravel, J.J. Rehr, H. Stragier, C.E. Bouldin, & J.C. Woicik, (1994) *Resonant Anomalous X-Ray Scattering: Theory and Applications*, Ed. G. Materlik, C. J. Sparks, and K. Fischer, North-Holland, pp. 389-420.

28 J.J. Rehr, & A.L. Ankudinov, (2001) New developments in the theory of X-ray absorption and core photoemission, *J. Electron Spectrosc. Relat. Phenom.*, 114, 1115–1121.

29 B. Ravel. ATOMS: crystallography for the X-ray absorption spectroscopist. *J. Synchrotron Radiat.*, 8, 314–316. (2001).

30 P.J. Ellis, & H.C. Freeman, (1995) Xfit: an Interactive EXAFS Analysis Program. *J. Synchrotron Radiat.*, 2, 190–195.

31 J.L. Hodeau, V. Favre-Nicolin, S. Bos, H. Renevier, E. Lorenzo, & J.-F. Berar, (2001) Resonant diffraction, *Chem. Rev.*, 101, 1843–1867.

[32] J.L. Hodeau, & R. Guinebretiere. (2007) Crystallography: past and present, *Appl. Phys. A - Mater. Sci. Process.* 89, 813–823.

[33] H. Stragier, J.O. Cross, J.J. Rehr, L.B. Sorensen, C.E. Bouldin, & J.C. Woicik, (1992) Diffraction anomalous fine-structure: a new X-ray structural technique, *Phys. Rev. Lett.* 69, 3064–3067.

[34] I.J. Pickering, M. Sansone, J. Marsch, & G.N. George, (1993) Diffraction anomalous fine-structure: a new technique for probing local atomic environment, *J. Am. Chem. Soc.* 115, 6302–6311.

[35] H. Renevier, & M.G. Proietti, (2013) Grazing incidence diffraction anomalous fine structure in the study of structural properties of nanostructures, in Characterization of Semiconductor Heterostructures and Nanostructures II (eds G. Agostini and C. Lamberti), Elsevier, Amsterdam.

[36] E.A. Stern, (1993) Number of relevant independent points in X-ray-absorption fine-structure spectra. *Phys. Rev. B* 48, 9825.

[37] D.C. Meyer, K. Richter, P. Paufler, & G. Wagner, (1999) X-ray analysis of the short-range order in the ordered-alloy domains of epitaxial (Ga,In)P layers by diffraction anomalous fine structure of superlattice reflections. *Phys. Rev. B* 59, 15253-15260.

[38] J. Röhler, A. Larisch, & R. Schafer, (1992) Evidence for charge redistribution below Tc in $YBa_2Cu_3O_{7-\delta}$. *Physica C* 191, 57.

[39] G.M. Luo, Z.H. Mai, T.P.A. Hase, B.D. Fulthorpe, B.K. Tanner, C.H. Marrows, & B.J. Hickey, (2001) Variable wavelength grazing incidence X-ray reflectivity measurements of structural changes on annealing Cu/NiFe multilayers. *Phys. Rev. B* 64, 245404.

[40] H. Renevier, J.L. Hodeau, P. Wolfers, S. Andrieu, J. Weigelt, & R. Frahm, (1997) Selective study of Fe atoms at the interfaces of an Fe/Ir(100) superlattice by means of diffraction anomalous fine structure. *Phys. Rev. Lett.* 78, 2775–2778.

[41] O. Ersen, V. Pierron-Bohnes, M.H. Tuilier, C. Pirri, L. Khouchaf, & M. Gailhanou, (2003) Short- and long-range order in iron and cobalt disilicides thin films investigated by the diffraction anomalous fine structure technique, *Phys. Rev. B* 67, 094116.

[42] O. Ersen, C. Ulhaq-Bouillet, V. Pierron-Bohnes, M.H. Tuilier, D. Berling, P. Bertoncini, C. Pirri, M. Gailhanou, & D. Thiaudière, (2002) Evidence of a ternary $Co_{1-x}Fe_xSi_2$ phase with a CaF_2-type structure: high-resolution transmission electron microscopy and diffraction anomalous fine structure study. *Appl. Phys. Lett.* 81, 2346–2348.

[43] D. Bazin, & D. Sayers, (1993) Comparison between XAS, AWAXS and DAFS applied to nanometer-scale supported metallic clusters 2. Bimetallic clusters. *Jpn. J. Appl. Phys.* 1, 32, 252–254.

[44] D.C. Bazin, D.A. Sayers, & J.J. Rehr, (1997) Comparison between X-ray absorption spectroscopy, anomalous wide angle X-ray scattering, anomalous small angle X-ray scattering, and diffraction anomalous fine structure techniques applied to nanometer-scale metallic clusters, *J. Phys. Chem. B* 101, 11040–11050.

[45] C. Meneghini, F. Boscherini, L. Pasquini, & H. Renevier, (2009) Diffraction anomalous fine structure study of iron/iron oxide nanoparticles, *J. Appl. Crystallogr.*, 42, 642–648.

[46] A.Y. Khodakov, J. Lynch, D. Bazin, B. Rebours, N. Zanier, B. Moisson, P. Chaumette, (1997) Reducibility of cobalt species in silica-supported Fischer-Tropsch catalysts, *J. Catal.* 168, 16–25.

[47] W. Welnic, A. Pamungkas, R. Detemple, Ch. Steimer, S. Blügel, & M. Wuttig, (2006) Unravelling the interplay of local structure and physical properties in phase-change materials, *Nat. Mater.* 5, 56–62.

[48] G. Milano, & G. Guerra, (2009) Understanding at molecular level of nanoporous and co-crystalline materials based on syndiotactic polystyrene, *Prog. Mater. Sci.* 54, 68–88.

[49] C. Suryanarayana, & F.H. Froes, (1993) Mechanical alloying of titanium-base alloys, *Adv. Mater.* 5, 96–106.

[50] J.P. Zheng, P.J. Cygan, & T.R. Jow, (1995) Hydrous ruthenium oxide as an electrode material for electrochemical capacitors, *J. Electrochem. Soc.* 142, 2699–2703.

[51] V.V. Kochervinskii, (2003) Piezoelectricity in crystallizing ferroelectric polymers: Poly(vinylidene fluoride) and its copolymers (a review), *Crystallogr. Rep.* 48, 649–675.

[52] M. Wuttig, & S. Raoux, (2012) The science and technology of phase change materials. *Z. Anorg. Allg. Chem.*,638, 2455–2465.

[53] H. Ikeda, Y. Qi, T. Cagin, K. Samwer, W.L. Johnson, & W.A. Goddard, III, (1999) Strain rate induced amorphization in metallic nanowires, *Phys. Rev. Lett.* 82, 2900–2903.

[54] B.S. Lee, J.R. Abelson, S.G. Bishop, D.-H. Kang, & B.-k. Cheong, (2005) Investigation of the optical and electronic properties of $Ge_2Sb_2Te_5$ phase change material in its amorphous, cubic, and hexagonal phases, *J. Appl. Phys.* 97, 093509.

[55] A.A. Dabous, (2002) Uranium isotopic evidence for the origin of the Bahariya iron deposits, *Egypt, Ore Geol. Rev.*, 19, 165–186.

[56] E. Mikhailov, S. Vlasenko, S.T. Martin, T. Koop, U. Poeschl, (2009) Amorphous and crystalline aerosol particles interacting with water vapor: conceptual framework and experimental evidence for restructuring, phase transitions and kinetic limitations, *Atmos. Chem. Phys.* 9, 9491–9522.

[57] J. Aizenberg, G. Lambert, L. Addadi, & S. Weiner, (1996) Stabilization of amorphous calcium carbonate by specialized macromolecules in biological and synthetic precipitates. *Adv. Mater.* 8, 222–226.

[58] A.I. Frenkel, A.V. Kolobov, I.K. Robinson, J.O. Cross, Y. Maeda, & Ch.E. Bouldin, (2002) Direct separation of short range order in intermixed nanocrystalline and amorphous phases. *Phys. Rev. Lett.* 89, 285503.

[59] A.V. Kolobov, Y. Maeda, & K. Tanaka, (2000) Raman spectra of Ge nanocrystals embedded into SiO_2, *J. Appl. Phys.* 88, 3285–3289.

[60] A.V. Kolobov, H. Oyanagi, Y. Maeda, & K. Tanaka, (2001) Local structure of Ge nanocrystals embedded in SiO_2 studied by X-ray absorption fine structure. *J. Synchrotron Radiat.*, 8, 511–513.
[61] A.L. Patterson, (1939) The Scherrer formula for X-ray particle size determination. *Phys. Rev.* 56, 978–982.
[62] J.I. Langford, & A.J.C.Wilson, (1978) Scherrer after sixty years: a survey and some newresults in the determination of crystallite size. *J. Appl. Crystallogr.* 11, 102–113.
[63] P. Scherrer, (1918) Bestimmung der Grosse und der inneren Struktur von Kolloidteilchen mittels Roentgenstrahlen. *Nachr. Ges. Wiss. Goettingen*, 26, 98–100.
[64] N. Seljakow, (1925) Eine roentgenographische Methode zur Messung der absoluten Dimensionen einzelner Kristalle in Korpern von fein-kristallinischem Bau, *Z. Phys.* 31, 439–444.
[65] M. von Laue, (1926) Lorentz-Factor und Intensitaetsverteilung in Debye-Scherrer ringen. *Z. Krystallogr.* 64, 115–142.
[66] J.O. Cross, M. Newville, L.B. Sorensen, *et al.* (1997) Separated anomalous scattering amplitudes for the inequivalent Cu sites in $YBa_2Cu_3O_7$-d using DAFS, *J. Phys. IV* 7, 745–747.
[67] A. Yang, Z.H. Chen, A.L. Geiler, *et al.* (2008) Element- and site-specific oxidation state and cation distribution in manganese ferrite films by diffraction anomalous fine structure. *Appl. Phys. Lett.* 93, 052504.
[68] C. Lamberti, E. Borfecchia, J. A. van Bokhoven, & M. Fernandez-Garcia, (2016) XAS Spectroscopy: Related Techniques and Combination with Other Spectroscopic and Scattering Methods, in *X-ray absorption and X-ray emission spectroscopy: theory and applications* / edited by Jeroen A. van Bokhoven and Carlo Lamberti, ISBN 978-1-118-84423-6.
[69] A.I. Frenkel, J.O. Cross, D.M. Fanning, & I.K. Robinson, (1999) DAFS analysis of magnetite. *J. Synchotron Radiat.* 6, 332–334.
[70] S. Sasaki, (1995) Fe^{2+} and Fe^{3+} ions distinguishable by x-ray anomalous scattering: Method and its application to magnetite. *Rev. Sci. Instrum.* 66, 1573.
[71] K. Kobayashi, H. Kawata, & K. Mori, (1998) Site specification on normal and magnetic XANES of ferrimagnetic Fe_3O_4 by means of resonant magnetic Bragg scattering. *J. Synchrotron Rad.* 5, 972-974.
[72] Y. Joly, J. E. Lorenzo, E. Nazarenko, J.-L. Hodeau, D. Mannix, & C. Marin, (2008) Low-temperature structure of magnetite studied using resonant X-ray scattering. Phys. Rev. B 78, 134110.
[73] E. Kravtsov, D. Haskel, A. Cady, A. Yang, C. Vittoria, X. Zuo, & V.G. Harris, (2006) Site-specific local structure of Mn in artificial manganese ferrite films, *Phys. Rev. B* 74, 104114.
[74] A.M. Brabers, (1995) Progress in spinel ferrite research, in *Handbook of Magnetic Materials*, vol. 8, (ed. K.H.J. Buschow), Elsevier, New York.

[75] B. Ravel, S. Grenier, H. Renevier, & C.B. Eom, (2001) Valence selective DAFS measurements of Mn in $La_{1/3}Ca_{2/3}MnO_3$, *J. Synchrotron Radiat.*, 8, 384–386.

[76] J. Yano, Y. Pushkar, P. Glatzel, A. Lewis, K. Sauer, J. Messinger, U. Bergmann, & V. Yachandra, (2005) High-resolution Mn EXAFS of the oxygen-evolving complex in photosystem II: Structural implications for the Mn_4Ca cluster. *J. Am. Chem. Soc.* 127, 14974–14975.

[77] S. Bordiga, E. Groppo, G. Agostini, J.A. van Bokhoven, & C. Lamberti, (2013) Reactivity of surface species in heterogeneous catalysts probed by in situ X-ray absorption techniques. *Chem. Rev.* 113, 1736–1850.

[78] Y.R. Lee, Y.P. Stetsko, W.H. Sun, Sh.-Ch. Weng, Sh.-Y. Cheng, G.-G. Lin, Y.-L. Soo, & Sh.-L. Chang, (2006) Multiple wave diffraction anomalous fine structure. *Phys. Rev. Lett.* 97, 185502.

[79] C. Lamberti, (2004) The use of synchrotron radiation techniques in the characterization of strained semiconductor heterostructures and thin films. *Surf. Sci. Rep.* 53, 1–197.

[80] A. Antolini, P. J. Bradley, C. Cacciatore, D. Campi, L. Gastaldi, F. Genova, M. Iori, C. Lamberti, C. Papuzza & C. Rigo, (1992) Investigations on the interface abruptness in CBE-grown InGaAs InP QW structures. *J. Electron. Mater.* 21, 233–238.

[81] L. Mino, D. Gianolio, G. Agostini, A. Piovano, M. Truccato, A. Agostino, S. Cagliero, G. Martinez-Criado, S. Codato, & C. Lamberti, (2010) Structural characterization of multi-quantum wells in electroabsorption-modulated lasers by using synchrotron radiation micrometer-beams. *Adv. Mater.* 22, 2050–2054.

[82] L. Alagna, S. Turchini, & T. Prosperi, (2001) Diffraction anomalous fine structure of forbidden reflection of super-ordered GaInP. *J. Synchrotron Radiat.* 8, 387–389.

[83] C. Leclere, V. Fellmann, C. Bougerol, D. Cooper, B. Gayral, M. G. Proietti, H. Renevier, & B. Daudin, (2013) Strain assisted inter-diffusion in GaN/AlN quantum dots. *J. Appl. Phys.* 113, 10.

[84] S. Grenier, M.G. Proietti, H. Renevier, L. Gonzalez, J.M. García, J.M. Gérard, J. García, (2001) Glancing-angle diffraction anomalous fine structure of InAs quantum dots and quantum wires, *J. Synchrotron Radiat.* 8, 536–538.

[85] J.C. Woicik, J.O. Cross, C.E. Bouldin, B. Ravel, J. G. Pellegrino, B. Steiner, S. G. Bompadre, L. B. Sorensen, K. E. Miyano, & J. P. Kirkland, (1998) Diffraction anomalous fine-structure study of strained $Ga_{1-x}In_xAs$ on GaAs(001). *Phys. Rev. B* 58, R4215–R4218.

[86] J.C. Woicik, J.G. Pellegrino, B. Steiner, K. E. Miyano, S. G. Bompadre, L. B. Sorensen, T.-L. Lee, & S. Khalid, (1997) Bond-length distortions in strained semiconductor alloys, *Phys. Rev. Lett.* 79, 5026–5029.

[87] M.G. Proietti, H. Renevier, J. L. Hodeau, J. García, J. F. Bérar, & P. Wolfers, (1999) Diffraction-anomalous-fine-structure spectroscopy applied to the study of III-V strained semiconductors. *Phys. Rev. B* 59, 5479–5492.

[88] J. Coraux, M.G. Proietti, V. Favre-Nicolin, H. Renevier, & B. Daudin, (2006) Step-by-step capping and strain state of GaN/AlN quantum dots studied by grazing-incidence diffraction anomalous fine structure. *Phys. Rev. B* 73, 205343.

[89] J. Coraux, V. Favre-Nicolin, M.G. Proietti, B. Daudin, & H. Renevier, (2007) Grazing-incidence diffraction anomalous fine structure: application to the structural investigation of group-III nitride quantum dots. *Phys. Rev. B* 75, 235312.

[90] D. Denley, P. Perfetti, R.S. Williams, D. A. Shirley, & J. Stöhr, (1980) Carbon K-edge fine structure in graphite foils and in thin-film contaminants on metal surfaces. *Phys. Rev. B* 21, 2267–2273.

[91] G. Comelli, & J. Stoehr, (1988) SEXAFS studies of C-C bond lengths in condensed and chemisorbed molecules. *Surf. Sci.* 200, 35–44.

[92] P. Luches, S. D'Addato, S. Valeri, E. Groppo, C. Prestipino, C. Lamberti, & F. Boscherini, (2004) X-ray absorption study at the Mg and O K-edges of ultrathin MgO epilayers on Ag(001). *Phys. Rev. B* 69, 045412.

[93] E. Groppo, C. Prestipino, C. Lamberti, R. Carboni, F. Boscherini, P. Luches, S. Valeri, & S. D'Addato, (2004) O K-edge X-ray absorption study of ultrathin NiO epilayers deposited in situ on Ag(001). *Phys. Rev. B* 70, 165408.

[94] K. Hamalainen, D. P. Siddons, J. B. Hastings, & L. E. Berman, (1991) Elimination of the inner-shell lifetime broadening in X-ray-absorption spectroscopy. *Phys. Rev. Lett.* 67, 2850.

[95] F.M.F de Groot, M. H. Krisch, & J. Vogel, (2000) Site-selective XAFS: a new tool for catalysis research. *Top. Catal.*, 10, 179.

[96] J. van Bokhoven, C. Louis, J. Miller, M. Tromp, O. Safonova, & P. Glatzel, (2006) Activation of oxygen on gold/alumina catalysts: in situ high-energy-resolution fluorescence and time-resolved X-ray spectroscopy. *Angew Chem Int Ed Engl*, 45, 4651.

[97] (a) K. Hamalainen, C. C. Kao, B. Hastings, D. P. Siddons, L.E. Berman, V. Stojanoff, & S. P. Cramer, (1992) Spin-dependent X-ray absorption of MnO and MnF_2. Phys. Rev. B, 46, 14274.

[98] Q. Qian, T. A. Tyson, S. Savrassov, C. C. Kao, & M. Croft, (2003) Electronic structure of $La_{1-x}Ca_xMnO_3$ determined by spin-polarized X-ray absorption spectroscopy: Comparison of experiments with band-structure computations. *Phys. Rev. B* 68, 014429.

[99] P. Glatzel, L. Jacquamet, U. Bergmann, F.M.F. de Groot, & S.P. Cramer, (2002) Site-selective EXAFS in mixed-valence compounds using high-resolution fluorescence detection: a study of iron in Prussian Blue. *Inorg. Chem.* 41, 3121.

[100] P. Glatzel, F.M.F. de Groot, O. Manoilova, D. Grandjean, B.M. Weckhuysen, U. Bergmann, & R. Barrea, (2005) Range-extended EXAFS at the L edge of rare earths using high-energy-resolution fluorescence detection: A study of La in LaOCl. *Phys. Rev. B* 72, 014117.

101 J. Fink, E. Schierle, E. Weschke, & J. Geck, (2013) Resonant elastic soft X-ray scattering. *Rep. Prog. Phys.* 76, 056502.

102 W.S. Lee, A.P. Sorini, M. Yi, Y.D. Chuang, B. Moritz, W.L. Yang, J.-H. Chu, H.H. Kuo, A.G. Cruz Gonzalez, I.R. Fisher, Z. Hussain, T.P. Devereaux, & Z.X. Shen, (2012) Resonant enhancement of charge density wave diffraction in the rare-earth tritellurides. *Phys. Rev. B* 85, 155142.

103 Y.D. Chuang, W.S. Lee, Y.F. Kung, A.P. Sorini, *et al.* (2013) Real-time manifestation of strongly coupled spin and charge order parameters in stripe-ordered $La_{1.75}Sr_{0.25}NiO_4$ nickelate crystals using time-resolved resonant X-ray diffraction. *Phys. Rev. Lett.* 110, 127404.

104 L.J.P. Ament, M. van Veenendaal, T.P. Devereaux, J.P. Hill, & J. van den Brink, (2011) Resonant inelastic X-ray scattering studies of elementary excitations. *Rev. Mod. Phys.* 83, 705.

Chapter 11

X-ray Magnetic Circular Dichroism

11.1 Introduction

X-ray magnetic circular dichroism (XMCD) is the difference spectrum of two X-ray absorption spectra obtained in a magnetic field parallel and antiparallel to the helicity of a circularly polarized X-ray beam.[1] In the case of absorption of circularly polarized X-rays, the excited core-level electrons are polarized regarding both their angular momentum and spin. This polarization is reversed for reversed polarization of X-rays. Spin-polarized electrons detect the spin polarization of the final state that is not occupied, and in the case of matched polarization, a large absorption intensity is generated, otherwise a small absorption intensity is generated, so-called circular dichroism. The intensity of XMCD is directly proportional to the spin polarization (magnetization) of the unoccupied final state. Careful analysis of the XMCD spectra can give the information about the magnetic properties of the atoms, such as the spin and orbital magnetic moment of the absorbing atoms.

The credit of the first theoretical investigation of the X-ray dichroism problem should be given to E.A. Stern *et al.* who performed a band structure calculation and predicted a magnetic circular dichroism at the $M_{2,3}$ absorption edges of ferromagnetic nickel in 1975.[2] The XMCD phenomenon was first observed by Schütz *et al.* in 1987 at the Fe K-edge.[3] The dichroism spectrum at the $L_{2,3}$-edges was first collected by Chen *et al.* for Ni using soft X-rays in 1990.[4] Derivation of magneto-optical sum rules has greatly strengthened the XMCD,[1] making this technique capable to disentangle orbital and spin contributions to the total magnetic moment carried by the absorbing atom. Also, in the beginning of 1990s, the advent of the third-generation synchrotron radiation facilities offered the scientific community the X-ray beamlines with full polarization control and greatly improved flux. As a result, polarization dependent X-ray spectroscopy has become a standard tool in modern magnetism research, leading to a deeper understanding of the microscopic origin of magnetic state of matter, as well as to major technological advances.

The first theoretical and experimental results excited the development of this powerful element-specific magnetometer technology. Especially, the spin and orbital moment sum rules for XMCD are very useful to understand the exchange bias and magnetic anisotropy in thin films and multilayers. To remain at the forefront in new emerging areas of magnetism research, however, requires continued improvements to both the instrumentation and optics in order to provide greater flux, improved polarization control, higher detection sensitivity and enhanced sample environments. With the further developments of circular dichroism in (resonant) photoemission and Auger, as well as X-ray detected optical activity, XMCD is increasingly applied in X-ray spectroscopy and imaging for the study of magnetic materials, including the magnetic behavior of bulk materials, thin films and multilayers, magnetic nanostructures and clusters, and spin transmission systems. Because of their inherent element and orbital specificity and ability to probe extremely small sample volumes, when used with X-ray absorption spectroscopy, it can provide supplementary information about the electronic structure, which is of interest to many disciplines such as physics, chemistry, materials science, surface science, life sciences, earth sciences, and catalytic research. XMCD is considered to be one of the most important discoveries in the field of magnetism in the last decades, and it is hard to imagine modern research into magnetism without the aid of polarized X-rays.

11.2 Theory of XMCD

The concept of XMCD spectroscopy was first proposed by Gisela Schütz *et al.* 1987's,[5] are illustrated in Figure 11.1. In a magnetic material, the breaking of time-reversal symmetry allows XMCD. The 2p core state of a 3d metal is split in a j = 3/2 level ($p_{3/2}$, L_3-edge) and j = 1/2 level ($p_{1/2}$, L_2-edge), where spin and orbit are coupled parallel ($l + s$) and antiparallel ($l - s$), respectively. The underlying physics of XMCD process is most easily understood in the following two-step picture.[6]

First, the interaction of circularly polarized X-rays with the p shell results in the excitation of spin-polarized electrons. The right or left circularly polarized photons transmit their angular momentum $\hbar$ and -$\hbar$, respectively, to the excited photoelectrons. If the photoelectron comes from the spin-orbit splitting level, for example at the $p_{3/2}$

level (L_3 edge) and $p_{1/2}$ level (L_2 edge), the angular momentum of the photon can be transferred in part to the spin through the spin-orbit coupling. The emission of an optical helicity vector parallel (antiparallel) to the 2p orbital moment will produce excited electrons with

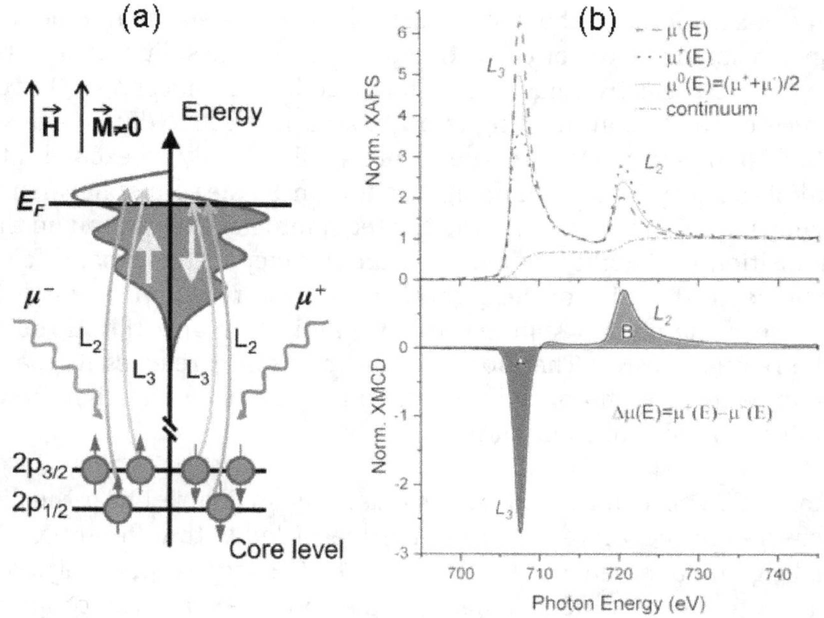

Figure 11.1 (a) Schematic representation of XMCD at $L_{2,3}$ edges for 3d transition metals. In the first step, a circularly polarized photon excites a spin polarized electron from the spin–orbit split 2p level (Fano-effect). The absorption of left (right) circularly polarized light leads to spin-up (spin-down) polarization of the photoelectrons at the $2p_{3/2}$ level (L_3-edge) via spin-orbital coupling (j=l+s) in the core level. According to the dipole selection rules, electrons of one spin direction will preferentially be promoted to the fermi level according to whether right or left circular polarization is used. In the second step, the d-valence band serves as a spin detector. Due to spin-conversation the left-circularly polarized light probes mainly the larger number of unoccupied spin-up d states with respect to the direction of the magnetization M. Therefore, the absorption cross sections for the two helicities differ due to the imbalance of the spin-up and spin-down d valence band. The effect reverses at the $2p_{1/2}$ level (L_2-edge) due to the opposite spin-orbital coupling the core level (j=l-s) the spin polarized electrons have to find a place in the unoccupied 3d band, resulting in a difference if the 3d electron are spin polarized. (b) XAFS at the Fe $L_{2,3}$ for right (μ^+) and left (μ^-) circular polarization together with the difference spectrum: the XMCD.

preferred spin up (down) direction. In the absorption step, the 'spin-up' and 'spin-down' are defined relative to the photon helicity or photon spin, which is parallel (right) or antiparallel (left) to the X-ray propagation direction.

In the second step, the spin-polarized electrons are analyzed by a spin-resolving detector consisting of the exchange split d final state. The excited electron has to find its place in the unoccupied 3d valence band. As spin flipping is prohibited in electric dipole transitions, photoelectrons from the p-core shell can only be excited into d-hole states with the same spin orientation. So the spin-split valence band acts as an analyzer for the excited photoelectron spins, and the transition probability depends on the vacancy in the d band for a given spin. The size of the dichroic effect also relies on the angle θ between the photon spin and the magnetization direction through the relation cos(θ). Therefore, the dichroic effect reaches its maximum when the photon spin direction and the magnetization direction are parallel and antiparallel.

An XMCD spectrum of Fe metal is illustrated in Figure 11.1b, the differences of the white line intensities recorded with different X-ray helicity are denoted as A (L_3 edge) and B (L_2 edge), respectively. The opposite sign of A and B stems from the opposite spin-orbit coupling of the $p_{3/2}$ and $p_{1/2}$ levels. According to the sum rule below, the spin moment is quantitatively related to the area A-2B, and the orbital moment in the d shell is linked to the dichroism intensity A-B. Therefore, the number of d holes and the magnitude of the spin and orbital magnetic moments determine the L_3 and L_2 resonance intensities and their difference between the parallel and antiparallel directions of the photon spin and magnetization directions. The anisotropy of spin density and orbital moment can be determined by the angle-dependent measurements the external magnetic field. It is easy to show that the XMCD intensity is the difference of the spectra obtained for the two X-ray helicities with a fixed magnetic field is equivalent to that obtained by switching the magnetization direction with a fixed X-ray helicity.

11.2.1 Dipole transition - the XMCD sum rules

The magnetic moment $\langle L_z \rangle$ of ground-state orbital is linked to the integrated signal of the XMCD through the famous sum rule derived by

Thole *et al.* in 1992.[7] This sum rule for single-electron wave functions (Slater determinants, eigenstates of L_z) was intuitively demonstrated by Altarelli [8] using the properties of the squared 3j symbols, while a pictorial demonstration was provided in Ref. [9] One year later, Carra *et al.* [10] derived a second sum rule for XMCD, which gives the relation between the branching ratio of the XMCD and the spin magnetic moment $\langle S_z \rangle$ and magnetic dipole term $\langle T_z \rangle$ in the ground state. Carra *et al.* first reported the quadrupole moment sum rule for XMLD and also extended the results to electric-quadrupole transitions.[11]

A sum rule links the ground-state expectation value of some operator to a function of the integral of the XMCD spectra due to a given term of the cross section. In the case of XMCD, these operators coincide with the contribution of spin and orbit to the magnetic moment, thus providing experimentalists with a simple interpretative paradigm. The total intensity of the L_3 and L_2 resonances is proportional to the number N of empty d states (holes) in valence band. The *d* valence shell can hold up to 10 electrons, therefore the number of filled states is 10-N which are filled into band states up to the Fermi level. For magnetic materials, the spin moment of the d-shell is given by the imbalance of the electrons in the spin-up and spin-down, or equivalently (except for the sign) the hole imbalance in the spin-up and spin-down. The difference in the number of *d* holes with up and down spin can be measured through a spin dependent X-ray absorption process by using right or left circularly polarized photons, which transfer their angular momentum to the excited photoelectron. This effect requires the breaking of time-reversal symmetry and the presence of a spin-orbit interaction. The transition of the electric dipole (E1) and in some cases the quadrupole (E2) is the cause of the inner shell excitation.

The atomic theory of sum rules

First, the expression for the transition probability is separated into a dynamic and a geometric part. The former gives the radial part, while the latter gives the angular dependent part. The radial part is assumed to be constant over the region of the spectrum and hence can be factored out. The transition probabilities of the polarized spectra are then obtained using Fermi's golden rule. The integrated intensities ρ_q for q polarized light are obtained by summing the $c \rightarrow l$ transition probability over the final states f and the components γ and m of c and l, respectively, where for simplicity diagonal basis states are assumed to be $m' = m$, $\sigma' = \sigma$ and $\gamma' = \gamma$, as in SO_2 symmetry,

$$\rho_q = \Sigma_{f\gamma m\sigma}\langle g|l_{m\sigma}c^{\dagger}_{\gamma\sigma}|f\rangle\langle f|l^{\dagger}_{m\sigma}c_{\gamma\sigma}|g\rangle \begin{pmatrix} l & 1 & c \\ -m & q & \gamma \end{pmatrix}^2, \qquad (11.1)$$

The values $q = -1, 0, +1$ correspond to right-circularly, z, and left circularly polarized radiation, respectively. The relative intensities $lm \rightarrow c\gamma$ are given by the squared 3j-symbol denoted by the parentheses, which is nonzero for $= q + \gamma$, and $l^{\dagger}_{m,\sigma}(c^{\dagger}_{\gamma}, \sigma)$ represents the creation of the valence electron, while $l_{m,\sigma}(c_{\gamma}, \sigma)$ denote the annihilation operator of core electron.

Final states $|f\rangle$ that cannot be reached in the dipole transition have no contribution to the intensity. The core-hole operators can be removed by extending the set of final states to the whole Hilbert space and using the completeness relation, so the integrated signals depend only on ground state operators. Since $|c^{\dagger}\rangle = \langle c| = 0$, then

$$\Sigma_{f\gamma m\sigma}\langle g|l_{m\sigma}c^{\dagger}_{\gamma\sigma}|f\rangle\langle f|l^{\dagger}_{m\sigma}c_{\gamma\sigma}|g\rangle = \Sigma_{m}\langle g|l_{m\sigma}l^{\dagger}_{m\sigma}|g\rangle = \Sigma_{m\sigma}\, n_{m\sigma} \qquad (11.2)$$

where $n_{m\sigma}$ is the number of holes with indices m and σ. Substitution of (13.2) into (13.1) and recoupling the angular momenta by rewriting the squared 3j-symbol using theorem 4 of Yutsis, Levinson, and Vanagas (YLV4) [12] leads to

$$\rho_q = \Sigma_{m\sigma x}\, n_{m,\sigma}(2x+1) \begin{Bmatrix} l & x & l \\ 1 & c & 1 \end{Bmatrix} \begin{pmatrix} 1 & x & 1 \\ -q & 0 & q \end{pmatrix} \begin{pmatrix} l & x & l \\ -m & 0 & m \end{pmatrix}. \qquad (11.3)$$

The integrated intensities ρ^x of the fundamental spectra can be defined as linear combinations of the integrated intensities ρ_q of the primitive spectra,

$$\rho^x \equiv \Sigma_q(-1)^{1-q} \begin{pmatrix} 1 & x & 1 \\ -1 & 0 & 1 \end{pmatrix}^{-1} \begin{pmatrix} 1 & x & 1 \\ -q & 0 & q \end{pmatrix} \rho_q, \qquad (11.4)$$

where the superscript $x = 0$, 1, and 2 denote the isotropic spectrum and the circular and linear dichroism, respectively. This simply gives $\rho^0 \equiv \rho_1 + \rho_0 + \rho_{-1}$, $\rho^1 \equiv \rho_1 - \rho_{-1}$, and $\rho^2 \equiv \rho_1 - 2\rho_0 + \rho_{-1}$, For the $l = c + 1$ case (*e.g.*, p → d or d → f), substitution of (13.3) into (13.4) results in

$$\rho^0 = \frac{1}{2l+1} \sum_{m\sigma} n_{m\sigma} = \frac{n_l}{2l+1}, \tag{11.5a}$$

$$\rho^1 = \frac{1}{l} \sum_{m\sigma} n_{m\sigma} m = \frac{\langle L_z \rangle}{l}, \tag{11.5b}$$

$$\rho^2 = \frac{3}{l(2l-1)} \sum_{m\sigma} n_{m\sigma} \left[m^2 - \frac{l(l+1)}{3} \right] = \frac{3\langle Q_{zz} \rangle}{l(2l-1)}. \tag{11.5c}$$

which gives the number of holes n_h in the l shell, its orbital moment and charge quadrupole moment, respectively. In intermediate coupling and in the presence of hybridization, the occupation numbers have of course non-integer values.

The sum rules can be demonstrated by using the dipole transition $p^6d^n \rightarrow p^5d^{n+1}$ in a transition metal. The transition probabilities of the primitive spectra with different polarization are calculated using the square of the 3j symbol in Eq.13.1. Left-circularly polarized light (L) with selection rule q = −Δm = +1 gives transition probabilities of 1/30, 1/10 and 1/5 for excitation into the m = 0, 1 and 2 sublevel, respectively of the ground state d state to the γ = −1, 0, +1 level of the *c* core-hole. Using time-reversal symmetry, {L,±m} ↔ {R,∓m}, gives the transition probabilities for right-circularly polarized light (R). Z-polarized light (perpendicular to the plane of L and R) gives transition probabilities of 2/15, 1/10 and 0 for excitation into the |m| = 0, 1, and 2 sublevels, respectively.

The fundamental spectra are linear combinations of the primitive spectra using Eq.12.4. In the isotropic spectrum, the transition probability is constant, *i.e.*, independent of *m*. In the XMCD, the transition probability is linear in *m*. In the linear dichroism, the transition probability has the shape of a parabola, and is thus quadratic in *m*. For an arbitrary initial state, the integrated intensity is the sum over *m* and spin σ of the occupation number n_m times the transition probability. For the isotropic spectrum, XMCD and XMLD this gives $\langle n \rangle$, $\langle L_z \rangle$, and $\langle Q_{zz} \rangle$, respectively. Applying the Wigner-Eckart theorem to obtain the reduced matrix elements in the desired point group,[13] the sum rules can be extended to crystal field symmetry. The spin-involving Sum rules requires a separate integration over the intensity of spin-orbit split core levels. This demands that the angular momentum *j* of the core-hole is a good quantum number, which is fulfilled when the core spin-orbit interaction is much larger than the core-valence interactions. Taking the weighted difference over the spin-orbit split core levels the sum rule leads to

$$\delta^1 \equiv \sum_{j=c+\frac{1}{2}} \rho^1 - \frac{c+1}{c} \sum_{j=c-\frac{1}{2}} \rho^1 = \frac{2}{3}\left(\langle S_z \rangle + \frac{(2l+3)\langle T_z \rangle}{l}\right). \quad (11.6)$$

In the limit of small 3d spin-orbit interaction, the magnetic dipole term, $\langle T_z \rangle = [\sum_i s - 3\hat{r}_i(\hat{r}_i \cdot \hat{s}_i)]_z$, corresponds to the asphericity of the charge density of the *l* valence shell, and describes the difference between the charge quadrupole moment for spin down and spin up electrons, $T = \frac{1}{2}(Q^{\downarrow} - Q^{\uparrow})\hat{S}$. Since the quadrupole moment is a second rank tensor with vanishing trace, the contributions of T over three orthogonal directions should be zero.[14] Alternatively, one can determine $\langle S_z \rangle$ at the magic angle (54.7°), where $\langle T_z \rangle$ has to vanish.[15] $\langle T_z \rangle$ is often small in cubic 3d transition metals, but this is not the case in 4f and 5f metals.[16] For instance, in the XMCD at the uranium $M_{4,5}$ edges of a ferromagnetic uranium monosulphide crystal measured by monitoring the fluorescence signal over,[17] the contribution of $\langle T_z \rangle$ to the dichroism signal in uranium monosulphide crystal was found to be larger than that of $\langle S_z \rangle$.

Using these *A*, *B*, and *C* to respectively denote the appropriate integrals of L_3 XMCD, L_2 XMCD signals, and XAFS signal, neglecting the $\langle T_z \rangle$ term, and assuming that $\mu_0 = (\mu^+ + \mu^-)/2$ and d-orbital hole number is n_h, yields the following simple expressions for the sum rules:

$$\langle L_z \rangle = \frac{2(A+B)}{3C} n_h \quad (11.7a)$$

$$\text{and } \langle S_z \rangle = \frac{A-2B}{2C} n_h \quad (11.7b).$$

The calculated absorption spectra of 3d transition ions and their related magnetic moments based on multiplet theory have verified the sum rules validity. However, the appropriate ab initial spin-polarized relativistic band-structure calculation for 3d transition metals indicate that the values obtained from the sum rule analysis differ from the related magnetic moments. The orbital sum rules were found to be accurate within 10%. The spin sum rule can deviate between 15% up to 40% for the bulk ferromagnets Fe and Ni, respectively. The origin of the failure of the sum rules for itinerant magnets is related to the ambiguity in the choice of the integration range and the definition of the number of the d holes. Therefore, Wu and coworkers proposed to use another form of sum rules that avoid these issues by taking the ratio of the orbital and spin sum rules.

$$\frac{\langle L_z\rangle}{2\langle S_z\rangle+7\langle T_z\rangle}=\frac{2}{3}\frac{(A+B)}{(A-2B)}. \qquad (11.7c)$$

Sum rules from multipole moment analysis

To describe the asymmetric XMCD line shape of 3d transition metals, van der Laan introduced the multipole moment analysis concept. The magneto-optical sum rules relate areas to certain ground-state moments, and the spectral distribution is linked to the certain ground-state properties of the 3d valence band. The latter can be characterized by the multipole moments$\langle w^{xyz}\rangle$, where the orbital moment x and the spin moment y are coupled to a total moment z.[18] Moments with even and odd x respectively describe the shape of the charge distribution and an orbital motion. An underscore signifies that the moment is taken not over the electrons but over the holes, which is required for X-ray absorption. Thus the number of holes n_h, the spin-orbit coupling $\sum_i l_h\cdot s_h$, the quadrupole moment Q_{zz}, the orbital magnetic moment L_z, the spin magnetic moment S_z, the 'magnetic dipole' term T_z can be expressed as

$$\langle \underline{w}^{000}\rangle = n_h,\ \langle \underline{w}^{110}\rangle = -\textstyle\sum_i l_h\cdot s_h\ ,\ \langle \underline{w}^{202}\rangle = -Q_{zz}/2\ ,$$
$$\langle \underline{w}^{101}\rangle = L_z/2,\quad \langle \underline{w}^{011}\rangle = 2S_z\,,\ \langle \underline{w}^{211}\rangle = 7T_z/2$$

where $\langle \underline{w}^{xyz}\rangle \equiv (-1)^z$ is normalized per hole. The isotropic XAFS (a=0) and dichroic (a=1) signal intensity of the jm final states can be expressed as a linear combination of ground-state moments $I^a_{jm} = \sum_{xyz}\langle \underline{w}^{xyz}\rangle \sum_r C_j^{xyzar}\,\mu^r_{jm}$, with

$$\mu^r_{jm} = (-1)^{j-m}(2r+1)\begin{pmatrix} j & r & j \\ -j & 0 & j\end{pmatrix}\begin{pmatrix} j & r & j \\ -m & 0 & m\end{pmatrix} \qquad (11.8).$$

So the coefficients C_j^{xyzar} is the probability of creating a core hole with multipole moment r in the j level for unit ground state moments $\langle \underline{w}^{xyz} \rangle$, can be derived from the transition matrix element.

Table 11.1 The coefficients $\sum_r C_j^{xyzar} \mu_{jm}^r$ of the core hole j for the unit ground state moments $\langle \underline{w}^{xyz} \rangle$ in the p-d isotropic and dichroic absorption [G. van der Laan, (1997) J. Phys.: Condens. Matter 9, L259–L265].

$\sum_r C_j^{xyzar} \mu_{jm}^r$	$\langle \underline{w}^{000} \rangle$	$\langle \underline{w}^{110} \rangle$	$\langle \underline{w}^{101} \rangle$	$\langle \underline{w}^{011} \rangle$	$\langle \underline{w}^{211} \rangle$	$\langle \underline{w}^{202} \rangle$
Isotropic a=0						
j=3/2	$2\mu^0$	μ^0	$5\mu^1/3$	$10\mu^1/9$	$2\mu^1/9$	μ^2
j=1/2	μ^0	$-\mu^0$	μ^1	$-\mu^1/3$	$-2\mu^1/3$	-
Dichroic a=1						
j=3/2	$5\mu^1/9$	$4\mu^1/9$	$2\mu^0 + 2\mu^2/5$	$\mu^0/3 + 2\mu^2/3$	$2\mu^0/3 + 2\mu^2/15$	$10\mu^1/9$
j=1/2	$\mu^1/3$	$-\mu^1/3$	μ^0	$-\mu^0/3$	$-2\mu^0/3$	$2\mu^1/3$

The integration both edges of the XAFS and XMCD spectrum gives the number of 3d holes and the orbital moment, respectively. As the other ground-state moments have either an asymmetric distribution or opposite contributions at the L_3 and L_2 edges, the spin moment is obtained by integrating over weighted $L_{3,2}$ differences. Accordingly, the orbital and spin sum rules can be derived using $\int \mu_{jm}^r = \delta_{r0}$ and C_j^{xyzar} from Table 11.1.

$$N = \int_{j^+ + j^-} I_{jm}^0 d\omega = 3 \langle \underline{w}^{000} \rangle = 3n_h \, ,$$

$$\frac{3n_h}{N} \int_{j^+ + j^-} I_{jm}^1 d\omega = 3 \langle \underline{w}^{101} \rangle = \frac{3}{2} \langle L_z \rangle \, ,$$

$$\frac{3n_h}{N} \left(\int_{j^+} I_{jm}^1 d\omega - 2 \int_{j^-} I_{jm}^1 d\omega \right) = \langle \underline{w}^{001} \rangle + 2 \langle \underline{w}^{211} \rangle = 2\langle S_z \rangle + 2\langle T_z \rangle . \tag{11.9}$$

As the $L_{3,2}$ edges partly overlap in the XMCD spectra of the early 3d transition metals, the spectral shape can be described with the multipole moment analysis in order to avoid the deconvolution issue. However, the strong core-hole correlation effects in XAFS and XMCD spectra of early 3d metals lead to the breakdown of the spin sum rule, which make the results of the multiple moment analysis highly suspicious.

Energy dependence of the ground state moments

The sum rules can also illustratively demonstrated from the perspective of the ground-state moments [19] The $L_{2,3}$ XMCD spectrum can be decomposed into its contributions from the separate ground-state moments.[20] The shape and magnitude of the XMCD are mainly determined by n_h and S_z. However, only for L_z, the XMCD has a non-zero intensity integrated over both edges. For S_z and T_z the XMCD does not vanish over each separate edge but integrated over both edges the intensity is zero. For n_h and the spin-orbit interaction, $l \cdot s$, the integrated intensity of XMCD vanishes for each edge, but these still contribute to the energy dependence across each edge. This is also the case for the higher moments, such as Q_{zz}. Hence, the integrated intensities over the L_3 and L_2 edges are related to the orbital and spin moments, but also that they do not depend on the other ground-state moments. Thus, a correct differential formulation of the sum rules needs to include all ground state moments!

Possible causes of concerns

In history, the validity of sum rules has been questioned in many ways. (1) As reinforced by some standard solid-state textbooks, the 3d metals are spin-only magnets which have no orbital magnetic moment. Actually, in 3d metals, the orbital part to the magnetic moment is small, it is manifestly present in, *e.g.*, the magnetic anisotropy.[21] (2) In intermediate coupling, there might be cross terms, leading to interference terms. However, the sum rules can also be derived for *jj*-coupled operators. By including cross terms between the $j = l + \frac{1}{2}$ levels, it could be shown that the sum rules are generally applicable in intermediate coupling.[22] (3) Furthermore, in an atomic model the sum rules are exact as long as *jj* coupling can be neglected. Large deviations due to *jj* coupling arise when the L_3 and L_2 start to overlap, such as in the Mn $L_{2,3}$ spectrum, which can be taken into account by a correction factor.[23] (4) The sum rules are valid for the atomic case but not for the band structure case, or vice versa, valid for the band structure case but not for the atomic case. However, several theorists successfully rederived the sum rules for the band structure case, such as Ankudinov and Rehr [24] using the independent electron approximation, Ebert [25] using first-principles spin-polarized relativistic multiple-scattering calculations, and Wu *et al.* [26] using local density band structure calculations. (5) Hybridization effects could ruin the sum rule. However, Benoist *et al.* [27] derived corrections to the atomic orbital sum rule for XMCD in solids using orthonormal linear muffin-tin orbitals (LMTOs) as a single-particle basis for electron band states. Nowadays the accuracy about the sum rules is repetitively discussed. Under the proper assumptions (such as the absence of *jj* coupling) the derivation of the sum rules is exact. The practical limitations of the sum rules are well known (see, e.g., Refs. [28,29]) and an error bar of 5-10% is nowadays commonly accepted.

The various sources for errors and deviations of this widely used XMCD tool were reviewed in details by Piamonteze *et al.*[30] The effective spin sum rule error is linearly related with the final state effects of the core level spin-orbit coupling and the core-valence exchange interactions (multiplet effects). In the ligand field multiplet method, despite the analysis of these effects, combined with the interactions that affect the magnetic ground state (including crystal field strength and 3d spin-orbit coupling), it is found that for the late transition metal systems, the error in the effective spin moment is between 5 and 10%. Due to the potentially large $\langle T_z \rangle$ value, it is impossible to reliably determine the spin moment of all systems except Ni. The error for $3d^4$ systems is very large, which means that in the absence of more information, the resulting effective spin and regular values of the $3d^4$ system have no meaning.

However, before applying the law to quantify orbital and spin moments, some facts must be kept in mind. In 3d transition metals, due to the Coulomb interactions in the final state, the spin-orbit splitting of the $L_{2,3}$ edges are not large enough to prevent their mixing. This makes it impossible to accurately separate the two partners, but the effect becomes smaller (~ 5%) on approaching the end of the 3d series. Therefore, in order to apply spin and rule, L_3 and L_2 edges have to be sufficiently separated. Background subtraction should separate 2p → 3d from other transitions. Additionally, $\langle T_z \rangle$ may be important in low-dimensional systems.

The *L*-edge X-ray absorption spectra contain the contributions from both p→d and p→s transitions, but in practice the p→d channel dominates by a factor >20.[31] The sum of the white line intensities, denoted I_{L3} and I_{L2}, respectively, is directly proportional to the number of d holes. According to the dipolar transition selection rule $\Delta l = \pm 1$, $L_{2,3}$ edges probe both unoccupied s as well as d states. The s orbital contribution is neglected for 3d metals. Similarly, $M_{4,5}$ edges for 4f metal probes the unoccupied f and d states, but one can neglect the d orbital contribution.

Concerning the application of the sum rules for the $L_{2,3}$-edges, the most important facts we should keep in mind are:

- the derivation of the sum rules is restricted to dipole allowed transitions;
- it ignores the exchange splitting for the core levels;
- it neglects the differences between $d_{3/2}$ and $d_{5/2}$ wavefunctions;
- it ignores any energy dependence of the wavefunctions.

11.2.2 Quadrupole transition

Electric quadrupole transitions (E2) are introduced to explain the structure below the 'white line' edge in the XMCD spectra, taken at the $L_{2,3}$ edges of rare-earth compounds,[32] in the energy range 7-10keV (hard X-rays). In these systems the quadrupole peak (2p→4f transitions) appears below the edge because of the strong Coulomb interaction, in the final state, between the 2p core hole and the 4f electrons (excited state). Given a single ion in an arbitrary ground state $|\Psi\rangle$, consider the $c \rightarrow l$ electric 2L-pole transition (EL) of a single partner, $j_{\pm} = c \pm 1/2$ of a spin-orbit split edge, the total intensity of the XMCD at rare earth $L_{3,2}$ edges can be written.

$$\int_{j_{\pm}} \{(\mu^+ - \mu^-)_{E1} - (\mu^+ - \mu^-)_{E2}\}\, d\omega \cong$$
$$\frac{4\pi^2 N_A \alpha \hbar\omega_0 \cos\theta}{45A}\left\{\left[\frac{3}{4}[j_{\pm}]\langle L_z\rangle_{5d} \pm 2\langle S_z\rangle_{5d} \pm 7\langle T_z\rangle_{5d}\right] P^2_{2p\rightarrow 5d} + \right.$$
$$\frac{1}{84}\left(\frac{\hbar\omega_0}{\hbar c}\right)^2\left[[j_{\pm}]\langle L_z\rangle_{4f} \pm 4\langle S_z\rangle_{4f} \pm 12\langle T_z\rangle_{4f} + \frac{(5\cos^2\theta - 3)}{5}\left(\frac{[j_{\pm}]\langle O_{zzz}\rangle_{4f}}{2} \pm \right.\right.$$
$$\left.\left.\left.\frac{12\langle M_{zzz}\rangle_{4f}}{7} \pm \frac{16\langle N_{zzz}\rangle_{4f}}{21}\right)\right] P^2_{2p\rightarrow 4f}\right\}. \tag{11.10}$$

The operators

$$O_{zzz} = \frac{1}{2}\Sigma_i[(3l^2 - 5l_z^2 - 1)l_z]_i\,,$$
$$M_{zzz} = \frac{1}{2}\Sigma_i[(5l_z^2 - l^2)s_z - [l_z, l\cdot s]_+]_i\,,$$
$$N_{zzz} = \frac{1}{8}\Sigma_i\left[\frac{35}{4}[l_z^2, [l_z, l\cdot s]_+]_+ + 3(l^2 - 2)l^2 s_z - \right.$$
$$\left.\frac{5}{2}(6l^2 - 5)\left(l_{zs_z}^2 + \frac{1}{2}[l_z, l\cdot s]_+\right)\right]_i\,,$$

where $[\cdots]_+$ presents the anticommutator. $\cos\theta = \vec{k}\cdot\vec{z}$, where $\vec{k}$ and $\vec{z}$ are the unit vectors in the direction, respectively, of the photon momentum and local magnetization. These operators describe octupolar moments, with angular dependence $(5\cos^2\theta - 3)\cos\theta$; details of their derivation please refer to ref.[33] The action of a small crystal field on the 4f electrons will result in a scaling of the expectation values of the operators; in general, with a different scaling factor for each class of operators of total symmetry z. Experimentally, the angular dependence can be utilized to separate dipolar and octupolar operators: for θ=39.23°, only the dipolar contributions survive in the above expression.

At the $j_\pm$- edge and for l =c+2 quadruple transitions, let Q_3 and Q_2 be the integral arera of quadrople components at L_3 and L_2 edges, respectively, then the expected values of $\langle L_z\rangle_{4f}$ and $\langle S_z\rangle_{4f}$ can be expressed as

$$\langle L_z\rangle_{4f} = \frac{1}{K_f}\left[Q_3 + \left(\frac{\omega_3}{\omega_2}\right)^3 Q_2\right] - \frac{1}{10}\langle O_{zzz}\rangle_{4f}(5\cos^2\theta - 3), \quad (11.11a)$$

$$\langle S_z\rangle_{4f} = \frac{1}{2K_f}\left[Q_3 - 2\left(\frac{\omega_3}{\omega_2}\right)^3 Q_2\right] - 3\langle T_z\rangle_{4f} - \frac{1}{35}\left(3\langle M_{zzz}\rangle_{4f} + \frac{4}{3}\langle N_{zzz}\rangle_{4f}\right), \quad (11.11b)$$

with

$$K_f = \frac{2N_A}{A}\frac{\pi^2\alpha\hbar\omega_3}{315}\left(\frac{\hbar\omega_3}{\hbar c}\right)^2 P^2_{2p\to4f}\cos\theta$$

$$P^2_{2p\to4f} = \left|\langle 4f|C^L|2p\rangle R_{2p\to4f}\right|^2.$$

where C^L is the normalized spherical harmonic operator, $R_{2p\to4f}$ the radial matrix element of the $c\to l$ transition. N_A is the Avogadro number, A the atomic weight and α=1/137. In atomic units, $\hbar = 1$, c=2/α and $\hbar\omega_0$ is the energy of the edge in Rydbergs. This result can be employed to interpret the electric quadruple contribution in $L_{2,3}$-XMCD spectra of rare-earths. However, it still cannot be directly applied to the experiment, as the mixing of E1 and E2 spectra makes the normalization of the data (the removal of the radial parts) impossible. This drawback can be overcome with the help of numerical calculations. The E2 contribution is determined by the highly localized 4f states, which could be well described by atomic theory, for example, calculated with Cowan's multiplet program. These results fully determine the magnitude of the quadrupole part, which can be subtracted from the integrated experimental data. As described in the previous section, the remaining El contribution can then be normalized to determine $\langle L_z \rangle$, $\langle S_z \rangle$ and $\langle T_z \rangle$ of the 5d electrons.

11.2.3 K- and L_1- edge XMCD

Through the magneto-optical sum rules, the spin and orbital contribution to the magnetic moment can be obtained directly from the integral of the XMCD spectra, which has become an indispensable technique for studying the magnetism of matter.[34] These sum rules are widely and successfully applied at spin-orbit split $L_{2,3}$ edges of transition metals [35,36,37] and $M_{4,5}$ edges of actinides.[38] On the other hand, in the absence of spin-orbit splitting of the core state like for the K or L_1 edge, only the orbital magnetization sum rule applies, [39,40] and a quantitative analysis of the spectra is far from being straightforward.

At the K edge, the sum rule for the electric dipole-electric dipole term [41,42] relates the integral of the XMCD spectrum to the orbital magnetic moment of occupied p states that is proportional to $\langle L_z \rangle p$. Consider an ion with an incomplete outer shell with angular momentum l, and let l^n be the ground-state configuration, any state of the final configuration cl^n, the integral of the XMCD spectra to the orbital magnetic moment of occupied p states that is proportional is related to $\langle L_z \rangle$ through the sum rule for the electric dipole-electric dipole term,

$$\rho \equiv \frac{\int_{edge}(\mu^+ - \mu^-)d\omega}{\int_{edge}(\mu^+ + \mu^0 + \mu^-)d\omega} = \frac{c(c+1) - l(l+1) - 2}{2l(l+1)(4l+2-n)} \langle L_z \rangle. \qquad (11.12)$$

Here μ^+, μ^- and μ^0 denotes the absorption coefficient obtained with right, left circular and linear polarized light, respectively. As ρ is a quantity directly measurable from these polarization-dependent core-level spectra, the ground-state expectation value $\langle L_z \rangle$ per hole can be determined though the above expression. However, since the upper limit of the integral is not well defined and, in the case of 3d transition elements, the 4p states are almost unoccupied, $\langle L_z \rangle p$ is very small and has a minor impact on the magnetic moment of the material, it is almost impossible to apply this sum rule in practice.[43] However, for 3d transition elements, measuring XMCD at the K-edge is the main method of detecting magnetism under pressure. Despite the difficulty of interpretation, XMCD is a widely used technique.[44,45,46]

11.2.4 The $M_{4,5}$- edges of rare earths

The rare earth systems feature an almost pure *LSJ* coupling of Hund's rule ground state. In this case $\langle T_z \rangle$ can be evaluated analytically.[47] For $n \leq 2l+1$,

$$\langle T_z \rangle = \langle M \rangle (l-n+1/2) \frac{3(S-J)^2(S+J+1)^2 - L(L+1)[L(L+1)+2S(S+1)+2J(J+1)]}{2(2l+3)(2l-1)(2L-1)SJ(J+1)}. \tag{11.13a}$$

The case $n \geq 2l+1$ is accounted for by $n \rightarrow 24l+2-n$. For $l=0$ configurations (n =0, $2l$+ 1, and $4l$ + 2): $\langle T_z \rangle = 0$, as S=J. Also, Lande

$$\langle S_z \rangle = \langle M \rangle \frac{J(J+1)+S(S+1)-L(L+1)}{2J(J+1)}. \tag{11.13b}$$

Within an *LSJ* term, the $\langle S_z \rangle$-to-$\langle T_z \rangle$ ratio is constant and the two contributions can be separated. In the 5d ions, a strong spin-orbit coupling (the spin-orbit parameter $\xi_d = -2\ eV$, HF) makes $\langle T_z \rangle$ rather large. In this case, further experimental information and/or calculation are required to separate $\langle S_z \rangle$ and $\langle T_z \rangle$.

11.3 Experimental Setup

X-ray Magnetic Circular Dichroism (XMCD) uses the differential absorption of left and right circularly polarized light in a magnetic field to examine magnetic materials and properties. Synchrotron radiation can offer a number of unique advantages in XMCD measurement based on the following principles: (a) Synchrotron radiation can provide highly circularly polarized monochromatic light source; (b) X-ray spectroscopy and scattering can be made sensitive to the magnetic moments by using the strong polarization dependence of the electric dipole transitions; (c) Element-specificity is obtained by tuning the X-rays to the resonance energy of the core-to-valence transition; (d) Electric-dipole transitions from the ground state reach only a limited subset of final states, thereby providing a fingerprint for the specific ground state; (e) Spin and orbital properties can be separated, as the light only interacts on the orbital part of the wave function; (f) Instrumental resolution is of a similar order of magnitude as the core-hole lifetime (typically a few hundred meV) allowing resolution of the multiplet structure and charge-transfer satellites.

With the third-generation synchrotron radiation sources, it is feasible to perform magnetic imaging using X-ray techniques at a sub-micrometer level and record spectrum at nanoseconds. By combining X-ray microscopy or X-ray photoelectron microscopy with spectroscopy, magnetic domain structures can be imaged with a lateral spatial resolution of a few 10 nm. In a new development of lens-less imaging by Fourier transform X-ray holography, the diffraction pattern of a coherently illuminated specimen can be recorded in Fourier space. These methods provide a key technique for research on small structures important in microelectronics, which are often heterogeneous and composed of different elements. With the temporal structure of the synchrotron radiation, the dynamics of domain-wall displacements and transformation or dynamics of the magnetization of mesoscopic magnetic structures can be investigated in element-sensitive time-domain ways, which are of prime importance for magnetic recording.[48] The underlying processes occur at times in the nano- to femto-second range.

A further advantage of insertion devices is the possibility to produce X-rays with variable linear, elliptical and circular polarization. This allows for the analysis of the magnetic behaviors of the samples, *e.g.* with the XMCD technique. The APPLE II (Advanced Planar Polarized Light Emitter-2) type helical undulator is an example of such an insertion device (Figure 11.1). An APPLE-II undulator consists of four identical arrays of permanent magnets arranged in four quadrants around the axis with two rows of planar permanent magnets above and below the electron orbit plane. Two arrays (upper right and lower left) are movable while two others are fixed. In the APPLE-II undulator, the variation of the phase is accomplished by translating movable magnetic rows longitudinally in the same direction. By shifting the opposing magnet rows relative to each other the magnetic field can be modified, producing various types of electron orbits such as vertical sinusoidal, horizontal sinusoidal or helical motion. The gap-scan technique also for these undulator types ensures a constancy of the degree of circular polarization in combination with a high photon flux. For zero shift the APPLE-II provides linear horizontal polarization and resembles a usual planar pure magnet undulator. Moving longitudinally in parallel mode two arrays it is possible to change the magnetic field from horizontal to vertical. With a quarter wavelength shift the 1st harmonic provides 100% circular polarized radiation. Note that not only the helicity but also the spectrum of APPLE-II undulator depends strongly on the value of phase. Any changes of the undulator gaps and/or phases do not affect the position and the angles of the X-ray beam emitted by undulators. Additionally, the closed orbit and the dynamics of the electron beam in the storage ring are also unperturbed by the gaps or phase changes and the use of the undulators is fully transparent for any other beamline around the storage ring.

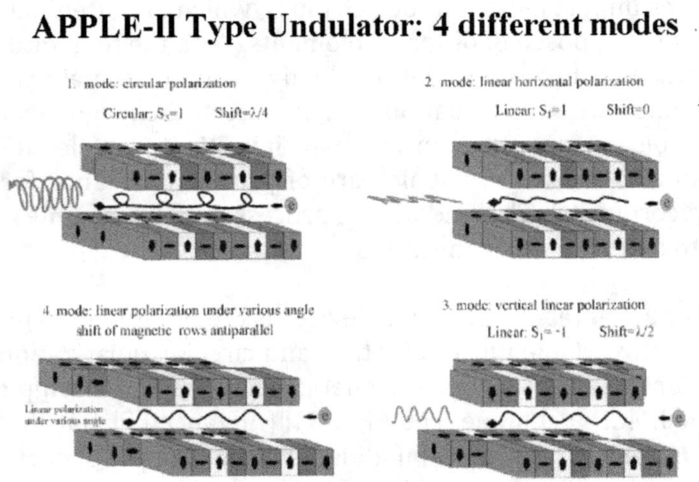

Figure 11.2 Apple-II type undulator: 4 different modes.

As one of the main tools to study magnetism, XMCD benefits from the rapidly variable helicity of circularly polarized X-ray photons. Recently, the alternating frequency between the right-handed and left-handed photons can be increased to the MHz level, which is three orders of magnitude faster than the existing technology.[49] The method is based on a twin elliptical undulator installed in an electron storage ring. The undulator is operated in a novel mode in which the electron optics is tuned close to a resonance with electrons captured in transverse resonance island buckets. Propagating through double undulators, electrons from different islands emit photons of the same wavelength but opposite helicity. Switching between ultrafast emissions of monochromatic circularly polarized X-rays with opposite helicity is shown within a few nanoseconds. In a proof-of-principle experiment at BESSY II, XMCD at the $L_{2,3}$ absorption edges of Ni was demonstrated with an 800 ns helicity flip. Rapidly switching the helicity of polarized X-rays is desirable for studying magnetic dynamics via XMCD spectroscopy.

The monochromator is the most critical component of any beamline designed for spectroscopy applications. This is because the quality of the spectra can be dramatically spoiled by a poor energy resolution, by the transmission of unwanted harmonics or by instabilities of the exit beam during energy scans, *etc.* The monochromator often equipped with a pair of Si < 111 > crystals is utilized for hard X-ray, which the grating monochromator is often used for soft X-ray. Note that monochromator is often kept at a fixed temperature using a cryogenic cooling system developed in-house and which proved to be totally free of undesirable vibrations.

A major concern in circular dichroism experiments is the determination of the circular polarization rate of the monochromatic beam, which depends not only on the polarization of the undulator radiation but also on the polarization transfer by all optical components including the monochromator. In fact, the monochromator does not preserve the initial degree of polarization of the undulator beam, except for pure σ or π linearly polarized beams. In the case of Si < 111 > monochromator, high circular polarization rate of the X-ray undulator beam (≥ 97%) drops down to very low values at photon energies around ca. 2.8 keV: this implies that XMCD measurements should be most difficult in the corresponding energy range. Moreover, when the helicity of the incident X-ray beam is reverted, indeed the circular polarization rate of the monochromatic X-ray beam is altered, but there is also a quite significant change affecting the Stokes-Poincare component P'_2, *i.e.* the linear polarization rates at 45° and 135° respectively, whereas Stokes-Poincare component P'_1 remains totally unaffected. This can have dramatic consequences for all types of circular dichroism measurements (*i.e.* XMCD as well as XNCD) on biaxial crystals because the weak XNCD signal can often be masked by a much stronger linear dichroism signal. Similar difficulty is also to be expected for circular dichroism measurements on uniaxial crystals, *i.e.*

whenever the optical axis of the sample is not strictly parallel to the direction of propagation of the X-rays.[50] Therefore, it is highly desirable to perform X-ray polarimetry measurements by combining a quarter-wave plate with a linear polarization analysis. X-ray phase plates exploit the birefringence of perfect crystals under the conditions of Bragg diffraction, the refraction indices are different for the σ and π polarization components.[51] To fully characterize the polarization state of the monochromatic beam, it is most convenient to insert downstream with respect to the monochromator a UHV compatible quarter wave plate chamber equipped with a diamond single crystal.

Fluorescence detectors such as Si drift detector are often utilized for hard X-ray XMCD data acquisition, single- or multi-anode ion implanted Si P^+NN^+ photodiodes can be used for all beam intensity monitors or beam position monitors. The very high efficiency, the fast response time and the excellent linearity make photodiodes the best detectors for X-ray spectroscopy. For the soft X-ray XMCD signals are often collected in electron yield mode, for example, using microchannel-plate multiplier near the sample.

The simplest samples for XMCD experiments are permanent magnets, such as the domains in disk storage devices. These samples can be magnetized separately from the XMCD measurement.[52] Some thin film samples can be magnetized in situ with a pulsed magnetic field[53,54] Ferromagnets are also easy to study, since only a small field need be applied. NdFeB alloy permanent magnets are still used in some experiments to polarize the sample. The field is switched by manually reversing the external magnet. Electromagnets make this process easier to automate, and several groups have used them in a variety of ways.[55] For example, an eight-pole electromagnet system recently built by Arenholz,[56] in combination with a sample manipulator that allows 180$_\circ$ polar and 360$_\circ$ azimuthal rotation of the specimen, permits any geometry of magnetic field, sample surface, and incident photon direction. Fields of 1-2 T are generally employed for conventional XMCD experiments. However, a high magnetic field could be the essential parameter to explore new phenomena in materials with important magnetic degrees of freedom. For such a purpose, the external magnetic field has to be comparable with the magnitude of the basic magnetic interaction energies. This typically requires magnetic fields of the order of several 10 T. For instance,

ID12@ESRF XMCD beamline is equipped with a high-vacuum (< 10^{-7} mbar) 52 mm cold bore solenoid superconducting magnet producing a horizontal magnetic field of 17 teslas.

Lower specimen temperature can help reduce the thermal disorder, which can benefit the XMCD signal. Full magnetization of the sample at 2–3 K can be readily achieved with pumped ^{4}He cryostats, and a temperature of ~0.5K can be reached by using a ^{3}He insert.[57] Some instruments also employ a ^{3}He–^{4}He dilution refrigerator.[58] Since XMCD is essentially a measurement of relative absorption coefficients, all of the methods used for conventional XAFS can in principle be applied. In practice, the three most important modes are transmission, fluorescence, and electron yield. The preferred mode depends on the concentration and spatial distribution of the element under investigation, as well as the fluorescence yield and the matrix absorption coefficient.

In order to effectively extend the capability of the XMCD technique, the current developments in the XMCD technique at the third-generation synchrotron radiation facility mainly focus on (a) significant improvements in the accuracy and sensitivity of XMCD measurements with a quick helicity switching of circularly polarized X-rays; (b) introduction of extreme conditions such as strong pulsed magnetic field and high pressures; and (c) new experimental setup using the high-brilliance X-ray source including microfocusing, reflection geometry, and high-pressure cells. With these achievements, as a unique magnetic probe with an element-specificity, XMCD will provide fundamental scientific insights in magnetic nanoparticles/moleculars, spintronics and magnetic recording devices, and magnetism under high-magnetic field and high pressures.

11.4 XMCD Applications

XMCD has had a major impact on our understanding of the physics of 3d transition metal, lanthanide and actinide systems. It is the only element-specific technique that can distinguish between the spin and orbital part of the magnetic moment. XMCD is used in many areas, such as engineering (magneto-electronics), physics (magnetic topological insulators, dilute magnetic semiconductors) chemistry (organo-metallic compounds), earth sciences (spinels, geomagnetism) and life sciences (metalloproteins, biomagnets).

11.4.1 3d transition metal

In the case of transition metals such as iron, cobalt, and nickel, the absorption spectra for XMCD are usually measured at the *L*-edge. For transition metal $L_{2,3}$-edges, which have an initial 2p level, the transitions are primarily to states of 3d-symmetry, since the 2p → 4s transitions are ~20-fold weaker than the 2p → 3d transitions.[59] Because the 3d electron states are the origin of the magnetic properties of the elements, the spectra contain information on the magnetic properties. From a spintronic perspective, the 3d transition metal $L_{2,3}$ absorption edges are probably the most important edges, however, their dichroism was only first measured in 1990 mainly due to the poor accessibility of the soft X-ray range below 800 eV. Fortunately, improvements in the performance of grating monochromators, especially the design of the double-headed Dragon monochromator for soft X-ray circular dichroism studies, opened up this new arena.[60]

Kao *et al.*[61] measured the specular reflectivity of p-polarized light near the Fe $L_{2,3}$ absorption edges from a single-crystal iron film with an external magnetic field perpendicular to the scattering plane. Large changes in reflectivity were observed upon reversal of the direction of the magnetic field. This resonant magnetization-sensitive effect can be attributed to the interference between magnetic and nonmagnetic contribution to the resonant scattering. Similar effects can be expected in other ferromagnetic and ferrimagnetic systems. The reflectivity and scattering spectra at resonance contain both the real imaginary part of the scattering cross section, which makes the analysis more complicated compared to absorption spectra.[62,63] Chen *et al.* measured the XMCD at the $L_{2,3}$ edges (2p → 3d transitions) of Ni;[64] followed by that of Co and Fe ferrites and garnets.[65] The measurements were performed using the Dragon monochromator at the National Synchrotron Light Source in Brookhaven. The transitions from the 2p core level take place directly to the empty 3d states, which are magnetically polarized, hence giving a strong dichroism. In the Ni L_3 edge, the XAFS shows a satellite at 6.3 eV, whereas the XMCD shows a satellite at 3.7 eV. This could be explained with multielectron calculations including configuration interaction on the basis of the Anderson impurity model, which shows that the spectrum contains a main peak p^5d^{10} and a satellite peak p^5d^9. [66] Koide *et al.* [67] measured the XMCD in the 3d-metal $M_{2,3}$ edges (2p → 3d transitions) which shows a strong Fano line shape. The experimental

XMCD of Ni metal is quite different that predicted by the exchange-split-valence band model, but it could be explained by multielectron calculations.[68]

XMCD can also be employed to study the origin of magnetism and anisotropy in ferromagnetic semiconductors, such as in $(Ga_{1-x}Mn_x)As$.[69] While most applications have been in solid-state physics and technology, more recently the field of research has broadened considerably.

In the study of a nanoscale nonlocal spin injection structure consisting of a perpendicular polarizer and an in-plane free layer,[70] XMCD measurements were performed with an operating current to the sample. It was found that the free layer magnetization tilts to the perpendicular direction owing to injection of perpendicular spin. A photoelectron collection method utilizing an order-sorting-aperture is developed for this measurement. This order-sorting-aperture collection method enables XMCD measurement on a sample under current application, and its measurement quality is comparable to that recorded with the conventional total electron yield method.

The interface between magnetic materials and heavy metals with large spin orbit interactions is the source of various spin-related phenomena. Recently, through the nonlinear anomalous Hall Effect in Pt/Cr_2O_3 bilayers, the special spin-dependent transport at a Pt/Cr_2O_3 interface was investigated.[71] The XMCD results show no appreciable magnetic moment at the interface originating from Cr 3d and Pt 5d orbitals, which could be associated with the anomalous Hall Effect response. Based on the detection limit of the XMCD measurements, a possible interfacial magnetic moment M at the Pt/Cr_2O_3 interface, yields an extraordinary large anomalous Hall conductivity (σ_{AHE}) per unit net magnetic moment (M) ($-\sigma_{AHE}/M$, of 0.57 V^{-1}), as compared with that for general magnetic materials. Together with first-principles calculations, the results suggest the intrinsic anomalous Hall Effect in the Pt/Cr_2O_3 interface might not rely on the net magnetic moment.

XMCD spectroscopy was utilized to investigate the isostructural correlation-driven metal–insulator transition of V_2O_3 doped with 1.1% Cr near room temperature in its paramagnetic state. [72] According to the magnetometer and literature, the magnetic moment per mass value is about 1 J/(T·kg), and a relative XMCD amplitude of about 2 permille is observed at the vanadium $L_{2,3}$ absorption edge. During the transition from metal to insulator, the spectral shape of vanadium XMCD changed significantly. According to atomic multiplet simulations, these changes are caused by a changing orbital occupation, which indicates a changing phase composition. According to the estimates used in this study, the dipole moment of the spin density distribution 7<Tz> in the bulk increases, so that the effective vanadium spin moment increases by a few percent with temperature in the two-phase region. Therefore, it partially compensates for the reduction in relative XMCD amplitude due to the reduction in paramagnetic moment alignment. After several smaller temperature cycles, the sample is in a two-phase state, where the XMCD and X-ray linear dichroic spectra are almost independent of temperature, and the specific electric resistance is in the intermediate state, showing only the weak signal of the metal-insulator transition.

Cautions must be taken in the XMCD analysis for the light transition metals. The 2p spin–orbit splitting is significantly reduced for the early 3d transition metals, and the decreased spin–orbit splitting results in stronger electron core–hole correlations. Additionally, quantum mechanical mixing of $j_{3/2}$ and $j_{1/2}$ excitations emerges and reduces the observed XMCD related spin and magnetic dipole term contributions. All these result in dramatic changes in the spectral intensities and prevent the multipole-moment analysis for XMCD spectra and the direct application of XMCD spin sum rules. The induced magnetic moments in ultrathin films of the light 3d elements (Ti, V and Cr) at the interface to Fe cannot be directly determined by the XMCD sum rule or multipole-moment analysis at the $L_{2,3}$ edges.[73] This is due to the correlation effects which cause the intensity ratio (branching ratio) deviate from its statistical value. The fine structures for the Fe/V system can be modeled theoretically by means of fully relativistic *ab initio* calculations that also provide corresponding magnetic moments for vanadium. A large number of 2p→3d absorption spectra have been fitted nearly perfectly by a simple and phenomenological model, which takes into account lifetime effects and provides quantitative information about *jj*-mixing at the light transition metals. On the basis of this mixing coefficient, sum rule correction factors have been determined. The proposed model results in renormalized magnetic projected XMCD spin moments, verified for different compounds of V, Cr, and Mn.[74]

11.4.2 4f rare earth

A class of materials of particular scientific interest are the actinide metals and their compounds. Their physical properties, deriving from the 5f electron states, show many similarities with the lanthanides, such as electron correlations, superconductivity, or ordered magnetism. But compared to the 4f metals, their properties and their magnetic structure, in particular, remain poorly understood. This is due to experimental complications and the exotic behavior of the 5f states that appear to be delocalized for the light actinide metals, however, becomes localized in the latter part of the series. Considerable insight into the electronic ground state can be obtained from core-level XAFS spectroscopy and electron energy loss spectroscopy, together with recent theoretical simulations.

Rare-earth (R)-doped $Bi_{2-y}R_yX_3$ (R = Ce, Gd; X = Se, Te) single crystals are promising candidates for magnetic topological insulators. The XAFS and XMCD measurements at R $M_{4,5}$–edges provide evidence that R ions are trivalent and magnetic, confirming that doped R ions have purely magnetic effects. ARPES measurements of $Bi_{2-y}Ce_ySe_3$ show a finite energy gap for $y>0$, arising from the broken time reversal symmetry due to magnetic R impurities. CD-ARPES measurements of $Bi_{2-y}Ce_ySe_3$ reveal finite CD effects in the topological surface states. This work demonstrates that $Bi_{2-y}R_yX_3$ (R = Ce, Gd) are magnetic topological insulators.[75]

GdNi is a ferrimagnetic material with a Curie temperature of $T_C = 69$ K and has a large magnetocaloric effect, so it can be used in magnetic refrigerator applications. The electronic structure of GdNi was investigated with XAFS and XMCD at T =25 K in the ferrimagnetic phase.[76] Gd $M_{4,5}$-edge (3d-4f) XAS atomic multiple calculations show that Gd is trivalent in GdNi, which is consistent with the local 4f state. On the other hand, a model clustering calculation performed on Ni $L_{2,3}$-edge (2p-3d) XAFS show that Ni is effectively divalent in GdNi and strongly hybridizes with nearest-neighbor Gd states, resulting in a d electron number of 8.57. The Gd $M_{4,5}$-edge XMCD spectrum is consistent with a ground-state configuration of S = 7/2 and L = 0. The Ni $L_{2,3}$-edge XMCD results indicate that the antiferromagnetically aligned Ni moments exhibit a small but finite total magnetic moment ($m_{tot} \sim 0.12\mu B$) with the ratio $m_o/m_s \sim 0.11$. Valence band hard X-ray photoemission spectroscopy shows Ni 3d features at the Fermi level, confirming a partially filled 3d band, while the Gd 4f states are at high binding energies away from the Fermi level. The results indicate that the Ni 3d band is not fully occupied and contradicts the charge-transfer model for rare-earth based alloys. The obtained electronic parameters indicate that GdNi is a strongly correlated charge-transfer metal with the Ni on-site Coulomb energy being much larger than the effective charge-transfer energy between the Ni 3d and Gd 4f states.

High-pressure induced magnetic behavior is of interest. The XANES and XMCD measurements at the Yb L_3 edge on $YbCu_2Si_2$ reveals that this intermetallic is a well-known intermediate valence compound which exhibits a pressure-induced paramagnetic-ferromagnetic phase transition at about 8 GPa. The pressure dependence of the Yb magnetization was determined by tracking the evolution of the maximum of the electric quadrupole ($2p_{3/2}$-4f) transition dichroic signal as a function of pressure. The XMCD-deduced magnetization curves confirm the ferromagnetic property of the high-pressure phase as well as the first-order nature of the transition. The order moment of Yb derived from XMCD is about 1.25 μB at 8.7 GPa and 2.7 K, which is in very good agreement with that obtained by Mössbauer spectroscopy for the magnetically ordered component of the spectra. This observation indicates that, in the XMCD experiment, the paramagnetic and ferromagnetic phases coexist only in a narrow pressure range. This is not the case for the Mössbauer measurements, which show that the paramagnetic component disappears only at 20 GPa. These results clarify the existing discrepancy with dc magnetization measurements that reported values of the Yb moment more than two times smaller. These apparent contradictory results are attributed to different experimental conditions, which influence the range of coexistence of the paramagnetic and ferromagnetic phases, as found in many similar situations and frequently addressed in the literature.[77]

XMCD studies at the Er $L_{2,3}$ edges of $Er_3Fe_5O_{12}$ exhibit a change of the spectral shape as a function of temperature and magnetic field. Using singular value decomposition, this variation is understood as a linear combination of two components. The dominating component is associated with the Er magnetization, while the second contribution is identified as an induced signal from the Fe sites. XMCD at either of the L-edges in $Er_3Fe_5O_{12}$ provides information on the net magnetization of both sublattices. Their evolution in fields up to 30 T reveals details of the ferrimagnetic interactions on two very different scales.[78]

The analysis of the $L_{2,3}$ XMCD of rare earth elements is not just hampered by the partial overlap of electric quadrupolar contributions (E2) with the dipolar contributions (El). Even if the two are separated, the direct application of the sum rules leads to the wrong sign of the induced 5d moment. This stems from the spin dependence of the transition matrix elements. To solve these difficulties, the experimental spectra need to be compared with initio calculations of the entire isotropic XANES and the XMCD signal. The localized magnetic 4f electrons spin polarize the 5d final states E1 transition, but also have a strong radial overlap with the 2p initial states. So the core hole effects, p→d, and p→f transitions are all intercorrelated. Standard branching ratios, like in XAS and XMCD of 3d transition elements, and simple proportionality of bulk magnetization and XMCD intensity are not expected. On the other hand, these experiments will open a new challenge for example to measure in parallel the shell-specific magnetization of 5d and the 4f band by means of dipolar and quadrupolar contributions. The complex field- and temperature dependent magnetic properties encountered in rare earths may then be resolved. A first step towards this goal is to combine the details of the experimental fine structure with ab initio calculations. The results of our fully relativistic multiple scattering or KKR formalism is in fair agreement with all details of the experimental spectra.[79] By introducing spin asymmetry parameters, the X-ray absorption and rules of spin and orbital moments are generalized to consider the spin dependence of the dipole matrix elements. This generalization is important for the $L_{2,3}$-edge XMCD analysis of rare-earth elements, because the original sum rules erroneously predict an antiparallel alignment of the 4f and 5d moments. The corrections to the sum rules are of the same order as the extracted moments for these systems. For this case the extracted 5d spin moment was corrected from the antiparallel alignment with 4f spin $\mu_S = -0.27$ μB to a parallel alignment $\mu_S = +0.37$ μB.[80] This correction brings XMCD data in agreement with the total magnetization measurement, which predicts that the 5d contribution $\mu_S+\mu_L > +0.34$ μB. The full test requires XMCD data at Tb $M_{4,5}$ edges to obtain spin and orbital moments from the 4f states. The inclusion of these spin-dependent corrections into standard sum rule analysis thus permits a more reliable determination of magnetic moments with the XMCD technique, and therefore makes the technique more robust and more generally useful. Thus, with the help of theory the XMCD technique yields valuable information for the spin-dependent band-structure which can be related to the magnetic ground state properties.

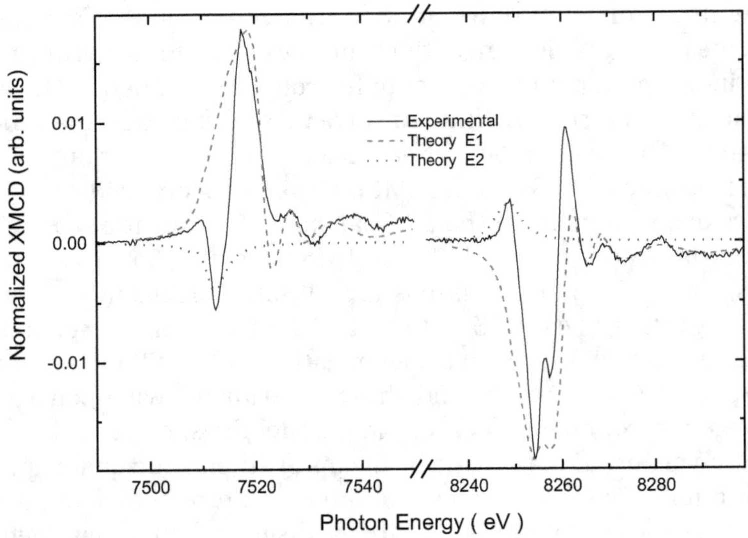

Figure 11.3 Comparison of experimental (black) and theoretical XMCD spectra at Tb $L_{2,3}$ edges: dipolar (red) and quadrupolar (blue) contributions.

11.5 The X-Ray Measurement of Ferromagnetic Resonance

Ferromagnetic resonance (FMR) can be explored using XMCD with time resolution in pump-probe experiments. The resonant precession of the magnetization can be pumped by microwave radiation. The frequency of oscillation of an applied magnetic field is set in phase with the X-ray pulses of the synchrotron source. The oscillation of moments under resonance conditions can be probed by XMCD. The precession of the moments determines a periodic alignment with the XMCD probing direction, so that the dynamics properties of the ferromagnetic specimen can be probed in the frequency domain. The relative phase of precession in individual magnetic layers of a spin valve structure can be understood by exploiting the pulsed time structure of the storage ring synchrotron radiation.[81]

The time-resolved XMCD offers ferromagnetic resonance FMR element- and site-specific.[82] This pump-probe technique is the only spectroscopy which can be utilized to explore dynamical aspects of orbital and spin magnetism separately, and therefore, can be viewed as the dynamic extension of XMCD.[83] In favorable cases, X-ray detection of magnetic resonance (XDMR) can be even site selective probe.[84] Particularly interesting, XDMR experiments carried out at K-edges can be a unique source of information regarding dynamical aspects of orbital magnetism, a subject that could not be explored experimentally before. Over the past decade, X-ray detection of magnetic resonance has emerged as a powerful experimental tool to study magnetization dynamics [85,86] and is expected to become a state-of-the-art technique at XFEL facilities. XDMR turns out to offer unique capabilities to probe spin transfer torque [87] and spin currents. [88]

Experimentally, XDMR experiments can be classified into two distinct categories: time-averaged and time-resolved methods. Time-averaged measurements are typically carried out in a longitudinal geometry, where the magnetization M is oriented along the incident X-ray beam. In this configuration, the precession of the magnetic moments is excited at the ferromagnetic resonance and the change in the time averaged longitudinal component of the magnetization is recorded via resonant XMCD. To enhance sensitivity, it is desirable to increase the microwave power at the expense of running into the non-linear fold over regime of ferromagnetic resonance. [89] Since the collected signal is independent of the phase of precession of the magnetic moments, this method can be used at all excitation frequencies. In general, the time-averaged approach can provide information on the amplitude, resonance field, and linewidth of the ferromagnetic resonance with element, site, and valence state specificity, yet it cannot provide any details about the phase of the magnetic precession. Time-resolved measurements are performed in a transverse geometry, where the wavevector of the incident X-ray pulses is oriented perpendicular to the external bias field (magnetization). In this configuration, the magnetic moments are continuously excited by the RF field while the response from the phase-dependent magnetization components along the X-ray beam is probed stroboscopically. This pump-probe technique requires the microwave excitation to be synchronized to integer multiples of the X-ray pulse frequency and is typically limited to frequencies in the low GHz regime, due to the finite pulse length of the X-ray bunches.

11.6 Outlook

After its discovery a quarter century ago, XMCD quickly attracted the attention of the wider magnetism community working with synchrotron radiation. The derivation of the sum rules enormously simplified the analysis of XMCD, leading to an explosion in the use of this technique for wider applications. The sum rules are independent of the theory, because they apply to both atomic and band structure models. Today, the applications of XMCD are vast and provide almost endless possibilities to study magnetic materials.[90] XMCD can be used for magnetic imaging on the nanoscale using X-ray transmission microscopy and for X-ray holography, using the coherent properties of synchrotron radiation.[91,92] New applications have emerged in areas such as chemistry, biology and earth sciences. For instance, in environmental sciences and microbiotechnology, XMCD has become a unique and powerful technique to determine the relative site occupations in (bio)spinels and other ternary oxides.[93] Rapidly increasing information density of modern magnetic data storage devices requires an answer to the question of the fundamental limits in bit size and writing speed. Time-resolved XMCD with femtosecond laser excitation is used to study magnetic switching, giving information of the driving force behind femtosecond spin-lattice relaxation.[94] In soft X-ray resonant emission the core-hole clock effect can be used to study the spin dependent screening in ferromagnetic metals on a femtosecond timescale.[95]

In ideal cases, XMCD should be able to reveal: (a) the distribution of spin and orbital angular momentum in transition metal complexes from sum rule analysis, (b) the strength of magnetic coupling between different centers from the field dependence, and (c) the total magnetic moment from the magnetization curve. In practice, intelligent use of the technique requires some caution—caveat emptor. For example, for systems more complicated than d^9 ions, the XMCD effect at a given edge is often bipolar or even more complex (see the calculations of van der Laan and Thole [96] and witness the detailed analysis of an Mn_{12} cluster). For the analysis of coupled systems, one is often faced with the question—is the XMCD bipolar because of the antiferromagnetically coupled ions, or because of side lobes in a single component spectrum? In the extreme cases of Fe–S clusters, it seems that individual features can completely overlap, leaving a relatively featureless XMCD with multiple interpretations.

As shown above, the standard XMCD analysis can fail for magnetic systems such as $L_{2,3}$ edge of early 3d series and $L_{2,3}$ of rare-earth elements. A systematic study along the 3d series revealed that the sum rule analysis as well as the multipole-moment analysis breaks down for light 3d elements. This failure could be related to the change of the branching ratio. The change of $L_{2,3}$ X-ray absorption spectra can be described by using a double-pole approximation within the TDDFT framework. Within this model the matrix elements of the unknown XC kernel can be determined. These values reveal that the shift of the spectral weight from the L_3 to the L_2 edge is primarily due to the reduced spin-orbit coupling. In the 5d magnetism of rare earth metals, the direct application of the integral sum rule analysis of the $L_{2,3}$ edges XMCD would lead to the wrong sign of the 5d moment, due to the spin-dependence of the transition matrix elements. The inclusion of this effect in the analysis yields the correct 5d moments in agreement with other magnetometries. Both examples reveal that only by comparison of the experimental spectra to calculations of the X-ray absorption coefficients and the corresponding dichroic spectra the magnetic ground state properties can be determined.

Experimental artifacts are also a concern, particularly radiation damage. Many samples are photoreduced in seconds on modern beamlines, and precautions such as sample motion or rapid scanning are often essential. Other potential problems include surface oxidation of reactive samples, beam heating at cryogenic temperatures, and surface anisotropy of magnetic properties. As discussed above, each detection scheme has the potential for mistakes as well. Despite these potential difficulties, XMCD opens a window into electronic and magnetic structure and provides information that is often difficult to obtain by other techniques. For bioinorganic samples, it is likely that there will be considerable improvement in experimental sensitivity. This can be obtained with larger arrays of faster and/or higher resolution detectors, such as superconducting tunnel junction detectors [97] or perhaps improved Si devices such as 'drift detectors'.[98,99] The more rapid circular polarization switching will also yield better signal-to-noise in XMCD spectra. Higher field magnets will allow experiments on wider ranges of nested magnetization curves, while faster magnetic field scans will yield more precise magnetization curves. All of these improvements are underway at the synchrotron sources. Overall, one can be optimistic about the future of this technique.

---

[1] G. van der Laan, A.I. Figueroa, (2014) X-ray magnetic circular dichroism—A versatile tool to study magnetism", *Coordination Chemistry Reviews* 277–278, 95–129.

[2] J.L. Erskine, E.A. Stern, (1975) Calculation of the $M_{2,3}$ magneto-optical absorption spectrum of ferromagnetic nickel. *Phys. Rev. B* 12, 5016.

[3] G. Schutz, W. Wagner, W. Wilhelm, P. Kienle, R. Zeller, R. Frahm, & G. Materlik, (1987) Absorption of circularly polarized x rays in iron. *Phys. Rev. Let.* 58, 737-740.

[4] C.T. Chen, F. Sette, Y. Ma, S. Modesti, (1990) Soft-X-ray magnetic circular dichroism at the $L_{2,3}$ edges of nickel. *Phys. Rev. B* 42, 7262.

[5] G. Schütz, W. Wagner, W. Wilhelm, P. Kienle, R. Zeller, R. Frahm, & G. Materlik, (1987) Absorption of circularly polarized x rays in iron. *Phys. Rev. Lett.* 58,737 - 740.

[6] J. Stohr, (1999) Exploring the microscopic origin of magnetic anisotropies with X-ray magnetic circular dichroism (XMCD) spectroscopy. *J. Magn. Magn. Mater.* 200, 470.

[7] B.T. Thole, P. Carra, F. Sette, G. van der Laan, (1992) X-ray circular dichroism as a probe of orbital magnetization. *Phys. Rev. Lett.* 68, 1943.

[8] M. Altarelli, (1993) Orbital-magnetization sum rule for X-ray circular dichroism: A simple proof. *Phys. Rev. B* 47, 597.

[9] G. van der Laan, (1994) Sum rules and fundamental spectra of magnetic X-ray dichroism in crystal field symmetry. *J. Phys. Soc. Jpn* 63, 2393.

[10] P. Carra, B.T. Thole, M. Altarelli, X. Wang, (1993) X-ray circular dichroism and local magnetic fields. *Phys. Rev. Lett.* 70, 694.

[11] P. Carra, H. Konig, B.T. Thole, M. Altarelli, (1993) Magnetic X-ray dichroism: General features of dipolar and quadrupolar spectra. *Physica B* 192, 182.

[12] D.A. Varshalovich, A.N. Moskalev, V.K. Khersonskii, (1988) *Quantum Theory of Angular Momentum* (World Scientific, Singapore), ISBN: 978-9971-5-0107-5 (hardcover).

[13] R.D. Cowan, *The theory of atomic structure and spectra* (University of California Press, Berkeley, 1981), ISBN: 9780520038219.

[14] J. Stohr, H. Koenig, (1995) Determination of spin-and orbital-moment anisotropies in transition metals by angle-dependent X-ray magnetic circular dichroism. *Phys. Rev. Lett.* 75, 3748.

[15] H.A. Durr, G. van der Laan, (1996) Magnetic circular X-ray dichroism in transverse geometry: Importance of noncollinear ground state moments. *Phys. Rev. B* 54, R760.

[16] G. van der Laan, B.T. Thole, (1996) X-ray-absorption sum rules in -coupled operators and ground-state moments of actinide ions. *Phys. Rev. B* 53, 14458.

[17] S.P. Collins, D. Laundy, C.C. Tang, G. van der Laan, (1995) An investigation of uranium $M_{4,5}$ edge magnetic X-ray circular dichroism in US. *J. Phys.: Condens. Matter* 7, 9325.

[18] G. van der Laan, & B.T. Thole, (1995) Core hole polarization in resonant photoemission. *J. Phys.: Condens. Matter* 7, 9947.

[19] B.T. Thole, G. van der Laan, M. Fabrizio, (1994) Magnetic ground-state properties and spectral distributions. I. X-ray-absorption spectra. *Phys. Rev. B* 50, 11466.

[20] G. van der Laan, (1997) Line shape of 2p magnetic-X-ray-dichroism spectra in 3d metallic systems. *Phys. Rev. B* 55, 8086.

[21] G. van der Laan, (1998) Microscopic origin of magnetocrystalline anisotropy in transition metal thin films. *J. Phys.: Condens. Matter* 10, 3239.

[22] G. van der Laan, (1998) Angular momentum sum rules for X-ray absorption. *Phys. Rev. B* 57, 112.

[23] K.W. Edmonds, N.R.S. Farley, T.K. Johal, G. van der Laan, R. P. Campion, B.L. Gallagher, C.T. Foxon, (2005) Ferromagnetic moment and antiferromagnetic coupling in (Ga,Mn)As thin films. *Phys. Rev. B* 71, 064418.

[24] A. Ankudinov, J.J. Rehr, (1995) Sum rules for polarization-dependent X-ray absorption. *Phys. Rev. B* 51, 1282.

[25] H. Ebert, (1996) Magneto-optical effects in transition metal systems. *Rep. Prog. Phys.* 59, 1665.

[26] R. Wu, D. Wang, A.J. Freeman, (1993) First principles investigation of the validity and range of applicability of the X-ray magnetic circular dichroism sum rule. *Phys. Rev. Lett.* 71, 3581.

[27] R. Benoist, P. Carra, O.K. Andersen, (2000) Band structure and atomic sum rules for X-ray dichroism. *Eur. Phys. J. B* 18, 193.

[28] M. Altarelli, P. Sainctavit, (1996) in *Magnetism and synchrotron radiation*, ed. by E. Beaurepaire, B. Carriere, J.P. Kappler (Les Editions de Physique, Les Ulis), p. 65.

[29] G. van der Laan, (1999) Sum rule practice. *J. Synch. Rad.* 6, 694.

[30] Cinthia Piamonteze, Piter Miedema and Frank M. F. de Groot, (2009) The accuracy of the spin sum rule in XMCD. *Journal of Physics: Conference Series* 190, 012015.

[31] H. Ebert, J. Stoehr, S.S.P. Parkin, M. Samant, A. Nilsson, (1996) L-edge X-ray absorption in fcc and bcc Cu metal: Comparison of experimental and first-principles theoretical results. *Phys. Rev. B* 53, 16067.

[32] P. Fischer, G. Schiitz, & G. Wiesinger, (1990) Spin-dependence of the rare earth $L_{2,3}$ absorption in $(RE)_2Co_{17}$. *Solid State Commun.* 76, 777.

[33] P. Carra, H. König, B.T. Thole, & M. Altarelli, (1993) Magnetic X-ray dichroism: General features of dipolar and quadrupolar spectra. *Physica B* 192, 182.

34 P. Carra, B. T. Thole, M. Altarelli, & X. Wang, (1993) X-ray circular dichroism and local magnetic fields. *Phys. Rev. Lett.* 70, 694.

35 J. Vogel, A. Fontaine, V. Cros, F. Petroff, J.-P. Kappler, G. Krill, A. Rogalev, & J. Goulon, (1997) Structure and magnetism of Pd in Pd/Fe multilayers studied by X-ray magnetic circular dichroism at the Pd sedges. *Phys. Rev. B* 55, 3663.

36 J. Stöhr, (1999) Exploring the microscopic origin of magnetic anisotropies with X-ray magnetic circular dichroism (XMCD) spectroscopy. *J. Magn. Magn. Mater.* 200, 470.

37 K. W. Edmonds, N. R. S. Farley, T. K. Johal, G. van der Laan, R. P. Campion, B. L. Gallagher, & C. T. Foxon, (2005) Ferromagnetic moment and antiferromagnetic coupling in (Ga, Mn) As thin films. *Phys. Rev. B* 71, 064418.

38 F. Wilhelm, R. Eloirdi, J. Rusz, R. Springell, E. Colineau, J.-C. Griveau, P. M. Oppeneer, R. Caciuffo, A. Rogalev, & G. H. Lander, (2013) X-ray magnetic circular dichroism experiments and theory of transuranium Laves phase compounds. *Phys. Rev. B* 88, 024424.

39 B. T. Thole, P. Carra, F. Sette, & G. van der Laan, (1992) X-ray circular dichroism as a probe of orbital magnetization. *Phys. Rev. Lett.* 68, 1943.

40 M. Altarelli, (1993) Orbital-magnetization sum rule for X-ray circular dichroism: A simple proof. *Phys. Rev. B* 47, 597.

41 J.I. Igarashi, & K. Hirai, (1996) Orbital moment and magnetic circular dichroism at the K edge in ferromagnetic cobalt. *Phys. Rev. B* 53, 6442.

42 A. Ankudinov, & J.J. Rehr, (1995) Sum rules for polarization-dependent X-ray absorption. *Phys. Rev. B* 51, 1282.

43 N. Bouldi, N.J. Vollmers, C.G. Delpy-Laplanche, Y. Joly, A. Juhin, Ph. Sainctavit, Ch. Brouder, M. Calandra, L. Paulatto, F. Mauri, & U. Gerstmann, (2017) X-ray magnetic and natural circular dichroism from first principles: Calculation of K- and L_1-edge spectra. *Phys. Rev. B* 96, 085123.

44 F. Baudelet, S. Pascarelli, O. Mathon, J.P. Iti, A. Polian, M. d'Astuto, & J.C. Chervin, (2005) X-ray absorption spectroscopy and X-ray magnetic circular dichroism simultaneous measurements under high pressure: The iron bcc–hcp transition case. *J. Phys. Condens. Matter* 17, S957.

45 R. Torchio, A. Monza, F. Baudelet, S. Pascarelli, O. Mathon, E. Pugh, D. Antonangeli, & J. P. Itié, (2011) Pressure-induced collapse of ferromagnetism in cobalt up to 120 GPa as seen via X-ray magnetic circular dichroism. *Phys. Rev. B* 84, 060403.

46 J.-D. Cafun, J. Lejeune, J.-P. Itié, F. Baudelet, & A. Bleuzen, (2013) XMCD at the transition metal k-edges as a probe of small pressure-induced structural distortions in Prussian blue analogues. *J. Phys. Chem. C* 117, 19645.

47 J. Kanamori, (1963) in *Anisotropy and Magnetostriction of Ferromagnetic and Antiferromagnetic Materials, Magnetism* Vol. I, edited by G.T. Rado, & H. Suhl (Academic, New York), Section 12.8.

48 D.J. Higley, K. Hirsch, G.L. Dakovski, E. Jal, *et al.* (2016) Femtosecond X-ray magnetic circular dichroism absorption spectroscopy at an X-ray free electron laser. *Rev. Sci. Instrum.* 87, 033110.

[49] K. Holldack, C. Schüssler-Langeheine, P. Goslawski, N. Pontius, T. Kachel, F. Armborst, M. Ries, A. Schälicke, M. Scheer, W. Frentrup & J. Bahrdt, (2020) Flipping the helicity of X-rays from an undulator at unprecedented speed, *Commun. Phys.* 3, 61.

[50] A. Rogalev, F. Wilhelm, J. Goulon, & G. Goujon, (2013) *Magnetism and Synchrotron Radiation: Towards the Fourth Generation Light Sources*, chapter 10, Springer.

[51] V.E. Dmitrienko, & V.A. Belyakov, (1989) *Sov. Phys. Uspekhi.* 32, 697.

[52] J. Stöhr, Y. Wu, B.D. Hermsmeier, M.G. Samant, G.R. Harp, S. Koranda, D. Dunham, & B.P. Tonner, (1993) Element-specific magnetic microscopy with circularly polarized X-rays. *Science* 259, 658.

[53] J.G. Tobin, G.D. Waddill, A.F. Jankowski, P.A. Sterne, & D.P. Pappas, (1995) Comparison of branching ratio and sum-rule analyses of magnetic circular dichroism in X-ray-absorption spectroscopy. *Phys. Rev. B* 52, 6530.

[54] H.A. Dürr, G. van der Laan, D. Spanke, F.U. Hillebrecht, & N.B. Brookes, (1997) Electron-correlation-induced magnetic order of ultrathin Mn films. *Phys. Rev. B* 56, 8156.

[55] R. Nakajima, (1997) Ph.D. Thesis, Stanford University.

[56] E. Arenholz, (2005) Design and performance of an eight-pole resistive magnet for soft X-ray magnetic dichroism measurements, *Review of Scientific Instruments* 76, 083908.

[57] J. Christiansen, G. Peng, A.T. Young, L.B. LaCroix, E.I. Solomon, & S.P. Cramer, (1996) X-Ray magnetic circular dichroism at temperatures< 1 K: demonstration with the blue copper site in plastocyanin. *Inorg. Chim. Acta* 118, 229.

[58] C. Bryant, (1998) M.S. Thesis, University of California, Davis.

[59] J. Stöhr, (1995) X-ray magnetic circular dichroism spectroscopy of transition metal thin films. *J. Electron Spectrosc. Relat. Phenom.* 75, 253.

[60] C.T. Chen, F. Sette, & N.V. Smith, (1990) Double-headed Dragon monochromator for soft X-ray circular dichroism studies. *Appl. Optics* 29, 4535.

[61] C. Kao, J.B. Hastings, E.D. Johnson, D.P. Siddons, G.C. Smith, & G.A. Prinz, (1990) Magnetic-resonance exchange scattering at the iron L_2 and L_3 edges. *Phys. Rev. Lett.* 65, 373.

[62] J.M. Tonnerre, L. Seve, D. Raoux, G. Soullie, B. Rodmacq, & P. Wolfers, (1995) Soft X-Ray Resonant Magnetic Scattering from a Magnetically Coupled Ag Ni Multilayer. *Phys. Rev. Lett.* 75, 740.

[63] G. van der Laan, Soft X-ray resonant magnetic scattering of magnetic nanostructures. *C. R. Physique* 9, 570 (2008)

[64] C.T. Chen, F. Sette, Y. Ma, & S. Modesti, (1990) Soft-X-ray magnetic circular dichroism at the $L_{2,3}$ edges of nickel. *Phys. Rev. B* 42, 7262.

[65] F. Sette, C.T. Chen, Y. Ma, S. Modesti, & N.V. Smith, (1991) in *X-ray absorption fine structure*, ed. by S.S. Hasnain (Ellis Horwood, Chichester, UK, p. 96.

[66] T. Jo, & G.A. Sawatzky, (1991) Ground state of ferromagnetic nickel and magnetic circular dichroism in Ni 2p core X-ray-absorption spectroscopy. *Phys. Rev. B* 43, 8771.

[67] T. Koide, T. Shidara, H. Fukutani, K. Yamaguchi, A. Fujimori, & S. Kimura, (1991) Strong magnetic circular dichroism at the $M_{2,3}$ edges in ferromagnetic Ni and ferromagnetic Fe_3O_4. *Phys. Rev. B* 44, 4697.

[68] G. van der Laan, & B.T. Thole, (1992) Electronic correlations in Ni 2p and 3p magnetic X-ray dichroism and X-ray photoemission of ferromagnetic nickel. *J. Phys.: Condens. Matter* 4, 4181.

[69] K.W. Edmonds, G. van der Laan, A. Freeman, N.R.S. Farley, T.K. Johal, R.P. Campion, C.T. Foxon, B.L. Gallagher, & E. Arenholz, (2006) Angle-Dependent X-ray magnetic circular dichroism from (Ga, Mn) As: anisotropy and identification of hybridized states. *Phys. Rev. Lett.* 96, 117207.

[70] H. Suto, N. Kikuchi, S. Okamoto, K. Toyoki, Y. Kotani, & T. Nakamura, (2020) Magnetization manipulation induced by nonlocal spin injection from a perpendicular spin polarizer: nano-beam X-ray MCD study using an order-sorting-aperture collection method. *Appl. Phys. Express* 13, 043002.

[71] T. Moriyama, Y. Shiratsuchi, T. Iino, H. Aono, M. Suzuki, T. Nakamura, Y. Kotani, R. Nakatani, K. Nakamura, & T. Ono, (2020) Giant Anomalous Hall Conductivity at the Pt/Al_2O_3 Interface. *Phys. Rev. Applied* 13, 034052.

[72] D. Schmitz, C. Schmitz-Antoniak, F. Radu, H. Ryll, Ch. Luo, S. Bhandary, S. Biermann, K. Siemensmeyer, H. Wende, S. Ivanov, & O. Eriksson, (2020) Soft X-ray Magnetic Circular Dichroism of Vanadium in the Metal–Insulator Two-Phase Region of Paramagnetic V_2O_3 Doped with 1.1% Chromium. *Phys. Status Solidi B* 257, 1900456

[73] A. Scherz, H. Wende, & K. Baberschke, (2004) Fine structure of X-ray magnetic circular dichroism for early 3d transition metals. *Appl. Phys. A* 78, 843.

[74] E. Goering, (2005) X-ray magnetic circular dichroism sum rule correction for the light transition metals. *Philosophical Magazine*, 85(25), 2895–2911.

[75] J.-S. Kang, E. Lee, S. Seong, M. Y. Yang, J. Kim, M.-H. Jung, B.-G. Park, Y. Kim, G. Kim, & Jeehoon Kim, (2020) Rare-earth-doped Bi_2X_3 (X = Se, Te) as candidates for magnetic topological insulators. *Philosophical Magazine*, 100, 1258-1267.

[76] C. W. Chuang , H. J. Lin, F. M. F. de Groot, F. H. Chang, C. T. Chen, Y. Y. Chin, Y. F. Liao, K. D. Tsuei, J.A. Chelvane, R. Nirmala, & A. Chainani, (2020) Electronic structure investigation of GdNi using X-ray absorption, magnetic circular dichroism, and hard X-ray photoemission spectroscopy. *Phys. Rev. B* 101, 115137.

[77] F. Wilhelm, J.-P. Sanchez, D. Braithwaite, S. M. Ramos, E. N. Hering, G. Lapertot, & A. Rogalev, (2019) Pressure-induced first-order magnetic phase transition and Yb ordered moment in $YbCu_2Si_2$. *Phys. Rev. B* 99, 180409(R).

[78] C. Strohm, P. van der Linden, O. Mathon, & S. Pascarelli, (2019) Simultaneous Observation of the Er- and Fe-Sublattice Magnetization of Ferrimagnetic $Er_3Fe_5O_{12}$ in High Magnetic Fields Using X-Ray Magnetic Circular Dichroism at the Er $L_{2,3}$ Edges. *Phys. Rev. Lett.* 122, 127204.

[79] C. Sorg, A. Scherz, K. Baberschke, H. Wende, F. Wilhelm, A. Rogalev, S. Chadov, J. Minár, & H. Ebert, (2007) Detailed fine structure of X-ray magnetic circular dichroism spectra: Systematics for heavy rare-earth magnets. *Phys. Rev. B* 75, 064428.

[80] A. L. Ankudinov, J. J. Rehr, H. Wende, A. Scherz, & K. Baberschke, (2004) Spin-dependent sum rules for X-ray absorption spectra. *Europhys. Lett.*, 66 (3), pp. 441–447.

[81] M.K. Marcham, L.R. Shelford, S.A. Cavill, P.S. Keatley, W. Yu, P. Shafer, A. Neudert, J.R. Childress, J.A. Katine, E. Arenholz, N.D. Telling, G. van der Laan, & R.J. Hicken, (2013) Phase-resolved X-ray ferromagnetic resonance measurements of spin pumping in spin valve structures. *Phys. Rev. B* 87, 180403.

[82] J. Goulon, A. Rogalev, F. Wilhelm, N. Jaouen, C. Goulon-Ginet, G. Goujon, J.B. Youssef, & M.V. Indendom, (2005) X-ray detected magnetic resonance at the Fe K-edge in YIG: Forced precession of magnetically polarized orbital components. *J. Exp. Theor. Phys. Lett.* 82, 696.

[83] Ch. Klewe, Q. Li, M. Yang, A. T. N'Diaye, D. M. Burn, Th. Hesjedal, A. I. Figueroa, Ch.g Hwang, J. Li, R. J. Hicken, P. Shafer, E. Arenholz, G. van der Laan & Z. Qiu (2020) Element- and Time-Resolved Measurements of Spin Dynamics Using X-ray Detected Ferromagnetic Resonance, *Synchrotron Radiation News*, 33(2), 12-19

[84] J. Goulon, A. Rogalev, F. Wilhelm, Ch. Goulon-Ginet, & G. Goujon, (2007) Element-selective X-ray detected magnetic resonance: a novel application of synchrotron radiation. *J. Synch. Rad.* 14, 257.

[85] D.A. Arena, Y. Ding, E. Vescovo, S. Zohar, Y. Guan, & W.E. Bailey, (2009) A compact apparatus for studies of element and phase-resolved ferromagnetic resonance. *Rev. Sci. Instrum.* 80, 083903.

[86] J. Goulon, A. Rogalev, F. Wilhelm, G. Goujon, A. Yaresko, Ch. Brouder, & J. Ben Youssef, (2012) Site-selective couplings in X-ray-detected magnetic resonance spectra of rare-earth-substituted yttrium iron garnets. *New J. Phys.* 14, 063001.

[87] J. C. Slonczewski, (1996) Current-driven excitation of magnetic multilayers. *J. Mag. Mag. Mater.* 159, L1–L7.

[88] G. van der Laan, (2017) Time-resolved X-ray detected ferromagnetic resonance of spin currents. *J. Electron Spectrosc. Relat. Phenom.* 220, 137–146.

[89] J. Goulon, A. Rogalev, F. Wilhelm, N. Jaouen, Ch. Goulon-Ginet, & Ch. Brouder, (2006) X-ray detected ferromagnetic resonance in thin films. *Eur. Phys. J. B* 53, 169.

[90] G. van der Laan, (2013) Applications of soft X-ray magnetic dichroism. *J. Phys.: Conf. Ser.* 430, 012127.

91 S. Eisebitt, J. Luning, W.F. Schlotter, M. Luergen, O. Hellwig, W. Eberhardt, & J. Stoehr, (2004) Lensless imaging of magnetic nanostructures by X-ray spectro-holography. *Nature* 432, 885.

92 T. Duckworth, F. Ogrin, S.S. Dhesi, S. Langridge, A. Whiteside, T. Moore, G. Beutier, & G. van der Laan, (2011) Magnetic imaging by X-ray holography using extended references. *Opt. Express* 19, 16223.

93 V.S. Coker, A.G. Gault, C.I. Pearce, G. van der Laan, N.D. Telling, J.M. Charnock, D.A. Polya, & J.R. Lloyd, (2006) XAS and XMCD evidence for species-dependent partitioning of arsenic during microbial reduction of ferrihydrite to magnetite. *Environ. Sci. Technol.* 40, 7745.

94 C. Stamm, T. Kachel, N. Pontius, R. Mitzer, T. Quast, K. Holldack, S. Khan, C. Lupulesco, E.F. Aziz, N. Wietstruk, H.A. Durr, & W. Eberhardt, (2007) Femtosecond modification of electron localization and transfer of angular momentum in nickel. *Nature Mater.* 6, 740.

95 L. Braicovich, G. Ghiringhelli, A. Tagliaferri, G. van der Laan, E. Annese, & N.B. Brookes, (2005) Femtosecond dynamics in ferromagnetic metals investigated with soft X-ray resonant emission. *Phys. Rev. Lett.* 95, 267402.

96 G. van der Laan, & B.T. Thole, (1991) Strong magnetic X-ray dichroism in 2p absorption spectra of 3d transition-metal ions. *Phys. Rev. B* 43, 13401.

97 S. Friedrich, T. Funk, O.S.E. Drury, S.P. Labov, & S.P. Cramer, (2002) A multichannel superconducting soft X-ray spectrometer for high-resolution spectroscopy of dilute samples. *Rev. Sci. Inst.* 73, 1629.

98 E. Gatti, & P. Rehak, (1984) Semiconductor Drift Chamber: an application of a novel charge transport scheme. *Nucl. Instrum. Methods A* 225, 608.

99 L. Strüder, P. Lechner, & P. Leutenegger, (1998) Silicon drift detector–the key to new experiments. *Naturwiss* 85, 539.

Chapter 12

Magnetic Extended X-ray Absorption Fine Structure

12.1 Introduction

The oscillation structure present in the absorption coefficient far above the edge, is called Extended X-ray absorption fine structure (EXAFS), this well-established spectroscopy provides a very useful probe to investigate local structure of specific element inside various solid materials, surfaces and molecules. Near the core absorption edges, the resonance absorption effect is manifested in a difference of absorption coefficients for the right and left circularly polarized quanta in ferromagnetic media and is called X-ray magnetic circular dichroism (XMCD).[1] It stems from the interplay of spin-orbit interaction in the initial state and angular momentum conservation in the absorption process, and reflects the local spin and orbital polarization of the final states. The beauty of XMCD is the applicability of the so-called 'sum rules' which can be utilized to determine spin and orbital contributions to the local magnetic moments separately. Its inherent element selectivity and high orbital sensitivity makes it a powerful tool to get element-specific insight into the ferro(i)magnetism of solids on an atomic or microscopic scale.

Magnetic contributions occur also in the EXAFS range, and the magnetic circular dichroism EXAFS performed with circularly polarized radiation are called magnetic or spin-dependent EXAFS (MEXAFS or SPEXAFS). The increased (decreased) absorption coefficients originate from the constructive (destructive) interference of the outgoing polarized photoelectron wave with the wave backscattered by the adjacent magnetic atoms. As the spin-dependent counterpart of EXAFS, MEXAFS experiments give insight into the spin dependent scattering of the photoelectron for ferromagnetic and ferrimagnetic materials.[2] The backscattering amplitude gives a quantitative measure of the local magnetic spin moments of the nearest neighbors. This scattering information yields the distances to the magnetic neighbors and results in a picture of the local magnetic structure.[3] One of the most enticing aspects concerning the structural information is the possibility of distinguishing the electronic and magnetic neighborhood of the absorbing atom by comparing EXAFS to MEXAFS.[4]

The potential of XMCD and MEXAFS has been experimentally illustrated in both hard and soft X-ray regimes for the rare earth-3d element intermetallics, magnetic thin films, multilayers and oxide compounds. Nowadays MEXAFS measurements can be utilized as a routine technique to investigate the local magnetic structure in comparison to the local crystallographic structure.[5] The theoretical descriptions of the MEXAFS effect are quite advanced; most experimental data can be well explained with a theoretical description by introducing an additive-exchange contribution to the complex Coulomb scattering and multiple-scattering expansion formalism.[6,7,8]

In the MEXAFS history, Schuetz and coworkers pioneered the development of MEXAFS, both experiment and theory.[9] By using circularly polarized radiation they first observed and analyzed the MEXAFS spectra for 3d metals at *K*-edges and rare earths at *L*-edges, and pointed out the problems and unknown factors for the analyses. It was concluded that MEXAFS supplies information on the short-range magnetic ordering around the absorbing atom, and the correlation of the orientation of spin moments between the relevant atoms. Most of the experimental investigations performed so far in this field were done in the hard X-ray regime, *e.g.* *K*-edges of 3d-transition metals [10] and the $L_{2,3}$-edges of rare earth elements.[11,12] Typical EXAFS oscillations were observed also in the soft X-ray regime for the MEXAFS-spectra at the $L_{2,3}$-edges of pure Fe, Co and Ni.[13,14,15] The Fourier transforms of the experimental MEXAFS for Ni–Mn alloys at Mn *K*-edge give unexpected peaks at the distance to the nonmagnetic scatterers (the 2nd peaks in the Fourier transform attributed to Mn-Mn pairs in the ordered phase), [16] they are enhanced in comparison with those in spin-averaged EXAFS. Ni atoms with small magnetic moment ($\approx 0.66\mu_B$). This suggests that some Ni atoms in the 2nd shell are substituted by Mn atoms, which have larger magnetic moment than Ni atoms, due to the heat-treatment induced atomic ordering. Embedded atom contributions to the MEXAFS was observed at the $L_{2,3}$-edges of rare-earth metallic systems, showing a smooth low-frequent background on which the magnetic EXAFS are superimposed.[17] Such an oscillating contribution to the MEXAFS only depends on the spin-polarization of the neighboring atoms. The ratio of this low-frequent background at the L_2- and L_3-edges reflect the existence of an orbital momentum similar as in the corresponding near-edge XMCD. However, this background is absent at the L_1-edge, and also not observed at the *K*-edge of transition metals and only very weak in rare-earth oxides. Additionally, the enhanced multiple-scattering peaks were reported in the Fourier transform (FT) of the MEXAFS signal, while the corresponding maxima at higher r-values in the spin-averaged EXAFS FT are much smaller or invisible.[18] This sharper structure is associated with the difference in the elastic scattering cross sections for parallel and anti-parallel orientation of the outgoing photoelectron wave and the spins of the magnetic electrons, due to the occurrence of additional harmonic frequencies in the magnetic raw spectra. Similar phenomenon was also observed in the XMCD of oxides, showing characteristic oscillations superimposed on the dichroic effect.

A first phenomenological description of MEXAFS was also proposed by Schuetz and coworkers.[19] Considering that the physical origin of MEXAFS - as for XMCD - is the interplay of spin-polarization and spin-orbit coupling, they introduced a spin-dependent backscattering term to conventional EXAFS. The EXAFS region is sensitive to the spin polarization of emitted photoelectrons. When the neighboring atom is spin polarized, there will be an exchange contribution in addition to the Coulomb scattering potential. This will modify both the amplitude $f_0(\pi, k)$ and phase shift $\phi_0(k)$ for electron backscattering, so that the traditional formula for EXAFS for a single absorber-scatterer interaction requires modification. Schütz and coworkers proposed addition of corrective terms $f_c(\pi, k)$ and $\phi_c(k)$ for the magnetic backscattering amplitude and phase shift, scaled by the degree of photoelectron polarization σ_z:[20]

$$\chi(k) = \frac{e^{-\left(2\sigma^2 k^2 + r/\lambda_0(k)\right)}}{kr^2} f(k) \sin[2kr + \phi(k)], \qquad (12.1)$$

with the amplitude and phase functions

$$f(\pi, k) = f_0(\pi, k) \pm \langle\sigma_z\rangle f_c(\pi, k)$$

$$\phi(k) = \phi_0(k) \pm \sigma_z \phi_c(k)$$

The other terms have their conventional meaning: r is the absorber-scatterer distance, σ is the root mean square variation in r, k is the photoelectron wave number, and λ is its mean free path.

Later on, a variety of attempts to formulate a complete description of the experimental data have been performed. One of the main sources of the MEXAFS is the spin–orbit coupling for core electrons ($L_{2,3}$-edge excitations) or photoelectrons (K-edge excitations), which produces the spin polarization. Another important source is the exchange interaction between the spin polarized photoelectrons and the magnetically polarized neighbors around the excited atoms.[21] Near-edge XMCD has extensively been studied because of rather intense signals,[22,23] but large scale full multiple-scattering calculations are necessary for the detailed analyses. On the other hand, MEXAFS can be analyzed on the basis of single or low-order multiple-scatterings. These theories are based on the multiple-scattering expansion formalism commonly applied to normal EXAFS. The atomic potentials in the EXAFS case are generally assumed to be independent of the spin and spherically symmetric, *i.e.*, independent of the magnetic quantum number m_l. This simplification is not justified in case of ferromagnetically ordered systems. The aim of the theory is therefore to calculate accurate spin-dependent potentials and to include the scattering from asymmetric potentials in the multiple-scattering expansion.

Brouder *et al.* [24,25] and Ankudinov and Rehr [26] developed two approaches based on the non-relativistic multiple-scattering theory for a finite cluster of atoms. While Brouder *et al.* solve the corresponding multiple-scattering equation by matrix inversion, Ankudinov and Rehr apply the scattering path expansion technique. Both methods were applied to the K-edge of Fe and to the L-edges of Gd giving a rather satisfying agreement with the experimental data. However, when considering spin-orbit coupling and magnetic ordering, both approaches use certain approximations. Brouder *et al.* simulated XMCD and MEXAFS at the L_3 edges of rare-earths (RE) in some $REFe_2$ intermetallic alloys within the full multiple-scattering approach.[27] The up and down potentials were obtained by means of a linear-muffin-tin orbital method. To test the influence of 4f electrons, whose very large role at the L-edges of RE is now well known, Brouder *et al.* chose Lu having no 4f magnetic moment, Gd having no 4f orbital momentum and Er having a strong 4f orbital momentum. Brouder *et al. also* discussed the multiple-scattering theory of MEXAFS at the K-edge of 3d transition metals. A semi-relativistic expansion of the Dirac Green function was utilized to calculate MEXAFS which is related to orbital momentum. Moreover, the high-energy behavior of MEXAFS is related to the exchange energy through the rigid band model and generalized to the case of orbital polarization. In particular Brouder and Hikam[28] pointed out that MEXAFS depends on the spin polarization of not only the neighboring atoms but also of the absorbing atom. This hypothesis was verified by the calculations of Gd L_3-edge MEXAFS in Gd cluster of 19 atoms for the cases: [29] (1) the X-ray absorbing atom is magnetized and surrounded by three shells of nonmagnetic Gd neighbors, (2) the central atom is not spin-polarized and the surrounding atoms are. Even in the case (1) EXAFS oscillation is observed in the MEXAFS. In case of the Gd $L_{2,3}$ edge theoretical calculations of spin-dependent EXAFS by Ankudinov and Rehr,[30] the half-filled shell of Gd ensures that the spherical symmetry of the scattering potential is retained. For the l=1 initial states, the spin-orbit interaction only has to be taken into account for the core electrons, whereas it is neglected for the photoelectron. This procedure, however, makes it very difficult to access more complex magnetic systems and the magnetic quantities of interest. A special case is the K and L_1 absorption, where, despite the absence of spin-orbit splitting in the initial s state, significant dichroic contributions occur.

Both the spin–orbit interaction and the spin polarization play essential roles for the analysis of MEXAFS spectra, a relativistic theory should be adopted within *ab initio* treatment to account for the effect of the spin–orbit interaction. So far two different relativistic multiple-scattering MEXAFS theories have been proposed. In the first approach, completely relativistic one-electron Green's functions were introduced.[31,32] However, this method is difficult to consider the Debye–Waller factor in EXAFS, which leads to the amplitude of the calculated MEXAFS spectra always lager than the experimental ones. In contrast to previous theoretical work, a fully relativistic description of MEXAFS developed by Ebert *et al.*[33] treats spin-polarization and all relativistic effects on the same level, working with the Dirac formalism for magnetic systems. Applications were made so far for the K- and $L_{2,3}$-edges of Fe in *bcc*-Fe [34] - using the matrix inversion technique - and for the K-edge of Fe in *bcc*-Fe and L-edges of Gd in *hcp*-Gd [35] -using the cluster expansion technique. In the second approach, the relativistic effect on photoelectron is treated as the perturbation, while the relativistic effect for core electrons is treat correctly. With this method, it is easy to study the temperature effects. The use of the relativistic one-electron Green's functions expanded in terms of non-relativistic Green's functions is suggested by Brouder.[36,37] Fujikawa has extended this theoretical method to the relativistic many-particle problems by use quantum electrodynamics (QED).[38,39] These theories of course can be applied to MEXAFS analyses in the low-order scattering approximation. An alternative theoretical approach has been developed by Ankudinov and Rehr.[40] They solve the Dirac equation for limiting cases and arrive at an interpolation scheme.

12.2 Theory for MEXAFS

12.2.1 Elastic exchange scattering model of MEXAFS

A phenomenological model of the MEXAFS can be derived by assuming that the exchange effects can be treated as an additive contribution to the Coulomb scattering amplitude. In the conventional spin-averaged EXAFS, the Coulomb backscattering amplitude f_c depends on the scattering angle β and the energy or wave vector k of the photoelectron. In ferromagnets, the output spin-polarized photoelectrons emitted after absorbing circularly polarized light also undergo exchange interactions with spin-polarized electrons carrying the magnetic moments of adjacent atoms. Similar to the description of the magnetic Compton scattering for polarized photons by the Klein-Nishima formula,[41] an additional exchange scattering contribution f_c is introduce to the overall scattering amplitude for parallel (+) and antiparallel (-) orientation of the majority-like spins to the photon-helicity. For a single electron-scattering process the amplitude becomes $f(\beta, k) = f_0(\beta, k) \pm \sigma_z f_c(\beta, k)$, the two possible values for the spin of the photoelectron in the photon-beam direction $\hat{z}$ being $\sigma_z = 1$ and $\sigma_z = -1$. Since the absorption experiment averages over many scattering processes with a particular ratio of spin-up and spin-down electrons, one actually observes a mean spin polarization of the photoelectrons. So σ_z should be replaced by its mean value at the respective edge $\langle \sigma_z \rangle$ as given by the Fano factor. As the scattering angle β is π for single backscattering, the explicit angle dependence can be omitted for simplicity and then:

$$f(k) = f_0(k) \pm \langle \sigma_z \rangle f_c(k). \tag{12.2}$$

As the inverse of the inelastic scattering cross section, the electron mean free path λ might also has a dependence on the spin of the photoelectron:

$$\lambda(k) = \lambda_0(k) \pm \langle\sigma_z\rangle\lambda_c(k). \tag{12.3}$$

All structural values are considered to be spin independent in this model, including the coordination number N, the shell radius r, the Debye-Waller factor $D_j(k) = exp(-2\sigma^2k^2)$, representing the damping by lattice vibrations. The amplitude reductions factor $S_j(k)$ is also assumed to be independent of spin, since its treatment is still a problem in the conventional EXAFS theory as well.

$\chi^+(k)$ and $\chi^-(k)$ can be constructed by using Eqs. (12.2) and (12.3) with the conventional EXAFS formula[42]

$$\chi(k) = -\sum_j \frac{N_j}{kr_j^2} S_j(k)D_j(k)e^{-2r_j/\lambda_0(k)} \times Im\left[e^{-2kr_j+i2\delta_i(k)}f(\beta,k)\right]. \tag{12.4}$$

The term $\delta_i(k)$ describes the phase shift experienced by the photoelectron while leaving and returning to the absorbing atom. Neglecting second-order terms the spin-averaged EXAFS are given by

$$\chi_0(k) = \frac{\chi^+(k)+\chi^-(k)}{2} = -\sum_j \frac{N_j}{kr_j^2} S_j(k)D_j(k)e^{-\frac{2r_j}{\lambda_0(k)}}F_{0,j}(k)\sin\left(2kr_j + 2\delta_i(k) + \Phi_{0,j}(k)\right). \tag{12.5}$$

This corresponds to the conventional EXAFS formula in which the complex scattering amplitude was replaced by its amplitude $F_{0,j}(k)$ and phase $\Phi_{0,j}(k)$ with $f_{0,j}(k) = F_{0,j}(k)exp\left(i\Phi_{0,j}(k)\right)$. The corresponding equation for the spin-polarized EXAFS can be written as

$$\chi_c(k) = \frac{\chi^+(k)-\chi^-(k)}{2} = -\sum_j \frac{N_j}{kr_j^2} S_j(k)D_j(k)\langle\sigma_z\rangle e^{-2r_j/\lambda_0(k)} \times \left\{F_{c,j}(k)\sin\left(2kr_j + 2\delta_i(k) + \Phi_{c,j}(k)\right) + \frac{2r_j\lambda_c(k)}{(\lambda_0(k))^2}F_{0,j}(k)\sin\left(2kr_j + 2\delta_i(k) + \Phi_{0,j}(k)\right)\right\}, \tag{12.6}$$

the complex scattering amplitudes are replaced by their amplitude and phase. If the absorbing atom carries a spin moment as well, the outgoing photoelectron will be influenced by the exchange interaction also upon leaving the atom adding a magnetic contribution $\delta_{c,i}(k)$ to $\delta_i(k)$. Since the numerical procedure does not allow the separation of a magnetic contribution to $\delta_i(k)$ from $F_{c,j}(k)$, the entire spin dependence of the photoelectron's phase is included in the value of $\Phi_{c,j}(k) = \delta_{c,i}(k) + \delta_{c,j}(k)$. The spin-polarized EXAFS function includes a term proportional to $\lambda_c(k)F_{0,j}(k)$. This enables one to estimate the spin-dependent mean free path simply by identifying contributions in the MEXAFS spectrum which scale with $F_{0,j}(k)$.

For simplicity a few phenomenological clues about the shape and character of the spin-dependent quantities are introduced. Evidently the effective exchange interaction results from those electrons that are carrying the magnetic moment, *i.e.*, the 3d electrons in transition metals and the 4f electrons in the rare earth. Hence the shape of the scattering potential should be comparable to that of a single electronic shell. The Coulomb scattering amplitude for simple potentials (Z<35) can be approximated by a Lorentzian and the phase by a polynomial.[43] Simarlly the electron mean free path can be substituted as $\lambda_0(k) = k^n/\eta$ where n=1 or 2. Since the number of magnetic electrons is small, one can use a Lorentzian for the spin-dependent backscattering amplitude and a polynomial of degree 3 for the phase shift, and a straight line for the spin-dependent mean free paths.

$$F_{c,j}(k) = \frac{A}{1+\left(\frac{k-C}{B}\right)^2},$$

$$\Phi_{c,j}(k) = p_0 + p_1 k + p_2 k^2 + p_3 k^3,$$

$$\lambda_c(k) = \xi k. \qquad (12.7)$$

In $F_{c,j}(k)$, A represents the amplitude, the factor B^2 corresponds to the full width at half maximum, and C gives the location of the maximum amplitude. Using Eqs. (12.6) and (12.7), the k distributions of the magnetic backscattering amplitude and phase shift can be extracted from the experimental data. This is accomplished in three steps: (1) the values $F_{0,j}(k)$, $\Phi_{0,j}(k)$, $\delta_i(k)$, $\lambda_0(k)$ are calculated using the FEFF code;[44] (2) the factors $D_j(k)$ and $S_j(k)$ together with the energy reference E_0, (*i.e.*, the difference of the kinetic energy origin of the photoelectron between the theoretical calculation and the inflection point of the absorption profile) are determined by fitting the $\chi_0(k)$ spectra; and finally (3) with the magnetic backscattering amplitude $F_{c,j}(k)$ and phase $\Phi_{c,j}(k)$ as well as $\lambda_c(k)$ the remaining free parameters are deduced from the $\chi_c(k)$ profiles. The second step is useful to reduce the number of free parameters in the overall fitting procedure.

Beside the element-specific determination of the magnetic short-range order, highly accurate studies of magnetic EXAFS provide a new way to determine electron-atom exchange scattering cross sections inside the solid, free of spin orbit and surface effects. This also laid an important foundation for testing the theory of spin scattering.

12.2.2 Fully relativistic theory for MEXAFS

Starting from Fermi's golden rule and representing the manifold of possible final states by the electronic Green's function G+, the X-ray absorption coefficient $\mu^{q\lambda}(\omega)$ can be expressed by [45]

$$\mu^{q\lambda}(\omega) \propto \sum_{iocc}\langle\Phi_i|X_{q\lambda}^{\times}\Im G^{+}(\hbar\omega + E_i)X_{q\lambda}|\Phi_i\rangle\,\delta(\hbar\omega + E_i - E_f). \tag{12.8}$$

As for the XMCD in the XANES-region, the most reliable way to deal with MEXAFS is to treat all ingredients here in a fully relativistic way. In particular, this means that the wave function Φ_i of the involved initial core states i are solutions to the Dirac equation for a spin-dependent potential.[46] Accordingly, the electron-photon interaction operator $X_{q\lambda} = ec\boldsymbol{\alpha}\cdot\boldsymbol{A}_{q\lambda}$ describes the coupling of the electronic current density operator to the vector potential $\boldsymbol{A}_{q\lambda}$ corresponding to light with wave vector q, frequency ω and polarization λ. Finally, using relativistic multiple-scattering theory the Green's function $G^{+}(\boldsymbol{r}_n, \boldsymbol{r}_n, E)$ can straightforwardly be expressed in terms of the regular and irregular solutions to the Dirac equation for an isolated potential well and the so-called scattering path operator $\tau_{\Lambda\Lambda'}^{nm}(\Lambda = (\kappa,\mu))$. The latter quantity describes the scattering of a wave of spin angular character Λ incoming on site m to a wave of character Λ' outgoing from site n with all possible multiple-scattering events accounted for. Accordingly, $\tau_{\Lambda\Lambda'}^{nm}$ has to fulfill the self-consistency condition

$$\underline{\tau}_{\Lambda\Lambda'}^{nm} = \underline{t}^{n}\delta_{nm} + \underline{t}^{n}\sum_{k\neq n}\underline{G}^{nk}\underline{\tau}^{km}, \tag{12.9}$$

where $\underline{t}^n$ is the single-site scattering t-matrix for site n and $\underline{G}^{nk}$ is the free electron propagator connecting sites k and n. For the XANES-region this equation for an infinite array of scatterers can be solved exactly by means of Fourier transformation. For the EXAFS-region on the other hand, it is sufficient to account for multiple-scattering of the electrons in the final state only within a finite cluster of atoms around the absorbing atom. In that case, Eq. (2) can be inserted repeatedly into itself, leading for $\underline{\tau}^{nm}_{\Lambda\Lambda'}$ in the expression for the Green's function to a series of scattering paths of increasing length and complexity. As an alternative to this approximate solution one can get an exact solution accounting for all scattering events within the cluster by the inversion of a matrix with its dimension fixed by the number of atoms in the cluster and the maximum angular momentum l_{max} used.

Similar to the sum rules for XMCD, Ebert *et al.* [34] demonstrated that these sum rules also apply for the MEXAFS spectra when they are used in a differential form. In their original integral form, these sum rules are given for the K-edge [47] and the $L_{2,3}$-edges [48,49] by

$$\frac{\int \Delta\mu_K/EdE}{\int \bar{\mu}_K/EdE} = \frac{3}{N_{hp}}\langle l_z\rangle_p \,,$$

$$\int(\Delta\mu_{L_3} + \Delta\mu_{L_2})dE = \frac{N}{2N_{hd}}\langle l_z\rangle_d \,,$$

$$\int(\Delta\mu_{L_3} - 2\Delta\mu_{L_2})dE = \frac{N}{3N_{hd}}(\langle\sigma_z\rangle_d + 7\langle T_z\rangle_d). \qquad (12.10)$$

where $\Delta\mu_X$ (X=K, L_2 and L_3) are the MEXAFS spectra, *i.e.* give the difference in absorption for left and right circularly polarized radiation at the various edges. N is the sum of the energy integrals for all partial absorption spectra and N_{hl} is the number of holes in the orbital with angular momentum character l. The expectation values $\langle\sigma_z\rangle$ and $\langle l_z\rangle$ give the spin and angular momentum of the absorber atom, while $\langle T_z\rangle$, a measure for the asphericity of the spin magnetization, can be very small in many cases.[50]

So far, many codes have been developed to simulate MEXAFS. Among these are SPRKKR, [51] FEFF, [52] and FDMNES.[53]

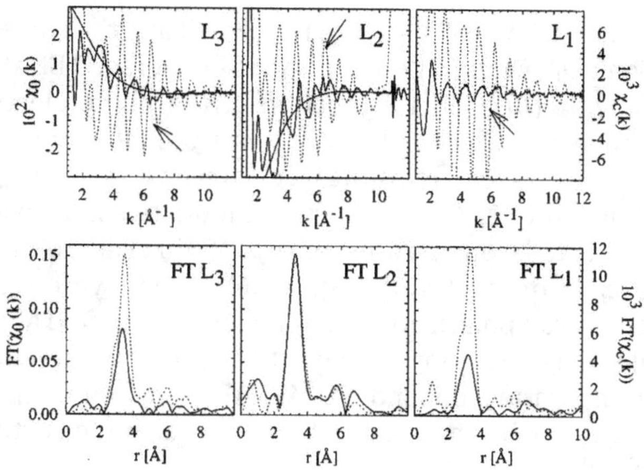

Figure 12.1 The spin-averaged $\chi_0(k)$(dots) and spin-dependent (solid) $\chi_c(k)$ spectra at the $L_{1,2,3}$ edges of Gd metal. The corresponding Fourier transforms after subtraction of the magnetic background (solid line in the $\chi_c(k)$ spectra) are shown in the lower panel. The arrows mark the location of multielectron transitions [Ahlers and Schuetz (1998) Phys. Rev. B 57(6), 3466-3473, reproduced with permission from American Physical Society].

12.3 Applications of MEXAFS

As the existence of a magnetic contribution in EXAFS region is a universal phenomenon, MEXAFS method provides a new attractive tool to study the short-range magnetism in ferromagnetic materials. One of the most important aspects concerning the structural information is the possibility of distinguishing the electronic and magnetic neighborhood of the absorbing atom by comparison of the EXAFS and spin-dependent MEXAFS (Figure 12.1). From a variety of systematic studies the relative amplitude of the dichroic contribution is found to be directly proportional to the average number of spin-polarized electrons of the first coordination shell, thus allowing a quantitative determination of the next-neighbor spin moment.

MEXAFS is defined as the difference between the EXAFS signal for right and left circular polarized X-rays. This signal is typically very small, and one needs 3rd generation synchrotron sources to determine its temperature dependence. From the measured spectra for parallel μ^+ and anti-parallel μ^- orientation of the photon helicity and magnetic field applied to the sample, one can obtain spin-averaged absorption coefficient $\mu_0(E) = (\mu^+(E)+\mu^-(E))/2$ and the corresponding dichroic signal $\mu_m(E)=\mu^+(E) - \mu^-(E)$. The spin-averaged EXAFS spectra $\chi_0(k)=(\mu_0(k)-\mu_{atom}(k))/\mu_0$ as a function of the photo-electron wave number k have been deduced in the conventional way from the absorption profile by subtracting the free atom absorption μ_{atom}. For the investigation of the MEXAFS spectra, a smooth magnetic background μ_{m0} was subtracted from the difference $\mu_m(E)$ in order to calculate the MEXAFS oscillation $\chi_m=(\mu_m-\mu_{m0})/(\mu_0.p_c.M_z)$. They were rescaled for full circular photon polarization $P_c=1$ and for completed alignment of the sample magnetization along photon beam direction $M_z=1$. Through temperature-dependent MEXAFS, not only can the local magnetic structure be analyzed, but also the magnetic fluctuation can be understood.

Reduction of XAFS data is conventionally based on least-squares fitting methods. However, if there are more parameters than the data alone allow to determine, this leads to an ill-conditioned system of equations. Beyond the least square fit analysis, Bayes–Turchin analysis [54] provides a priori for all parameters, is similar to a fit in which all parameters are restrained, as opposed to free or constrained. Bayes–Turchin analysis mathematically selects the parameters that most strongly affect the fit, in effect automating the process of choosing free and constrained parameters traditionally undertaken by the scientist doing the analysis. In contrast to a standard χ^2 fit, the Bayes-Turchin method allows to start with very large model parameter $\mathcal{Q}$ space. It then separates this space into a generally smaller subspace $\mathcal{R}$, where the data predominantly determine the outcome of the fit and the complementary space, in which assumptions determine largely or even completely the fit. The Bayes–Turchin approach makes it possible to start the data analysis with a very large model-parameter space. The method itself, rather than a mere guess, yields the subspace where the data determine the outcome of the fit. This allows the simultaneous analysis of atomic-like background and structure parameters. In particular, the effect of kinks in the muffin-tin potential at interstitial places can be separated from geometrical parameters describing the environment of the absorbing atom. The Bayes–Turchin approach further enables to include a large number of scattering paths in the sum,[55] which is indeed very important for the low k values of the spin-averaged EXAFS and the spin-polarized MEXAFS signals.[54] The relation between the EXAFS observable L_3 and the MEXAFS observable $\mu_{L_2}^M + \mu_{L_3}^M$ in the case of overlapping L-

edges,[56] is a sensitive probe of the spin dependence in the exchange correlation term of the local spin-density functional, used in the FEFF code. The relation can only be sensibly analyzed with a very careful treatment of background subtractions.

Alternatively, an iterative Van Cittert deconvolution approach can also be utilized to isolate the L_3-edge from the L_2-edge EXAFS and MEXAFS signals for the 3d transition metals.[57] The deconvoluted L_3-edge EXAFS data is in agreement with theoretically generated L_3-edge data, thus demonstrating the validity of this approach. Results from the deconvolution of the MEXAFS data also qualitatively agree with previously published MEXAFS data on the same elements at the *K*-edge. In principle, the *K*-edge MEXAFS of 3d metals could circumvent this overlap difficulty and access a longer *k*-range. However, due to the lower spin-polarization of the photoelectron at the *K*-edge, the dichroic signal at the *K*-edge is about an order of magnitude smaller compared to the $L_{2,3}$-edge signals, which stems from the spin–orbit splitting in the final p states. Furthermore, the *K*-edge MEXAFS signal is only relevant to the orbital polarization.

Another way to solve the overlap issue at $L_{2,3}$-edge MEXAFS is to utilize the so-called rigid-band model.[58] According to the model, the spin-dependent coefficients can be roughly expressed as spin-averaged coefficients being shifted in energy by an exchange-related energy ΔE,

$$\mu_{\uparrow\downarrow}(E) \approx \mu_0\left(E \pm \frac{\Delta E}{2}\right)\left[1 + \chi\left(E \pm \frac{\Delta E}{2}\right)\right], \tag{12.11}$$

where the positive (negative) sign of ΔE is used for spin-up (spin-down) electrons. The MEXAFS signal of the overlapping L_2 (j=1/2,l=1) and L_3 (j=3/2,l=1) components is now approximated by

$$\mu_{L_3}^M(E) + \mu_{L_2}^M(E) \approx \frac{1}{4}\left[\mu_{L_3}\left(E + \frac{\Delta E}{2}\right) - \mu_{L_3}\left(E - \frac{\Delta E}{2}\right)\right] - \frac{1}{2}\left[\mu_{L_2}\left(E + \frac{\Delta E}{2}\right) - \mu_2\left(E - \frac{\Delta E}{2}\right)\right]. \tag{12.12}$$

A relation between MEXAFS and the first derivative of EXAFS is obtained, when (1) the energy shift ΔE is neglected in the smoothly varying central atom absorption coefficient μ_0, (2) the difference in χ is approximated by $(d\chi/dE)\Delta E$, and (3) $\mu_{0,L2}(E)$ is replaced by $\mu_{0,L3}(E)/2$, assuming that the spin-orbit splitting E_{SO} is much smaller than the energy range over which $\mu_{0,L2}(E)$ changes significantly. Then the MEXAFS signal can be written as

$$\mu_{L_3}^{M}(E) + \mu_{L_2}^{M}(E) \approx \frac{\mu_{0,L_3}(E)}{4}\left[\frac{d\chi_{L_3}(E)}{dE} - \frac{d\chi_{L_2}(E)}{dE}\right]\Delta E \approx E_{SO}\frac{\mu_{0,L_3}(E)}{4}\left[\frac{d^2\chi_{L_3}\left(E-\frac{E_{SO}}{2}\right)}{dE^2}\right]\Delta E. \tag{12.13}$$

where in $\chi_{L2}(E)$ has been replaced by $\chi_{L3}(E-E_{SO})$. Within these approximations the energy dependence of MEXAFS is described by the energy dependence of EXAFS and the spin dependence is expressed by the exchange-related energy ΔE and the spin-orbit splitting E_{SO}.

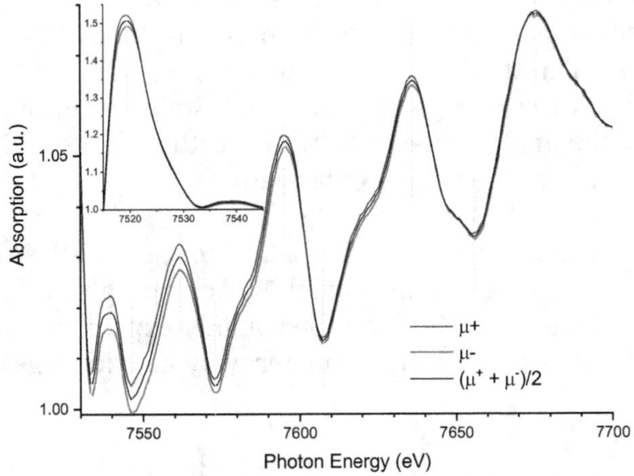

Figure 12.2 A typical spin-polarized XAFS spectra. The vertical amplitude variations of the polarized spectra (u^+ and u^-) **cannot** be simulated by shifting the spin-averaged spectrum ($u^0=(u^+ + u^-)/2$) horizontally.

However, the rigid-band model was improperly utilized to simulate MEXAFS by some researchers. They naively claimed that the spin fluctuation and lattice vibration contribution in magnetic EXAFS can be separated by modeling the absorption coefficients for right and left circularly polarized X-ray by the energy shifted experimental spin-averaged spectra,[59] and the temperature dependence of the MEXAFS can then be described by a temperature-dependent energy shift ΔE(T) which is proportional to the temperature-dependent magnetization M(T). However, the fundamental fact is the spin-polarized absorption spectra (μ^+ and μ^-) do not shift with respect to the spin-averaged one (μ_0= (μ^+ +μ^-)/2, EXAFS) along the photon energy-axis (see Figure 12.2). The oscillation peaks/valleys in spin-polarized spectra (μ^+ and μ^-) are exactly at the same positions as those in EXAFS, however, have different amplitudes. The vertical amplitude changes in the spin-polarized spectra apparently cannot be simulated by *horizontally* shifting a spin-averaged spectrum. There is no way to reproduce a vertical change by a horizontal shifting. Then, the difference between the shifted spectra ($\mu_{sim}{}^M(E) = \mu(E - ½\Delta E) - \mu(E + ½\Delta E)$) cannot be the true MEXAFS ($\mu^M = \mu^+ - \mu^-$) because $\mu(E - ½\Delta E) \neq \mu^+$ and $\mu(E + ½\Delta E) \neq \mu^-$. In fact, the calculated spectra from this "rigid band model" apparently do not match the experimental MEXAFS results at all.[60] The calculated MEXAFS based on "rigid band model" even leads to the wrong sign in the case of MEXAFS of Fe film. Actually, as early as 1996, Chakarian *et al.* already pointed out that "*these magnetic EXAFS results indicate a nonlinear dependence of this effect on the magnetic polarization of the scattering atom. The data further indicate that the dominant scattering paths responsible for the effect are different from those responsible for EXAFS*". [61] Even larger deviation between $\mu_{sim}{}^M(E)$ and $\mu^M(E)$ can be found in any compound magnet (such as supermalloy, ferrite, NdFeB, $DyFe_2$ and $NdCo_2$ alloys). MEXAFS spectra of these compounds don't share any similarity of EXAFS data, especially for those which contain both magnetic and nonmagnetic scatterers where the magnetic and nonmagnetic scatterers contribute differently in MEXAFS spectra. In spin-averaged EXAFS spectrum, only Coulomb potentials of the scattering atoms contribute (the spin/magnetism does not get involved or cancelled out), while MEXAFS spectra contains the magnetism information of the scattering atoms. The magnetism information cannot be reproduced from a non-magnetism-related spectrum. Under a magnetic field, the electronic structures of both absorbing and scattering atoms are altered (not simply the energy shift!), which cannot to be simulated by a system under zero magnetic field. The invalidity of the "rigid band model" in MEXAFS simulation was already pointed out by other scientists, [56] see p.134107-10, *"Contrary to the predictions of the HL energy density functional in combination with the rigid-band model, the energy dependence of the measured MEXAFS oscillations can be described by a negative exchange-related energy"*.

12.4 Summary and Outlook

The occurrence of a magnetic contribution in the EXAFS regime is a complicated problem but interesting phenomenon since it is expected to contain unique information on the magnetic short-range order. Experimental studies of the small effects with sufficient high accuracy are required to provide the possibility of a reliable data analysis and interpretation. This is often hampered by the occurrence of strong magnetic multi-electron effects and low-frequent atomic contributions as well as the strongly enhanced magnetic multiple-scattering.[62] Concerning the theoretical treatment problems arise, because the spin-dependency has to be included correctly and spin-orbit and exchange interaction have to be treated on the same level especially for higher photo-electron energies. By introducing an additive spin-dependent part of the complex scattering amplitude for a spin-polarized outgoing photo-electron wave, the experimental data can be reasonably interpreted in a first approximation. This method can be utilized to study magnetic and crystallographic structures and the low-energy exchange scattering process even in a quantitative manner. The double-electron transitions have strong influence on the EXAFS L-edges of rare-earth systems. Magnetic multielectron excitations (MMEEs) in the absorption channel,[63] result in strong nonoscillatory contributions to the MEXAFS. It is essential to correct the data for MMEEs at high precision. These MMEEs need to be subtracted accurately prior to Fourier transformation.

Beside the element-specific determination of the magnetic short-range order, highly accurate studies of magnetic EXAFS provide a new possibility to determine electron-atom exchange scattering cross sections inside the solid, free of spin orbit and surface effects, creating an important basis to test the theory of spin scattering. By further improvement of the accuracy of the data and with a better insight into the strong MMEE and enhanced multiple-scattering contributions, the exchange scattering cross section even for much lower electron energies can be addressed by this spectroscopic method. MEXAFS gives the information on the pair distribution of magnetic scattering sites around an absorption atom. MEXAFS will be a more common and powerful method for the study of magnetism in the near future with the further development of experimental technique and theory.

[1] Cz. Kapusta, P. Fischer, & G. Schuetz, (1999) Magnetic X-ray absorption spectroscopy. *J. Alloys Compd.* 286, 37–46.

[2] G. Schuetz, R. Frahm, P. Mautner, R. Wienke, W. Wagner, W. Wilhelm, & P. Kienle. (1989) Spin-dependent extended X-ray-absorption fine structure: Probing magnetic short-range order. *Phys. Rev. Lett.* **62**, 2620.

[3] M. Knuelle, D. Ahlers, & G. Schuetz, (1995) Spin dependent EXAFS in pure metals and garnets. *Soild State Commun.* **94**, 267.

[4] E. Dartyge, F. Baudelet, C. Brouder, A. Fontaine, J.P. Kappler, G. Krill, C. Giorgetti, M.F. Lopez, & S. Pizzini, (1995) Hard X-rays magnetic EXAFS. *Physica B* **208-209**, 751.

[5] D. Ahlers, & G. Schuetz, (1998) Elastic exchange scattering amplitude studied using magnetic EXAFS. *Phys. Rev. B* **57**, 3466.

[6] Ch. Brouder, M. Alouani, & K.H. Bennemann, (1996) Multiple-scattering theory of X-ray magnetic circular dichroism: Implementation and results for the iron K edge. *Phys. Rev. B* **54**, 7334.

[7] A. Ankudinov, & J.J. Rehr, (1997) Relativistic calculations of spin-dependent X-ray-absorption spectra. *Phys. Rev. B* **56**, R1712.

[8] H. Ebert, V. Popescu, & D. Ahlers. A fully relativistic theory for magnetic EXAFS-formalism and applications. *Phys. Rev. B* **60**, 7156 (1999).

[9] G. Schütz, R. Frahm, P. Mautner, R. Wienke, W. Wagner, W. Wilhelm, & P. Kienle, (1989) Spin-dependent extended X-ray-absorption fine structure: Probing magnetic short-range order. *Phys. Rev. Lett.* 62, 2620.

[10] E. Dartyge, F. Baudelet, C. Brouder, A. Fontaine, C. Giorgetti, J.P. Kappler, G. Krill, M.F. Lopez, & S. Pizzini. (1995) Hard X-rays magnetic EXAFS. *Physica B* 208-209, 751.

[11] M. Knuelle, D. Ahlers, & G. Schuetz, (1995) Spin dependent EXAFS in pure metals and garnets. *Solid State Commun.* 94, 267.

[12] E. Dartyge, A. Fontaine, Ch. Giorgetti, S. Pizzini, F. Baudelet, G. Krill, Ch. Brouder, & J.-P. Kappler, (1992) Multielectron excitations in Rare-Earth compounds revealed by magnetic circular X-ray dichroism. *Phys. Rev. B* 46, 3155.

[13] V. Chakarian, Y.U. Idzerda, G. Meigs, & C.T. Chen, (1995) Applications of soft X-ray magnetic circular dichroism. *IEEE Trans. Magn.* 31, 3307.

[14] Y.U. Idzerda, V. Chakarian, & J.W. Freeland, (1997) Soft X-ray magnetic circular dichroism at NSLS beamline U4B *Synchrotron Radiation News* 10, 6.

[15] H. Ebert, V. Popescu, D. Ahlers, G. Schütz, L. Lemke, H. Wende, P. Srivastava, & K. Baberschke, (1998) Theoretical and experimental investigations on the magnetic EXAFS at the K-and $L_{2,3}$-edges of bcc-Fe. *Europhys. Lett.* 42, 295.

[16] T. Miyanaga, T. Okazaki, R. Maruko, K. Takagahara, S. Nagamatsu, T. Fujikawa, H. Kon, & Y. Sakisaka, (2003) Magnetic X-ray absorption fine structure for Ni–Mn alloys. *J. Synchrtron Rad.* 10, 113.

[17] D. Ahlers, & G. Schütz, (1997) Embedded atom contributions to the magnetic EXAFS. *J. Phys. IV France* 7(C2), 461-462.
[18] D. Ahlers, S. Stahler, G. Schutz, (1997) Enhanced multiple path lines in magnetic EXAFS. *J. Phys. IV France* 7 (1997) 7(C2), 455-456.
[19] G. Schuetz, & D. Ahlers, (1996) in *Spin-orbit influenced spectroscopies of magnetic solids*, edited by H. Ebert and G. Schuetz, Lect. Notes Phys., Vol. 466 (Springer, Berlin), p. 229.
[20] G. Schütz, P. Fischer, E. Goering, K. Attenkofer, D. Ahlers, & W.Rößl, (2008) X-ray magnetic circular dichroism. *Syn. Rad. News* 10, 13-26.
[21] G. Schuetz, D. Ahlers, in: H. Ebert, C. Schuetz (Eds.), (1995) *Spin–Orbit-influenced Spectroscopies of Magnetic Solids*, Springer, Berlin, p. 229, Lecture Notes in Physics 466.
[22] I. Yamamoto, S. Nagamatsu, T. Nakamura, T. Fujikawa, & S. Nanao, (2002) Multiple scattering approach to Co K-edge XMCD and XANES spectra in Gd–Co alloys. *J. Elect. Spec. Relat. Phenom.* 125, 89. K. Okamoto, S. Nagamatsu, E.V. Voronina, T. Miyanaga, & T. Fujikawa, (2004) Fe K-edge X-ray magnetic circular dichroism analyses of Fe–M (M= Si, Al, Sn) alloys. *J. Elect. Spec. Relat. Phenom.* 141, 5.
[23] K. Okamoto, S. Nagamatsu, E.V. Voronina, T. Miyanaga, T. Fujikawa. Analysis for Fe-M (M= Si, Al, Sn) alloys by multiple scattering X-ray magnetic circular dichroism of Fe K-edge. *Phys. Scr.* T115, 645 (2005).
[24] E. Dartyge, F. Baudelet, Ch. Brouder, A. Fontaine, C. Giorgetti, J.P. Kappler, G. Krill, M.F. Lopez, and S. Pizzini, (1995) Hard X-rays magnetic EXAFS. *Physica B* 208-209, 751.
[25] C. Brouder, (1997) Multiple-Scattering Theory of Magnetic EXAFS. *J. Phys.* (Paris) 7 (C2), 377.
[26] A. Ankudinov, & J.J. Rehr, (1997) Relativistic calculations of spin-dependent X-ray-absorption spectra. *Phys. Rev. B* 56, R1712.
[27] C. Giorgetti, E. Dartyge, C. Brouder, M. Alouani, F. Baudelet, (1997) Hard-XMCD and magnetic EXAFS in the multiple scattering approach, *Journal de Physique IV* 7(C2), 435-436.
[28] Ch. Brouder, & M. Hikam, (1991) Multiple-scattering theory of magnetic X-ray circular dichroism. *Phys. Rev. B* 43, 3809.
[29] Ch. Brouder, M. Alouani, C. Giorgetti, E. Dartyge, F. Baudelet, (1995) in: H. Ebert, G. Schuetz (Eds.), Spin–Orbit-influenced Spectroscopies of Magnetic Solids, Springer, Berlin, p. 259, Lecture Notes in Physics 466.
[30] A. Ankudinov, & J. J Rehr, (1995) Calculation of X-ray magnetic circular dichroism in Gd. *Phys. Rev. B* 52, 10 214.
[31] H. Ebert, V. Popescu, & D. Ahlers, (1997) Fully relativistic theory for magnetic EXAFS: Formalism and applications. *Phys. Rev. B* 56, 7156.
[32] V. Popescu, H. Ebert, & D. Ahlers, (1999) Magnetic EXAFS at the K-and $L_{2,3}$-edges of FCC-Ni. *J. Magn. Magn. Mater.* 191, 368.
[33] H. Ebert, V. Popescu, & D. Ahlers, (1997) MXD - model calculations and a fully relativistic theory for magnetic EXAFS. *J. Phys. IV* 7 (C2), 131.

[34] H. Ebert, V. Popescu, D. Ahlers, G. Schuetz, L. Lemke, H. Wende, P. Srivastava, & K. Baberschke. (1998) Theoretical and experimental investigations on the magnetic EXAFS at the K- and $L_{2,3}$-edges of bcc-Fe. *Europhys. Lett.* 42, 295.

[35] D. Ahlers, G. Schütz, V. Popescu, & H. Ebert, (1998) A fully relativistic theory for magnetic EXAFS. *J. Appl. Phys.* 83, 7082.

[36] Ch. Brouder, M. Alouani, & K.H. Bennemann, (1996) Multiple-scattering theory of X-ray magnetic circular dichroism: Implementation and results for the iron K edge. *Phys. Rev. B* 54, 7334.

[37] T. Fujikawa, S. Nagamatsu, (2003) Relativistic multiple scattering theory of K-edge X-ray magnetic circular dichroism. *J. Elect. Spect. Relat. Phenom.* 129, 55.

[38] T. Fujikawa, (2004) Relativistic QED approach to XPS and XAFS spectra. *J. Elect. Spect. Relat. Phenom.* 136, 85.

[39] T. Fujikawa, (2005) Relativistic Keldysh Green's function approach to XAFS theory. *Phys. Scr.* T115, 35.

[40] A. Ankudinov, & J.J. Rehr, (1997) Relativistic calculations of spin-dependent X-ray-absorption spectra. *Phys. Rev. B* 56, R1712.

[41] R. M. Steffen, & H. Frauenfelder, (1968) in Alpha-, Beta- and Gamma ray Spectroskopy, edited by K. Siegbahn (North-Holland, Amsterdam), Vol. 2, p. 1456

[42] J.J. Rehr, J.J. Kas, F.D. Vila, & M. Newville, (2017) Theory and Analysis of XAFS. In: Iwasawa Y., Asakura K., Tada M. (eds) XAFS Techniques for Catalysts, Nanomaterials, and Surfaces. Springer, Cham.

[43] B. K. Teo, (1986) EXAFS: Basic Principles and Data Analysis (Springer, Berlin).

[44] J. J. Rehr, (1993) Recent developments in multiple-scattering calculations of XAFS and XANES. *Jpn. J. Appl. Phys., Part 1* **32**, 8.

[45] H. Ebert, (1996) Magneto-optical effects in transition metal systems. *Rep. Prog. Phys.* 59, 1665.

[46] H. Ebert, (1989) Fully relativistic treatment of core states for spin-dependent potential. *J. Phys.: Condens. Matter.* 1, 9111.

[47] J. Igarashi, & K.M. Hirai, (1994) Magnetic circular dichroism at the K edge of nickel and iron. *Phys. Rev. B* 50, 17820.

[48] P. Carra, B.T. Thole, M. Altarelli, & X. Wang, (1993) X-ray circular dichroism and local magnetic fields. *Phys. Rev. Lett.* 70, 694.

[49] B.T. Thole, P. Carra, F. Sette, & G. van der Laan, (1992) X-ray circular dichroism as a probe of orbital magnetization. *Phys. Rev. Lett.* 68, 1943.

[50] G.Y. Guo, H. Ebert, W.M. Temmerman, & P.J. Durham, (1995) Band theoretical investigation of circular magnetic X-ray dichroism in Fe and Co multilayers. J. *Magn. Magn. Mater.* 148, 66.

[51] H. Ebert, (2000) *Electronic Structure and Physical Properties of Solids* (*Lecture Notes in Physics* vol 535) ed H Dreyss (Berlin: Springer) pp 191–246.

52 J.J. Rehr, J.J. Kas, M.P. Prange, A.P. Sorini, Y. Takimoto, & F. Vila, (2009) Ab initio theory and calculations of X-ray spectra. *Comptes Rendus Physique* 10, 548.

53 N. Bouldi, N. J. Vollmers, C. G. Delpy-Laplanche, Y. Joly, A. Juhin, Ph. Sainctavit, Ch. Brouder, M. Calandra, L. Paulatto, F. Mauri, & U. Gerstmann, (2017) X-ray magnetic and natural circular dichroism from first principles: Calculation of K- and L1-edge spectra. *Phys. Rev. B* 96, 085123.

54 H.J. Krappe, & H.H. Rossner, (2009) The Bayes-Turchin approach to the analysis of extended X-ray absorption fine-structure data. *Physica Scripta* 79, 048302(7).

55 A. L. Ankudinov, B. Ravel, J. J. Rehr, & S. D. Conradson, (1998) Real-space multiple-scattering calculation and interpretation of X-ray-absorption near-edge structure. *Phys. Rev. B* 58, 7565.

56 H. H. Rossner, D. Schmitz, P. Imperia, H. J. Krappe, & J. J. Rehr, (2006) Bayes-Turchin analysis of X-ray absorption data above the Fe $L_{2,3}$-edges. *Phys. Rev. B* 74, 134107.

57 K.M. Kemner, W.T. Elam, Y.U. Idzerda, & V. Chakarian, (1997) Deconvolution of transition metal L-III/L-II-edge EXAFS and MEXAFS (MEXAFS) data, *J. Phys. IV*, 7(C2) , 431-432.

58 C. Brouder, M. Alouani, C. Giorgetti, E. Dartyge, & F. Baudelet, (1995) in *Spin-Orbit-Influenced Spectroscopies of Magnetic Solids* (Ref. 14), p. 259.

59 H. Wende, A. Scherz, C. Sorg, *et al.*, (2005) Temperature dependence of magnetic EXAFS for rare earth elements. *Physica Scripta* T115, 600.

60 J. Kurde, N. Ponpandian, J. Luo, C. Weis, K. Baberschke, P. Srivastava, & H. Wende, (2007) Scattering-path analysis and magnetic scattering properties of Fe/Ag(100) films: A temperature-dependent magnetic EXAFS study. *Phys. Rev. B* 76, 224418.

61 V. Chakarian, Y. U. Idzerda, & K. M. Kemner, (1996) Giant magnetic effects in the L-edge extended x-ray absorption fine structure of 3d transition metals. *J. Appl. Phys.* 79, 6493.

62 D. Ahlers, & G. Schuetz, (1997) Strong Magnetic Multi-Electron Excitations at the L-Edges in Rare Earths. *J. Phys. IV* 7, C2-457; G. Schuetz & D. Ahlers, (1997) Magnetic EXAFS. *J. Phys. IV* 7, C2-59,

63 D. Ahlers, G. Schuetz, V. Popescu, & H. Ebert, (1998) Fully relativistic theory for the magnetic extended X-ray absorption fine structure. *J. Appl. Phys.* 83, 7082.

Chapter 13

X-ray Magnetic Linear Dichroism

13.1 Introduction

Magnetic dichroism phenomena make the near-edge, inner-shell absorption of polarized X-rays a promising technique for investigating the magnetic behavior of bulk materials, thin films and multilayers, magnetic nanostructures, clusters and molecules, and spin transport systems.[1] X-ray magnetic linear dichroism (XMLD) is the difference in the absorption cross sections of horizontal and vertical linear polarized light with respect to a magnetization field. It depends on the square of the magnetic moment and on the magneto-crystalline anisotropy, which makes it favorable for the study of phenomena such as exchange biasing and magnetic anisotropy. This element-specific and surface sensitive technique has been employed to study the magnetism of transition metal, rare-earth, and actinide materials. It is often utilized alongside XANES spectroscopy, and can provide complementary information about electronic structure, which is of importance to a broad range of disciplines covering physics, materials science, surface science, chemistry, life sciences, earth sciences, biomanufacturing and catalysis research.

XMLD was first theoretically predicted by B.T. Thole in 1985 [2] and the first XMLD spectrum was experimentally obtained by Gerrit van der Laan the year after. [3] In XMLD measurement, electrons are excited from a selected core shell to empty valence states, the electric field vector ***E*** of linearly polarized X-rays acts as a search light for the direction of the maximum and minimum number of empty valence states. In most cases, the anisotropy in the bonding (*i.e.* the electrostatic potential) leads to the anisotropy of the charge in the atomic volume. In the absence of spin polarization in systems where the absorbing atom has lower than cubic symmetry, linear dichroic XANES spectroscopy can only probe charge order. But in the presence of spin order, even in a cubic system, spin-orbit coupling will lead to a preferential charge orientation relative to the spin direction. Based on this effect, the spin axis can be determined by XMLD spectroscopy for ferromagnetic systems and especially antiferromagnetic systems.

XMLD has been established as a powerful probe to examine the magnetocrystalline anisotropy of antiferro- as well as ferro- and ferrimagnetic systems.[4,5] The $L_{2,3}$ XMLD spectra of localized 3d transition metal ions have been successfully calculated using multiplet theory for different crystal field strengths.[6] For itinerant 3d metallic systems the line shapes of the $L_{2,3}$ XMLD spectra can be explained using the effective exchange field splitting of the core hole.[7] The expectation values of the charge quadrupole moment and the anisotropic spin-orbit interaction can be calculated from the integrated intensity over the L_3 and L_2 edges by using the XMLD sum rules.[8] Although the integral of XMLD is smaller than that of XMCD, their research has provided important insights in the microscopic origin of the magnetocrystalline anisotropy.[9] A huge magnetocrystalline anisotropy of cubic Fe, Co, and Ni metal displayed an opposite sign along the different high-symmetry quantization axes in Kunes and Oppeneer's calculation of the $L_{2,3}$ edge XMLD spectra. [10] The sign reversal of the XMLD signal in $LaFeO_3$ can be explained using the model for the effective exchange field splitting of the core hole.[11] The multiplet structure calculations confirmed the large anisotropy in the XMLD of the ferrimagnet Fe_3O_4 by taking properly into account the core-valence Coulomb and exchange interactions.[12] A large anisotropy in the XMLD line shape was found experimentally at the Mn^{2+} $L_{2,3}$ edge in (Ga,Mn)As, [13] in good agreement with atomic multiplet calculations. Since the half-filled 3d shell of Mn has no orbital moment, the angular dependence cannot be attributed to the magnetocrystalline anisotropy, and the integrated intensities of the $L_{2,3}$ XMLD actually disappears. The calculations found that the 3d spin-orbit interaction almost has no effect on the XMLD line-shape and that the angular dependence disappears when the crystal-field interaction is turned off. The full angular dependence of the XMLD has been studied by using vector magnet magnetometry, and the results were compared to those of the multiplet simulations for the transition metal $L_{2,3}$ edges of $Ga_{1-x}Mn_xAs$, Fe_3O_4,[11] $NiFe_2O_4$, NiO,[14,15] $CoFe_2O_4$,[16] Co-based Heusler alloys,[17] $La_{0.7}Sr_{0.3}FeO_3$,[18] and $GaxFe_{1-x}$,[19] as well as the rare earth $M_{4,5}$ edges of EuO.[20] It was found that the angular dependent XMLD in cubic symmetry could be described by a linear combination of two independent fundamental spectra.

For antiferromagnetic materials, the spin orientation can be probed using the linear polarization dependence of the intensity ratio of the two peaks of the L_2 X-ray absorption edge. Alders *et al.* [21,22] have measured a NiO specimen with a statistical distribution of $\langle \pm 1 \pm 1 \pm 2 \rangle$ possible spin orientations. However, the spin orientation with respect to the crystallographic axes was not explicitly taken into account in their analysis, which is correct for axial symmetry. The same approach was later utilized to analyze the antiferromagnetic domain structures of $LaFeO_3$ [23,24,25] and NiO,[26,27,28,29] not realizing that the XMLD strongly depends on the crystal field orientation. It was first demonstrated by Arenholz *et al.* [30] that the relative orientation of light polarization and magnetization with respect to the crystal frame have to be taken into account for a correct interpretation of the X-ray absorption and XMLD spectra. This has subsequently been applied in several recent studies of various NiO surfaces and interfaces.[31,32,33]

Originally measured in absorption experiments, XMLD was found to work in other X-ray techniques, including reflection, scattering, diffraction, photoemission and imaging. It is also the only substitute of X-ray magnetic circular dichroism (XMCD) for studying antiferromagnets. It is especially important for magnetic domain imaging with photoemission electron microscopy (PEEM), and forms the basis for X-ray resonant processes, such as X-ray resonant scattering, resonant Raman, resonant photoemission, *etc.* However, XMLD is more complicated than previously thought. Until recently, the complexity of its angular dependence has not been fully realized.[34]

13.2 XMLD Theory and Sum Rules

By measuring the X-ray absorption of linearly polarized light for different angles of incidence, one can use the electric field vector of the X-rays as a search light for the maximum and minimum unoccupied states. Thus, anisotropies in the charge density can be probed with X-ray linear dichroism, that may result from the crystal structure (X-ray natural linear dichroism) or magnetic ordering (X-ray magnetic linear dichroism). Given as single ion in an arbitrary ground state $|\Psi\rangle$, consider the $c \rightarrow l$ electric 2^L-pole transition (EL) of a single partner, $j_\pm = c \pm 1/2$ of a spin orbit split edge. For the total intensity, $I_{EL}^{j_\pm}$, of the $j_\pm$-edge, the following expression can be obtained

$$I_{EL}^{j_\pm} = [j_\pm]\left(\frac{1}{[L,c]} + \Sigma_{xyz}[x,y,z]^{1/2}(-)^{j_\pm+c+1/2} \times \begin{Bmatrix} c & 1/2 & j_\pm \\ 1/2 & c & y \end{Bmatrix} \begin{Bmatrix} l & c & L \\ l & c & L \\ x & y & z \end{Bmatrix} (-)^{L-q} \times \begin{Bmatrix} L & L & z \\ q & -q & 0 \end{Bmatrix} \left\langle \Psi \middle| W_0^{(xy)z} \middle| \Psi \right\rangle \right) \left(P_{cl}^{\pm}\right)^2. \quad (13.1)$$

In this expression the final states, accessible in a $c \rightarrow l$ transition, have been removed using the closure relation [2]. Here, $x = 0 \ldots\ldots 2l$, $y = 0$ and 1, $z = 0, \ldots, 2L$, $q=-L,\ldots,L$ and $[a, \ldots, z] = (2a+1)\ldots(2z+1)$; $P_{cl}^{\pm} = \langle l | C^{(L)} | c \rangle R^{\pm}$, with $R^{\pm}$ the radial matrix element of the $c \rightarrow l$ transition. The 9j-symbol is zero unless $x + y + z$ is even. Expression 13.1 generalizes the dipolar sum rule discussed in ref. [35] It gives the total intensity, $I_{EL}^{j_\pm}$, as a linear combination of the ground state expectation value of coupled tensor operators $W_0^{(xy)z}$.[36] The operators are spin independent when $y = 0$, while they depend on spin for $y = 1$; furthermore, the two classes can be separated, as discussed below.

Various kinds of spectra are obtained by considering the intensity as a function of q and z: $I_{EL}^{j_\pm}(q,z)$. The total intensity of the isotropic spectrum (XAFS), multiplied by 2L + 1, is obtained by summing over q and contains the $x = y = z = 0$ term only. Generalizing the result of Thole and van der Laan,[37] we have

$$\int_{j_+ + j_-} (\mu^{-L} + \cdots + \mu^{L})_{EL} d\omega \propto \Sigma_{q=-L}^{L} I_{EL}^{j_\pm}(q,0) = \frac{4l+2-\langle n \rangle}{2l+1} P_{cl}^2. \quad (13.2)$$

At the $j_\pm$-edge and for E1 transitions, the total intensity of the linear dichroism is given by a linear combination of coupled-tensor operators of the form $W_0^{(xy)z}$. For $l = c + 1$ transitions, expression 13.1 gives

$$\int_{j_\pm} \left(\mu^{\parallel} - \mu^{\perp}\right)_{E1} d\omega \propto \left(2I^{j_\pm}(0,2) - I^{j_\pm}(1,2) - I^{j_\pm}(-1,2)\right)_{E1} = \frac{\left(P_{cl}^{\pm}\right)^2}{(2c+1)l(2l+1)(2l-1)} \left\{ [j_\pm]\langle \Psi | Q_{zz} | \Psi \rangle \mp \frac{4c(2l-1)\langle \Psi | P_{zz} | \Psi \rangle}{5} \mp \frac{12c \langle \Psi | R_{zz} | \Psi \rangle}{5(l-1)} \right\}, \quad (13.3)$$

with the quadrupole operators

$$Q_{zz} = \frac{1}{2}\Sigma_i(3l_z^2 - \boldsymbol{l}^2)_i\,,$$

$$P_{zz} = \frac{1}{2}\Sigma_i(3l_z s_z - \boldsymbol{l}\cdot\boldsymbol{s})_i\,, \text{ and}$$

$$R_{zz} = \frac{1}{4}\sum_i [5l_z(\boldsymbol{l}\cdot\boldsymbol{s})l_z - (\boldsymbol{l}^2 - 2)\boldsymbol{l}\cdot\boldsymbol{s} - (\boldsymbol{2l}^2 + 1)l_z s_z]_i$$

A few remarks on the physical meaning of the coupled-tensors $W_0^{(xy)z}$ are now in order. Consider the magnetic field of the valence electrons, given by the superposition of two contributions: orbital and spin. The effect of the spin-orbit interaction (after the polarization of the spin) and/or crystal field effects result in a deformation of the electronic cloud. A multipolar expansion of the magnetic field will contain the various moments (of order 2^z) characteristic of the symmetry of the deformation and described by the coupled-tensors $W_0^{(x0)z}$. Spin-orbit coupling introduces a correlation between the spin direction and the position (x even) and motion (x odd) of the electron. This is accounted for by the coupled-tensors $W_0^{(xy)z}$ (a crystal field can generate even values of x only).

The operator $W_0^{(20)2} \propto Q_{zz} = \frac{1}{2}\Sigma_i(3l_z^2 - \boldsymbol{l}^2)_i$, for exampie, accounts for the quadrupolar moment of the charge distribution of the l-electrons; it can be induced either by the combined effect of exchange and spin-orbit interaction (the exchange interaction aligns s along a preferred direction, say $\hat{z}$, and the spin-orbit interaction aligns l and s) or by a crystal field. In the latter case, the point-group symmetry has to be low enough for $J = 2$ to branch to the totally symmetric representation, to yield nonzero matrix elements.

The coupled-tensors $W_0^{(xy)z}$ are one-particle operators; the only correlation they describe are between l and s of the same electron. As an example, a nonzero value of $W_0^{(11)2} \propto P_{zz} = \frac{1}{2}\Sigma_i(3l_z s_z - \boldsymbol{l}\cdot\boldsymbol{s})_i$ implies that the probability of having l and s parallel is largest in the $\hat{z}$ direction.

Linear (even z) and circular (odd z) dichroism, together with the type of transition involved (a 2L-pole electric transition detects the values z = 0 , . . . , 2L), select different classes of operators.

The orbital and spin contributions to expression 13.3 can be separated as in the following. By neglecting relativistic corrections to the radial part and setting $P_{cl}^{+} = P_{cl}^{-}$; this approximation introduces errors which are generally small. Adding the two partners of a spin-orbit split edge and normalizing to the non-polarized spectrum, one obtains, yielding the expectation value of the quadrupole moment per hole of the charge distribution.

$$\xi = \frac{\int_{j_{+}+j_{-}}(\mu^{\parallel}-\mu^{\perp})_{E1}d\omega}{\int_{j_{+}+j_{-}}(\mu^{+}+\mu^{0}+-\mu^{-})_{E1}d\omega} = \frac{2\langle Q_{zz}\rangle}{l(2l-1)(4l+2-\langle n\rangle)}. \quad (13.4a)$$

The ground state expectation value of the spin dependent part (per hole) is given by

$$\eta = \frac{\int_{j_{+}}(\mu^{\parallel}-\mu^{\perp})_{E1}d\omega-\left(\frac{c+1}{c}\right)\int_{j_{-}}(\mu^{\parallel}-\mu^{\perp})_{E1}d\omega}{\int_{j_{+}+j_{-}}(\mu^{+}+\mu^{0}+-\mu^{-})_{E1}d\omega} = \frac{4}{5l(4l+2-\langle n\rangle)}\left[\langle P_{zz}\rangle + \frac{4}{(l-1)(2l-1)}\langle R_{zz}\rangle\right]. \quad (13.4b)$$

Alternatively, one can consider the dichroic spectrum only and determine the ratio

$$\gamma = \frac{\int_{j_{+}}(\mu^{\parallel}-\mu^{\perp})_{E1}d\omega-\left(\frac{c+1}{c}\right)\int_{j_{-}}(\mu^{\parallel}-\mu^{\perp})_{E1}d\omega}{\int_{j_{+}+j_{-}}(\mu^{\parallel}-\mu^{\perp})_{E1}d\omega} = -\frac{(l-1)(2l-1)\langle P_{zz}\rangle+6\langle R_{zz}\rangle}{5(l-1)\langle Q_{zz}\rangle}, \quad (13.4c)$$

which determines the magnitude of the spin contributions in relation to the purely orbital part of the moment.

In transition metals, $\langle Q_{zz}\rangle$, $\langle P_{zz}\rangle$ and $\langle R_{zz}\rangle$ can be determined by measuring the linear dichroism at the $L_{3,2}$, $M_{3,2}$ and $N_{3,2}$ edges. In the 3d series, the spin-orbit splitting of the $L_{3,2}$ edges is not large enough to prevent their mixing, caused by Coulomb interactions in the final state. This makes an exact separation of the two partners impossible; however, the effect is small (<5%) on approaching the end of the series. Similar considerations apply to the $N_{3,2}$ edges of the 4d elements.

The rare earths' ground states are in compliance with an almost pure LSJ Hund's rule. The quadrupole operators can be numerically calculated for a $4f^{n}$ ground state configuration. For the M = -J level, they change in the presence of a crystal field. However, provided that for a given ion, the crystal field is weak enough to split the ground state J level without mixing in other levels, the ratio of the expectation values of the operators remains constant (crystal field scaling). In this case, the effects of the crystal field can be fully determined by measuring the total intensity of the dichroic signal. On the contrary, when there is hybridization, the ground state is not given by a split single configuration J level, then the $j_{\pm}$ peak ratios is required to be determined.

13.3 Methods to Obtain XMLD

The μ_{XMLD} is defined as a difference of linearly polarized X-ray absorption between measurements in two geometries: one in which the ***E***-vector is parallel to the magnetization ***M*** and the other in which it is perpendicular: $\mu_{XMLD} = \mu^{\parallel} - \mu^{\perp}$. By assuming the spin-spin splitting of the initial core-hole level, the μ_{XMLD} is then the difference for all the individual transition intensities, which only partially overlap in the spectrum.[38] In systems of lower than cubic symmetry, both *XMLD* and *XNLD* (X-ray natural linear dichroism, defined as the difference between vertically σ^V, and horizontally σ^H, polarized light) are present and need to be disentangled by experiment. Again, proper geometries allow to do that by defining experiments of *XMLD* of first type where the light scattering geometry with respect to the surface normal and crystallographic axes is fixed and it is the magnetization that is rotated by 90° from the easy to the hard magnetization direction. Assuming that the applied magnetic field ***H*** is sufficient to saturate ***M*** in the hard direction, the only net effect will be magnetic, i.e. *XMLD*, since the *XNLD* will be canceled by the spectral difference. The second possible geometry keeps ***M*** fixed in the easy axis and rotates ***E*** parallel and perpendicular (Figure 13.1). Both experiments become equivalent for negligible magnetocrystalline anisotropy or cubic or higher symmetry. Magnetic fields must be strong enough to magnetically saturate the samples in both directions.

The XMLD sum rules proposed by Carra *et al.*[39] have been verified by Dhesi *et al.*, [40] concerning the XMLD (first type) and the magnetocrystalline anisotropy energy ΔE_{SO} via the μ_{XMLD} on L_3 and L_2 edges and the spin-orbit parameter.

A change in magnetization direction results in a different distribution over the magnetic sublevels m of the atom in a crystal field. A change in polarization direction gives different selection rules (Δm = 0, ±1) for the electric-dipole transitions from ground to the final state. Changing both the magnetization and polarization direction can result in the same spectrum under symmetry conditions. The ground state in octahedral crystal field splits into E and T_2 orbitals with charge densities along different directions. The E orbitals are aligned towards the neighboring atoms whereas the T_2 orbitals point in between. The energy splitting allows to spectroscopically distinguish between both types of orbitals. Generally, when the energy degeneracy is broken, each level set obtains a different angular dependence. Therefore, energy splitting and angular distribution are two different sides of the same coin.

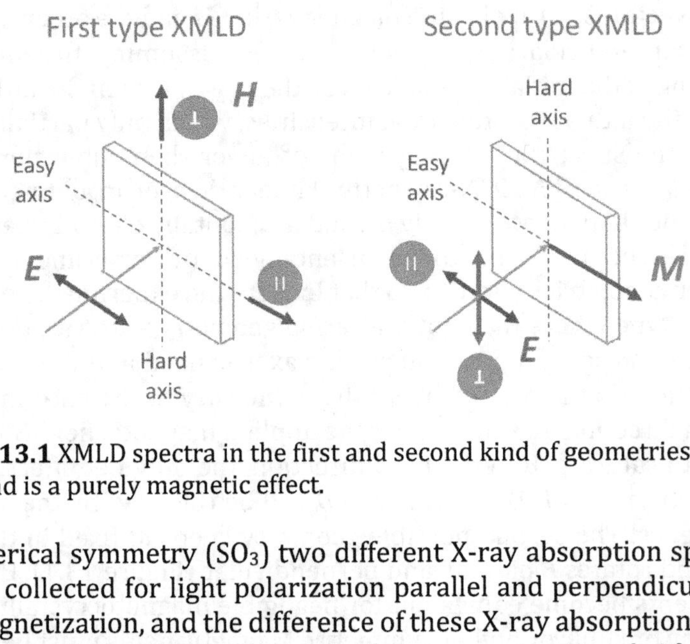

Figure 13.1 XMLD spectra in the first and second kind of geometries. The first kind is a purely magnetic effect.

In spherical symmetry (SO_3) two different X-ray absorption spectra can be collected for light polarization parallel and perpendicular to the magnetization, and the difference of these X-ray absorption spectra lead to single XMLD spectrum. If the magnetization (with symmetry C_∞) is along a high symmetry axis of a cubic crystal, there will be only one XMLD spectrum. However, when the magnetization is along arbitrary direction there should be more than one XMLD spectrum. The number of distinct X-ray absorption spectra is determined by the number of different irreducible representations in the point group of the crystal. The number of independent XMLD spectra is one less than the number of X-ray absorption spectra.

Taking the magnetization $\hat{M}$ and electric-dipole polarization $\hat{\boldsymbol{E}}$ and $\hat{\boldsymbol{E}}'$ for incident and scattered X-rays along arbitrary directions, the spectral intensity is separated in an angular part and a physical part. For electric dipole radiation the scattering tensor $\mathfrak{F}$ is composed of a scalar $\mathfrak{F}^{(0)}$, a pseudo-vector antisymmetric tensor $\mathfrak{F}^{(1)}(\hat{M})$ and a symmetric tensor $\mathfrak{F}^{(2)}(\hat{M})$, and the scattering intensity is generally given by

$$I(\hat{\boldsymbol{E}},\hat{\boldsymbol{E}}',\hat{M}) = (\hat{\boldsymbol{E}}^* \cdot \hat{\boldsymbol{E}}')\mathfrak{F}^{(0)} - i(\hat{\boldsymbol{E}}^* \times \hat{\boldsymbol{E}}')\mathfrak{F}^{(1)}(\hat{M}) + \hat{\boldsymbol{E}}^* \cdot \mathfrak{F}^{(2)}(\hat{M}) \cdot \hat{\boldsymbol{E}}', \tag{13.5}$$

with magnetization unit vector $\hat{\boldsymbol{M}} = (\hat{x}, \hat{y}, \hat{z})$ and polarization unit vector $\hat{\boldsymbol{E}} = (\hat{\varepsilon}_x, \hat{\varepsilon}_y, \hat{\varepsilon}_z)$, and $\hat{\boldsymbol{E}} = \hat{\boldsymbol{E}}'$ is assumed for X-ray absorption. $\mathfrak{F}^{(k)}(\hat{M},\omega) \equiv \sum_{k'\Gamma\gamma} C_{\Gamma'\gamma}^{(k')}(M) F_{\Gamma\gamma}^{(k')}(M)$, where $C_{\Gamma'\gamma}^{(k')}$ is a polynomial function of rank k' and irreducible representation $\Gamma'\gamma$, $F_{\Gamma\gamma}^{(k')}(M)$ gives the frequency dependent intensity as a scalar property.[41]

The scattering tensor provides a general framework for the X-ray absorption intensity, which is a function of the relative orientation of the magnetization M, the X-ray polarization E and the crystalline axes. In cubic crystal field, the XMLD spectrum is a linear combination of two independent spectra, which means that the maximum of the XMLD will be photon energy dependent and no longer at E ∥ M. The angular dependence of the XMLD signal can be divided into an isotropic part rotating with M and an anisotropic part depending on the crystal axis direction. At the Ni^{2+} L_2 edge,[29] where the isotropic part disappears, the angular dependence is completely determined by the anisotropic part. When E and M have equal but opposite angles with respect to the [100] direction, the anisotropy will be the largest.

There are many ways to obtain XMLD. Assume three orthogonal linear polarization directions $\mathcal{E}_1$, $\mathcal{E}_3$ and $\mathcal{E}_3$ and with magnetization direction $M_1 \parallel \mathcal{E}_1$. The isotropic X-ray absorption (XA) cross-section can be expressed as

$$\mu_{XA}^{iso} \equiv \frac{1}{3}[\mu_{XA}(\mathcal{E}_1, M_1) + \mu_{XA}(\mathcal{E}_2, M_1) + \mu_{XA}(\mathcal{E}_3, M_1)], \tag{13.6}$$

which is orientation independent. Ideally, the XMLD is defined as

$$\mu_{XMLD}(\mathcal{E}_1, M_1) \equiv \mu_{XA}(\mathcal{E}_1, M_1) - \tfrac{1}{2}\mu_{XA}(\mathcal{E}_2, M_1) - \tfrac{1}{2}\mu_{XA}(\mathcal{E}_3, M_1). \quad (13.7)$$

Such a geometry with three orthogonal vectors is not practical to measure, so that one normally settles for

$$\mu_{XMLD}(\mathcal{E}_1, M_1)_2 \equiv \mu_{XA}(\mathcal{E}_1, M_1) - \mu_{XA}(\mathcal{E}_2, M_1),$$

$$\mu_{XMLD}(\mathcal{E}_1, M_1)_3 \equiv \mu_{XA}(\mathcal{E}_1, M_1) - \mu_{XA}(\mathcal{E}_3, M_1).$$

Substitution gives

$$\mu_{XMLD}(\mathcal{E}_1, M_1) \equiv \tfrac{1}{2}[\mu_{XMLD}(\mathcal{E}_1, M_1)_2 + \mu_{XMLD}(\mathcal{E}_1, M_1)_3]. \quad (13.8)$$

Note that $\mu_{XMLD}(\mathcal{E}_1, M_1)$ is only unambiguously defined for $M_1 \parallel C_3$, C_4, C_6 and C_∞ axes, otherwise the $\mathcal{E}_2$ and $\mathcal{E}_3$ directions give different results. The XA and XMLD spectra can be readily obtained using multiplet calculations.[5] The angular dependence of the XMLD is governed by symmetry arguments. For simplicity, the energy dependence of the spectra is neglected, the X-ray absorption intensity in cubic symmetry along high symmetry axes is given in Table 13.1. It can be verified that for $\widehat{M} \parallel [100]$ and $\widehat{M} \parallel [111]$ the μ_{XA} in the perpendicular plane is constant due to the C_4 and C_3 symmetry axis, respectively. This is not true for other directions, as *e.g.* is seen for $\widehat{M} \parallel [110]$. By changing from the group symmetry notation to a spectroscopic notation and using the XMLD in the cubic symmetry, one can define

$$\mu_0 \equiv \mu\left(E_{[100]}, M_{[100]}\right) - \mu\left(E_{[100]}, M_{[100]}\right) = F_E^{(2)},$$

$$\mu_{45} \equiv \mu\left(E_{[110]}, M_{[110]}\right) - \mu\left(E_{[1\bar{1}0]}, M_{[110]}\right) = F_{T_2}^{(2)},$$

$$\mu_{[111]} \equiv \mu\left(E_{[111]}, M_{[111]}\right) - \mu\left(E_{[11\bar{2}]}, M_{[111]}\right) = F_{T_2}^{(2)},$$

$$\mu_{XA}^{iso} \equiv \tfrac{1}{3}\left[\mu\left(E_{[100]}, M_{[100]}\right) + \mu\left(E_{[010]}, M_{[100]}\right) + \mu\left(E_{[001]}, M_{[100]}\right)\right] = F_{A_1}^{(0)}, \quad (13.9)$$

where the results at the right-hand site are obtained from the XA given in Table 13.1. The μ_0 and μ_{45} are two independent spectra with different energy (ω) dependence, *i.e.*, different spectral shape, which are sometimes called 'fundamental spectra'. Since $\mu_{[111]}$ is equal to μ_{45}, it does not give an extra fundamental spectrum.

Table 13.1 Linear X-ray absorption in cubic crystal field for $\hat{\boldsymbol{E}}$ and $\widehat{M}$ along different high symmetry directions. The last column gives $\hat{\boldsymbol{E}} \cdot \mathfrak{F}^{(2)}(\widehat{M}) \cdot \hat{\boldsymbol{E}}$ in spectroscopic notation. $\hat{\boldsymbol{E}} \cdot \mathfrak{F}^{(2)}(\widehat{M}) \cdot \hat{\boldsymbol{E}} = \boldsymbol{F}_{A_1}^{(0)} - \frac{1}{3}\boldsymbol{F}_{E}^{(2)} = \mu_{XA}^{iso} - \frac{1}{3}\mu_0$. Adapted from Ref. 41.

M	E	μ_{XA}(O symmetry)	μ_{XA}(spectr. notation)
[100]	[100]	$\boldsymbol{F}_{A_1}^{(0)} + \frac{2}{3}\boldsymbol{F}_{E}^{(2)}$	μ_0
[100]	[010], [001]	$\boldsymbol{F}_{A_1}^{(0)} - \frac{1}{3}\boldsymbol{F}_{E}^{(2)}$	0
[110]	[110]	$\boldsymbol{F}_{A_1}^{(0)} + \frac{1}{6}\boldsymbol{F}_{E}^{(2)} + \frac{1}{2}\boldsymbol{F}_{T_2}^{(2)}$	$(\mu_0 + \mu_{45})/2$
[110]	[1$\bar{1}$0]	$\boldsymbol{F}_{A_1}^{(0)} + \frac{1}{6}\boldsymbol{F}_{E}^{(2)} - \frac{1}{2}\boldsymbol{F}_{T_2}^{(2)}$	$(\mu_0 - \mu_{45})/2$
[110]	[001]	$\boldsymbol{F}_{A_1}^{(0)} - \frac{1}{3}\boldsymbol{F}_{E}^{(2)}$	0
[111]	[111]	$\boldsymbol{F}_{A_1}^{(0)} + \frac{2}{3}\boldsymbol{F}_{T_2}^{(2)}$	$(\mu_0 + 2\mu_{45})/3$
[111]	[11$\bar{2}$], [$\bar{1}$10]	$\boldsymbol{F}_{A_1}^{(0)} - \frac{1}{3}\boldsymbol{F}_{T_2}^{(2)}$	$(\mu_0 - \mu_{45})/3$

To clearly reveal the dependence on the crystallographic axes, the (001) plane in cubic symmetry which is often studied experimentally, is chosen for the XMLD angular dependence study.

$$\mu_{XMLD}(\alpha,\beta) \equiv \mu_{XA}(\alpha,\beta) - \mu_{XA}(\alpha + 90°,\beta)$$

where α and β are the angles of $\hat{E}$ and $\widehat{M}$ with respect to the [100] axis, *i.e.*, $\hat{E} = (\varepsilon_x, \varepsilon_y) = (\cos\alpha, \sin\alpha)$ and $\hat{E} = (x, y) = (\cos\beta, \sin\beta)$, gives an angular dependence

$$\mu_{XMLD}(\alpha,\beta) = \mu_0 \cos 2\alpha \cos 2\beta + \mu_{45} \sin 2\alpha \sin 2\beta = \frac{1}{2}[(\mu_0 + \mu_{45})\cos(2\alpha - 2\beta) + (\mu_0 - \mu_{45})\cos(2\alpha + 2\beta)], \quad (13.10)$$

where μ_0 and μ_{45} are the $\mu_{XMLD}(\alpha,\beta)$ for $\alpha=\beta=0°$ and $\alpha=\beta=45°$, respectively. This expression sums over the terms containing an angular and an energy dependent factor. At $\beta = 0°$ or $45°$, no matter what the angle α between $\hat{E}$ and the [100] axis is, one has

$$\mu_{XMLD}(\alpha, 0°) = \mu_0 \cos 2\alpha$$

$$\mu_{XMLD}(\alpha, 45°) = \mu_{45} \sin 2\alpha$$

Only in the special cases with the magnetization direction $\widehat{M}$ along a high-symmetry axis $\langle 100\rangle$ or $\langle 110\rangle$ one measures a single spectrum (μ_0 or μ_{45}, respectively) where the direction of linear polarization $\hat{E}$ affects the magnitude of the XMLD but not its spectral shape. At other angle, the μ_{XMLD} is a linear combination of μ_0 and μ_{45}. Thus, in the frequently encountered geometry with $\widehat{M}$ along a high-symmetry axis, the full angular dependence of the XMLD remains hidden. In order to understand the angular dependence, the magnetization direction $\widehat{M}$ has to be forced along a non-symmetry direction using an applied magnetic field or bias field.

The angular dependence of the XMLD can be written as a sum over a rotational invariant term which is independent of the crystal axes, and a term which depends on the orientation with respect to the crystalline axes

$$\mu_{XMLD}(\alpha,\beta) = \frac{1}{2}\mu_{XMLD}^{iso}(\alpha,\beta) + \mu_{XMLD}^{ani}(\alpha,\beta), \tag{13.11}$$

with $\mu_{XMLD}^{iso}(\alpha,\beta) = (\mu_0 + \mu_{45})\cos(2\alpha - 2\beta)$, and

$$\mu_{XMLD}^{ani}(\alpha,\beta) = \frac{1}{2}(\mu_0 - \mu_{45})\cos(2\alpha + 2\beta).$$

The isotropic part, *i.e.*, the rotational invariant contribution, depends only on the difference $|\alpha - \beta|$ between the angles of $\hat{E}$ and $\widehat{M}$. Hence, the angular distribution is rigidly fixed to the direction $\widehat{M}$ and does not depend on the crystal frame orientation. The isotropic part of the XMLD behaves as what could be considered "normal" for a situation where the crystal field can be neglected. The isotropic part can be measured directly at $= \beta = 22.5°$, where the anisotropic part disappears. The anisotropic part, *i.e.*, the rotational dependent contribution depends on the directions of $\hat{E}$ and $\widehat{M}$ with respect to the crystal axes. The anisotropic part could lead to a situation where the maxima of the XMLD and XAFS do no longer coincide with the direction of M. Hence, the separation of the XMLD into its isotropic and anisotropic parts offers a powerful way to find the influence of the crystal-field interaction.

13.4 XMLD Applications

13.4.1 Determination of spin order through charge order

XMLD uses a uniaxial field to detect the anisotropy of the atomic charge. Since the spins are axially aligned along the magnetization field M through exchange, the spin-orbit interaction deforms the charge into a non-spherical shape, which can also be measured by magnetostriction. The emergence of XMLD requires unequal population of the m_j sublevels, and an energy splitting that can be resolved by the experiment.

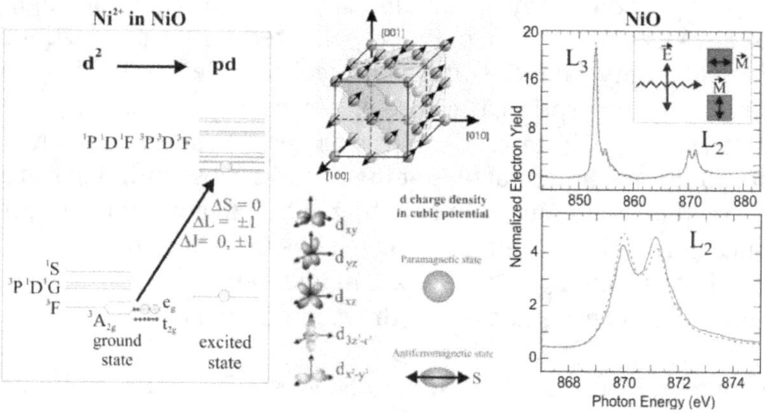

Figure 13.2 Origin of XMLD for cubic NiO (Fm-3m, 225). On the left displays the electronic configurations involved in the X-ray absorption process. In the ground state, there are two d holes (d^2 hole configuration), and their energy levels are determined by multiplet and crystal field effects. In the final state, a 2p hole will be formed by X-ray absorption, and the excited electron will fill a d hole. The resulting pd hole configuration again gives rise to multiplet splitting and the XAFS spectrum reflects the multiplet structure. The resulting pd pore configuration again causes multiplet splitting, and the XAFS spectrum reflects the multiplet structure. In the paramagnetic state, due to cubic symmetry, the absorption spectrum of NiO does not show polarization dependence. In the antiferromagnetic state, spin-orbit coupling causes charge density distortion, and the XMLD effect is observed.

A typical example is a cubic structured NiO (space group *Fm-3m, 225*). The cubic symmetry of the NiO lattice makes the charge distribution around the atoms almost spherical, and no linear dichroism effect exists in the absence of magnetic interaction, say, paramagnetic NiO above the Néel temperature (T_N=520 K). At room temperature, NiO is antiferromagnetic, and the antiferromagnetic ordering is caused by the super-exchange interaction along the Ni-O-Ni bond in the (100) direction, so the net moment at each O site disappears. The Ni spins are oriented in the (1,1,1) plane along three possible directions [$\bar{2}$, 1, 1], [1, $\bar{2}$, 1] or [1, 1, $\bar{2}$] in the fcc lattice (Figure 13.2). The equal number of spins pointing in opposite directions leads to no net magnetic moment, so only the preferred magnetic axis exists. The arrangement of local atomic spins along this axis breaks the cubic symmetry of the charge through spin-orbit coupling. As a result, the charge shows small anisotropy in the unit lattice, that is, the charge distribution is no longer spherical, but shows elliptical distortion along the magnetic direction. This charge anisotropy causes the asymmetry of the X-ray absorption signal by searching for light effects, *i.e.*, X-ray absorption of linear polarized X-rays proportional to density of empty valence states in direction of electric field vector E. During X-ray absorption, the core electrons are excited from the 2p shell to the empty d states. The X-ray absorption spectrum of NiO exhibits additional fine structure at the L_3 and L_2 absorption edges arising from final state multiplet coupling of the 2p core hole with the d valence hole ($2p^5 3d^9$ configuration). A linear dichroism effect is present at both L_2 and L_3- edges for alignment of the electric field vector parallel versus perpendicular to the magnetic axis.

As shown in Figure 13.2, for E parallel *versus* E perpendicular to the magnetic axis, the maximum XMLD difference is obtained. The spectra were collected by surface-sensitive electron yield for a 45 nm-thick NiO(100) film grown on MgO(100). Near the surface the magnetic axis is preferentially perpendicular to the surface. The green curve corresponds to E perpendicular to the surface and the red curve for E parallel to the surface. Contrary to the X-ray magnetic circular dichroism (XMCD) effect, the XMLD effect shows $\cos^2\theta$ dependence, where θ is the angle between electric field vector of linearly polarized X-ray beam and the magnetic axis of the specimen. The 3d metal has weak spin-orbit interaction and a large valence band width. When the d-state is summed over the Brillouin zone, it results in a small charge anisotropy. Therefore, in general, XMLD has little effect on 3d metal. However, in the case of multiplet splitting, a larger XMLD effect (about 10-30%) may be observed. Under certain multiple energies, by enhancing the matrix element effect of XMLD, only the selected d-valence state is detected. In a simple picture, each multiplet can be regarded as a non-spherical, strongly coupled spin-orbit state in space, resulting in a larger XMLD effect. In Figure 13.2, the XMLD effect is particularly pronounced at the Ni L_2 edge, and a considerable difference is observed between E parallel to the magnetic axis (green) and perpendicular to the magnetic axis (red). Since X-ray magnetic linear dichroism may be caused by electrical and magnetic asymmetry, the magnetic sequence effect and ligand field effect should be carefully distinguished. This is usually done through temperature-related measurements.

Theoretical XMLD spectra of NiO were simulated from atomic multiplet calculations [42,43] for the electric-dipole transition Ni^{2+} $2p^6 3d^8 \rightarrow 2p^5 3d^9$ in octahedral crystal field of 10Dq = 1.4 eV, which has a ground state $\left(t_2^6 e^2,\ {}^3A_2\right)T^2$. Multiplet calculations performed with $\hat{E}$ and $\widehat{M}$ separately incremented in steps of 5° steps over the entire angular range confirm the analytical relationship in Eq. 13.10 to high accuracy, which indicates that the possible higher order terms must be small. [44] For the entire L_2 edge, $\mu_{45} \approx -\mu_0$, however, the L_3 edge shows a strongly different μ_{45} and μ_0 as a function of photon energy. The angle correlation expression given in Eq.13.6 are valid for the entire $L_{2,3}$ spectrum. A more specific rule can be derived by restricting to the Ni^{2+} L_2 edge and using the fact that at this edge,[29] $\mu_{45} \approx -\mu_0$. This relation is a peculiarity of the Ni^{2+} L_2 edge and is not true for the L_3 edge,[45] but it holds to some extent for other L_2 edges, such as for Co^{2+} (Ref. [19]), Fe^{3+}, and Mn^{2+} (Ref. [14]). The two fundamental spectra for the L_2 XMLD have approximately the same shape due to the reduced number of allowed transitions for the j = 1/2 core-hole state. The electric-dipole selection rules restrict the set of final states that can be reached. Substitution of $\mu_{45} = -\mu_0$ reduces Eq.13.10 to which removes the isotropic part and retains only the anisotropic part. Therefore, the XMLD at the L_2 edge is strongly dependent on the orientation of $\widehat{M}$ and $\hat{E}$ with respect to the crystal axes. The maximal intensity is obtained when $\widehat{M}$ and $\hat{E}$ have equal angles of opposite sign with respect to the [100] axis.

Previously, the orientation of the Ni spins was determined by assuming that the second peak in the L_2 absorption edge is maximal when

$\hat{E} \parallel \widehat{M}$. However, this is only correct if the crystal field symmetry would play no role. It was found recently that a proper description of the angular dependent XMLD in cubic symmetry requires a linear combination of two fundamental spectra, μ_0 and μ_{45}. Taking into account the angular dependence, two extreme cases occur in the (001) plane: For $\widehat{M} \parallel [100]$ the XMLD is pure μ_0 while for $\widehat{M} \parallel [110]$ the XMLD is pure μ_{45}. The L_2 absorption edge consists of a well resolved double peak structure which is well suited to study the spin orientation. The relation $\mu_{45} \approx -\mu_0$ means that the second peak is maximal not only for $\hat{E} \parallel \widehat{M} \parallel [100]$ but also for $\hat{E} \perp \widehat{M} \parallel [110]$. While in the latter case the spins are perpendicular instead of parallel to $\hat{E}$, both orientations give the same Ni^{2+} L_2 X-ray absoprtion spectrum. Therefore, it is important in PEEM image analysis to consider the angular dependence of XMLD relative to the crystal frame in order to retrieve the spin axis of the antiferromagnetic NiO domain.

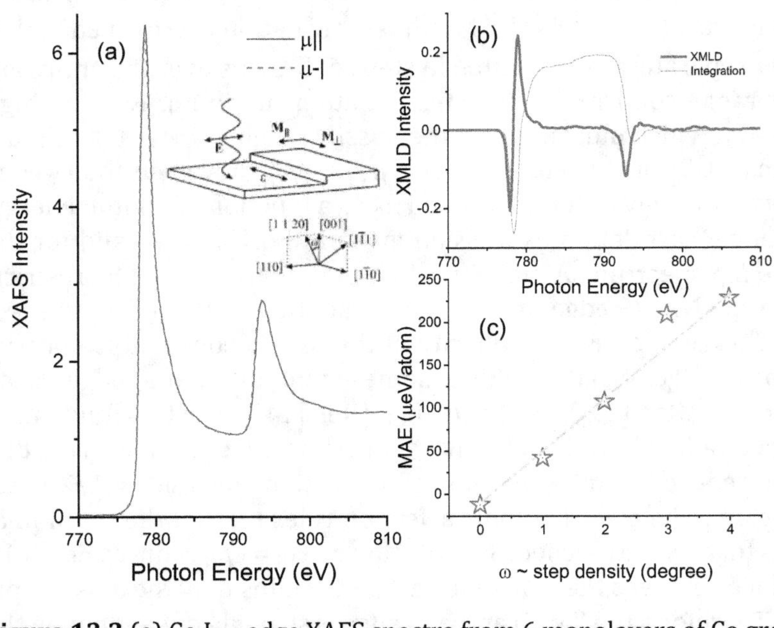

Figure 13.3 (a) Co $L_{2,3}$ edge XAFS spectra from 6 monolayers of Co grown on Cu (ω = 4°), measured with M perpendicular (solid blue) and M parallel (dashed red) to E. (b) The XMLD (solid black) is the difference between the two XAFS spectra and has an integrated intensity close to zero (dash red). (c) The Co MAE determined from the XMLD as a function of step density. The straight line is a linear fit to the data [Dhesi et al. (2001) Phys. Rev. B 87(6), 067201, reproduced with permission from American Physical Society].

13.4.2 Magnetocrystalline anisotropy energy

The magnetocrystalline anisotropy (MAE) refers to the energy necessary to rotate the magnetic moment in the single crystal from the easy direction to the hard direction. The easy and difficult directions come from the interaction of the spin magnetic moment with the lattice (spin-orbit coupling). MAE is an important parameter of magnetic storage media. In a magnetic multilayer, the interface magnetocrystalline anisotropy stabilizes the perpendicular magnetic anisotropy, which is important for high storage density and affects the oscillation exchange coupling, which controls the magnetic resistance in the reading device. Therefore, the microscopic understanding of the interface magnetocrystalline anisotropy is important, but a probe is required to specific to elements and locations. As X-ray absorption process involves a transition from the core level to a state above the Fermi level, X-ray absorption spectroscopy is a powerful element-specific probe for local electronic structures. For transition metals, the dipole selection rules ensure that the 2p core level electron is excited to the empty 3d level, which is the state of driving magnetism. In recent years, combining X-ray absorption spectroscopy with circularly polarized light, XMCD has become a unique element-specific tool for the separation of spin and orbital magnetic moments in ferrous and ferromagnetic systems. XMCD can then provide information about orbital moment anisotropy, which is related to magnetocrystalline anisotropy through the perturbation model. However, this method is not practical because orbital moment anisotropy and magnetocrystalline anisotropy are strictly disproportionate. Since magnetocrystalline anisotropy is caused by spin-orbit interactions, a technique to detect this quantity seems more appropriate. By combining X-ray absorption spectroscopy with linearly polarized light, XMLD allows to measure anisotropy in spin-orbit interactions, which can be related to the magnetocrystalline anisotropy using a straight-forward sum rule.[7] Additionally, XMLD is also sensitive to antiferromagnetic as well as ferrimagnetic and ferromagnetic.

To verify the proposed relationship between XMLD and magnetocrystalline anisotropy, ultra-thin Co films were grown on a Cu crystal substrate with 5 miscut surfaces (at a different angle ω to (001)). The surface magneto-optical Kerr measurement revealed that the energy of magnetocrystalline anisotropy of the stepped Co surfaces grow on

Cu increases linearly with the step density.[46] The XMLD spectra can be obtained by recording the Co *L*-edge absorption spectrum on a stepped Co film with M vertical and parallel M grown on Cu ($\omega = 4\,°$), using a strong linear polarized soft X-ray beam (Figure 13.3ab). The $L_{2,3}$ edges arise from the transitions from the spin-orbit split 2p core level to the unoccupied 3d states. As illustrated in Figure 13.3b, if no contribution comes from charge anisotropy, the integral of XMLD would be zero. The magnetocrystalline anisotropy of a stepped Co film determined using XMLD linearly changes as a function of step density (Figure 13.3c). The same linear trend of magnetocrystalline anisotropy suggests that XMLD is directly related to magnetocrystalline anisotropy. Thus, magnetocrystalline anisotropy can be investigated by using this experimentally established new element-specific tool XMLD. For example, a high-resolution photoelectron emission microscope can be used to image the magnetic anisotropy via XMLD on the symmetry broken surface and the exchange bias interface.

XNLD has a very wide range of applications in inherently low symmetry systems such as 2D interfaces and monolayer on surfaces. In that case, the anisotropy of the empty states parallel versus perpendicular to the surface / interface is directly probed by impinging the linearly polarized light at normal incidence in one experiment and at glancing incidence in a second experiment or exploring also intermediate incidence angles. When the electric field $\boldsymbol{E}$ is aligned along the orbit, it will provide maximum absorption, and when $\boldsymbol{E}$ is oriented along the nodal plane, it will give zero intensity. XNLD has a simple expression and definition as "search light" effect of $\boldsymbol{E}$ with respect to the empty orbitals in the low-D system, which can be equivalently described as probing the angular part of the matrix elements.

13.5 Summary

"Dichroism" refers to the absorption behavior related to polarization, which is caused by the anisotropy of the charge or spin distribution in the micrograph. If dichroism is only due to charge anisotropy, then it is called "charge dichroism" or "natural dichroism", such as X-ray natural linear dichroism (XNLD) and X-ray natural circular dichroism (XNCD). In the case of anisotropic spin distribution, it is called "magnetic dichroism" like XMLD and XMCD. The polarized X-ray beams play an increasingly important role in the exploration of the magnetic phenomena and magnetic materials. XMCD and XMLD allow insight into the underlying physics of basic effects such as exchange bias and magnetic anisotropy.

XMLD uses the differential absorption of horizontally and vertically linearly polarized light in a magnetic field to check magnetic materials and properties. In the X-ray absorption experiment, electrons are excited from the selected core shell to the empty state. For linearly polarized X-ray beams, the electric field vector acts like search light for the number of valence holes in different directions of the atomic volume. The transition intensity is proportional to the number of empty valence states in the direction of the X-ray electric field. $I_{XMLD} = I_{||} - I_{\perp} \propto \langle m^2 \rangle$, where $\langle m^2 \rangle$ is the expectation value of square of atomic magnetic moment. The anisotropy of charge in the atomic volume is mainly caused by the anisotropy of bonding, that is, by the electrostatic potential. In the absence of spin order, linear dichroic XAFS spectroscopy can only determine the order of charge in a system where the absorbing atoms has lower cubic symmetry. However, in the presence of spin order, even in a cubic system, spin-orbit coupling will result in a preferential charge order relative to the spin direction.

When atomic spins are aligned axially through exchange interactions, the XMLD effect is due to the non-spherical distortion of atomic charges caused by spin-orbit interactions. The shape and size of the XMLD spectrum are affected not only by the magnetic moment and the relative direction of the X-ray polarization, but also by their direction relative to the crystal axis. Therefore, for accurate interpretation of the XMLD data, all of these factors must be considered. XMLD is mainly employed to determine the spin order in ferri- and ferromagnets as well as antiferromagnets through charge order and magnetocrystalline anisotropy energy. By probing non-collinear magnetic state in ferrimagnetic single layer exchange bias systems, XMLD can also be utilized to discriminate between two main reversal mechanisms, namely perpendicular domain wall formation versus spin-flop transition.[47] Because the electric field vector oscillates along the time axis, and the radiation can be absorbed at any time, linearly polarized X-rays are only sensitive to the axial direction. Therefore, the direction of the antiferromagnetic axis or the ferromagnetic axis can be determined by using XMLD, but XMLD cannot provide information about the spin direction.

---

[1] P. Kuiper, B.G. Searle, P. Rudolf, L.H. Tjeng, & C.T. Chen, (1993) X-ray magnetic dichroism of antiferromagnet Fe_2O_3: The orientation of magnetic moments observed by Fe 2p X-ray absorption spectroscopy. *Phys. Rev. Lett.* 70, 1549.

[2] B.T. Thole, G. van der Laan, & G.A. Sawatzky, (1985) Strong Magnetic Dichroism Predicted in the X-Ray Absorption Spectra of Magnetic Rare-Earth Materials. *Phys. Rev. Lett.* 55, 2086;

[3] G. van der Laan, B.T. Thole, G.A. Sawatzky, J.B. Goekoop, J.C. Fuggle, J.-M. Esteva, R. Karnatak, J.P. Remeika, & H.A. Dabkowska, (1986) Experimental proof of magnetic X-ray dichroism. *Phys. Rev. B* 34, 6529.

[4] M.M. Schwickert, G.Y. Guo, M.A. Tomaz, W.L. O'Brien, & G.R. Harp, X-ray magnetic linear dichroism in absorption at the edge of metallic Co, Fe, Cr, and V. *Phys. Rev. B* 58, R4289 (1998).

[5] J.B. Kortright, & S.K. Kim, Resonant magneto-optical properties of Fe near its 2p levels: Measurement and applications. *Phys. Rev. B* 62, 12216 (2000).

[6] G. van der Laan, & B.T. Thole, Strong magnetic X-ray dichroism in 2p absorption spectra of 3d transition-metal ions. *Phys. Rev. B* 43, 13401 (1991).

[7] G. van der Laan, Line shape of 2p magnetic-X-ray-dichroism spectra in 3d metallic systems. *Phys. Rev. B* 55, 8086 (1997).

[8] G. van der Laan, Magnetic Linear X-Ray Dichroism as a Probe of the Magnetocrystalline Anisotropy. *Phys. Rev. Lett.* 82, 640 (1999).

[9] G. van der Laan, (1998) Microscopic origin of magnetocrystalline anisotropy in transition metal thin films. *J. Phys.:Condens. Matter* 10, 3239; E. Arenholz, G. van der Laan, A. McClure, &Y. Idzerda, (2010) Electronic and magnetic structure of Ga_xFe_{1-x} thin films. *Phys. Rev. B* 82, 180405(R).

[10] J. Kunes, & P.M. Oppeneer, (2003) Anisotropic X-ray magnetic linear dichroism at the $L_{2,3}$ edges of cubic Fe, Co, and Ni: ab initio calculations and model theory. *Phys. Rev. B* 67, 024431.

[11] S. Czekaj, F. Nolting, L.J. Heyderman, P.R. Willmott, & G. van der Laan, (2006) Sign dependence of the X-ray magnetic linear dichroism on the antiferromagnetic spin axis in $LaFeO_3$ thin films. *Phys. Rev. B* 73, 020401.

[12] E. Arenholz, G. van der Laan, R.V. Chopdekar, & Y. Suzuki, (2006) Anisotropic X-ray magnetic linear dichroism at the Fe $L_{2,3}$ edges in Fe_3O_4. *Phys. Rev. B* 74, 094407.

[13] A.A. Freeman, K.W. Edmonds, G. van der Laan, N.R.S. Farley, T.K. Johal, E. Arenholz, R.P. Campion, C.T. Foxon, & B.L. Gallagher, (2006) Giant anisotropy in X-ray magnetic linear dichroism in (Ga,Mn)As. *Phys. Rev. B* 73, 233303.

[14] E. Arenholz, G. van der Laan, & F. Nolting, (2008) Magnetic structure near the Co/NiO(001) interface. *Appl. Phys. Lett.* 93, 162506.

[15] H. Ohldag, G. van der Laan, & E. Arenholz, (2009) Correlation of crystallographic and magnetic domains at Co/NiO (001) interfaces. *Phys. Rev. B* 79, 052403.

[16] G. van der Laan, E. Arenholz, R.V. Chopdekar, & Y. Suzuki, (2008) Influence of crystal field on anisotropic X-ray magnetic linear dichroism at the Co^{2+} $L_{2,3}$ edges. *Phys. Rev. B* 77, 064407.

[17] N.D. Telling, P.S. Keatley, G. van der Laan, R.J. Hicken, E. Arenholz, Y. Sakuraba, M. Oogane, Y. Ando, K. Takanashi, A. Sakuma, & T. Miyazaki, (2008) Evidence of local moment formation in Co-based Heusler alloys. *Phys. Rev.* B 78, 184438.

[18] E. Arenholz, G. van der Laan, F. Yang, N. Kemik, M.D. Biegalski, H.M. Christen, & Y. Takamura, (2009) Magnetic structure of $La_{0.7}Sr_{0.3}MnO_3/La_{0.7}Sr_{0.3}FeO_3$ superlattices. *Appl. Phys. Lett.* 94, 072503.

[19] E. Arenholz, G. van der Laan, A. McClure, & Y. Idzerda, (2010) Electronic and magnetic structure of Ga_xFe_{1-x} thin films. *Phys. Rev. B* 82, 180405(R).

[20] G. van der Laan, E. Arenholz, A. Schmehl, & D.G. Schlom, (2008) Weak Anisotropic X-Ray Magnetic Linear Dichroism at the Eu $M_{4,5}$ Edges of Ferromagnetic EuO(001): Evidence for 4f-State Contributions. *Phys. Rev. Lett.* 100, 067403.

[21] D. Alders, J. Vogel, C. Levelut, S.D. Peacor, T. Hibma, M. Sacchi, L.H. Tjeng, C.T. Chen, G. van der Laan, B.T. Thole, & G.A. Sawatzky, (1995) Magnetic X-ray dichroism study of the nearest-neighbor spin-spin correlation function and long-range magnetic order parameter in antiferromagnetic NiO. *Europhys. Lett.* 32, 259.

[22] D. Alders, L.H. Tjeng, F.C. Voogt, T. Hibma, G.A. Sawatzky, C.T. Chen, J. Vogel, M. Sacchi, & S. Iacobucci, (1998) Temperature and thickness dependence of magnetic moments in NiO epitaxial films. *Phys. Rev. B* 57, 11623.

[23] A. Scholl, J. Stoehr, J. Luening, J.W. Seo, J. Fompeyrine, H. Siegwart, J.P. Locquet, F. Nolting, S. Anders, E.E. Fullerton, M.R. Scheinfein, & H.A. Padmore, (2000) Observation of antiferromagnetic domains in epitaxial thin films. *Science* 287, 1014.

[24] A. Scholl, F. Nolting, J. Stoehr, T. Regan, J. Luening, J.W. Seo, J.P. Locquet, J. Fompeyrine, S. Anders, H. Ohldag, & H.A. Padmore, (2001) Exploring the microscopic origin of exchange bias with photoelectron emission microscopy. *J. Appl. Phys.* 11, 7266.

[25] J. Luening, F. Nolting, A. Scholl, H. Ohldag, J.W. Seo, J. Fompeyrine, J.P. Locquet, & J. Stoehr, (2003) Determination of the antiferromagnetic spin axis in epitaxial $LaFeO_3$ films by X-ray magnetic linear dichroism spectroscopy. *Phys. Rev. B* 67, 214433.

[26] J. Stoehr, A. Scholl, T.J. Regan, S. Anders, J. Luening, M.R. Scheinfein, H.A. Padmore, & R.L. White, (1999) Images of the antiferromagnetic structure of a NiO (100) surface by means of X-ray magnetic linear dichroism spectromicroscopy. *Phys. Rev. Lett.* 83, 1862.

27 F.U. Hillebrecht, H. Ohldag, N.B. Weber, C. Bethke, U. Mick, M. Weiss, & J. Bahrdt, (2001) Magnetic moments at the surface of antiferromagnetic NiO (100). *Phys. Rev. Lett.* 86, 3419.

28 H. Ohldag, T.J. Regan, J. Stoehr, A. Scholl, F. Nolting, J. Luening, C. Stamm, S. Anders, & R.L. White, (2001) Spectroscopic identification and direct imaging of interfacial magnetic spins. *Phys. Rev. Lett.* 87, 247201.

29 H. Ohldag, A. Scholl, F. Nolting, S. Anders, F.U. Hillebrecht, & J. Stoehr, (2001) Spin reorientation at the antiferromagnetic NiO (001) surface in response to an adjacent ferromagnet. *Phys. Rev. Lett.* 86, 2878.

30 E. Arenholz, G. van der Laan, R.V. Chopdekar, & Y. Suzuki, (2007) Angle-Dependent Ni^{2+} X-Ray Magnetic Linear Dichroism: Interfacial Coupling Revisited. *Phys. Rev. Lett.* 98, 197201.

31 I.P. Krug, F.U. Hillebrecht, M.W. Haverkort, A. Tanaka, L.H. Tjeng, H. Gomonay, A. Fraile-Rodriguez, F. Nolting, S. Cramm, & C.M. Schneider, (2008) Impact of interface orientation on magnetic coupling in highly ordered systems: A case study of the low-indexed Fe_3O_4 interfaces. *Phys. Rev. B* 78, 064427.

32 G. van der Laan, & E. Arenholz, (2009) Anisotropic X-ray magnetic linear dichroism. *Eur. Phys. J. Special Topics* 169, 187.

33 G. van der Laan, N.D. Telling, A. Potenza, S.S. Dhesi, & E. Arenholz, (2011) Anisotropic X-ray magnetic linear dichroism and spectromicroscopy of interfacial Co/NiO (001). *Phys. Rev. B* 83, 064409.

34 J. Stoehr, & H.C. Siegmann, (2006) *Magnetism. From Fundamentals to Nanoscale Dynamics* (Springer Verlag, Heidelberg).

35 P. Carra, B.T. Thole, M. Altarelli, & X.D. Wang, (1993) X-ray circular dichroism and local magnetic fields. *Phys. Rev. Lett.* 70, 694.

36 B.R. Judd, (1967) Second Quantisation in Atomic Spectroscopy (Johns Hopkins University Press, Baltimore).

37 B.T. Thole, & G. van der Laan, (1988) Linear relation between X-ray absorption branching ratio and valence-band spin-orbit expectation value. *Phys. Rev, A* 38 , 1943.

38 J. Stohr, & H.C. Siegmann, (2006) *Magnetism From Fundamentals to Nanoscale Dynamics* (Springer). ISSN 0171–1873.

39 P. Carra, H. Koenig, B.T. Thole, & M. Altarelli, (1993) Magnetic X-ray dichroism: General features of dipolar and quadrupolar spectra. *Physica B* 192, 182.

40 S.S. Dhesi, G. van der Laan, E. Dudzik, & A.B. Shick, (2001) Anisotropic spin-orbit coupling and magnetocrystalline anisotropy in vicinal Co films. *Phys. Rev. Lett.* 87, 067201.

41 Gerrit van der Laan, (2013) chp 8. "Anisotropic X-ray magnetic linear dichroism", *Magnetism and Synchrotron Radiation: Towards the Fourth Generation Light Sources*, Springer.

42 G. van der Laan, & I.W. Kirkman, (1992) The 2p absorption spectra of 3d transition metal compounds in tetrahedral and octahedral symmetry. *J. Phys.:Condens. Matter* 4, 4189.

[43] G. van der Laan, (2006) Hitchhiker's guide to multiplet calculations. *Lect. Notes Phys.* 697, 143.

[44] G. van der Laan, N.D. Telling, A. Potenza, S.S. Dhesi, & E. Arenholz, (2011) Anisotropic X-ray magnetic linear dichroism and spectromicroscopy of interfacial Co/NiO (001). *Phys. Rev. B* 83, 064409.

[45] A.A. Freeman, K.W. Edmonds, G. van der Laan, N.R.S. Farley, T.K. Johal, E. Arenholz, R.P. Campion, C.T. Foxon, & B.L. Gallagher, (2006) Giant anisotropy in X-ray magnetic linear dichroism in (Ga,Mn)As. *Phys. Rev. B* 73, 233303.

[46] R.K. Kawakami, M.O. Bowen, Hyuk J. Choi, E.J. Escorcia-Aparicio, & Z.Q. Qiu, (1998) Effect of atomic steps on the magnetic anisotropy in vicinal Co/Cu (001). *Phys. Rev. B* 58, R5924.

[47] C. Luo, H. Ryll, C.H. Back, & F. Radu, (2019) X-ray magnetic linear dichroism as a probe for non-collinear magnetic state in ferrimagnetic single layer exchange bias systems. *Sci. Rep.* 9, 18169.

Chapter 14

X-ray Natural Dichroism (XNCD and XNLD)

14.1 Introduction

A polarization dependent absorption behavior is called "dichroism". X-ray dichroism refers to the difference in the X-ray absorption cross sections measured for two orthogonal polarization states of the incident X-ray beam: right- and left-handed polarizations for circular dichroism, and vertical and horizontal polarizations for linear dichroism measurements. In a microscopic view, this polarization dependence of X-ray absorption originates from an anisotropy of the charge or spin distribution. The dichroism caused by an anisotropic spin distribution is called "magnetic dichroism" like X-ray magnetic circular dichroism (XMCD) and X-ray magnetic linear dichroism (XMLD). In the case of a charge anisotropy only, it is called "charge dichroism" or "natural dichroism" like the X-ray natural circular dichroism (XNCD) and X-ray natural linear dichroism (XNLD). So far, compared to XMCD and XMLD, XNCD and XNLD are less used, but they have a fundamental meaning because they can access element-specific stereochemical information.[1] Renewed interest in this technique has recently grown [2] with the emergence of new materials that are both chiral and magnetic, especially in the domain of molecular magnetism.

XNCD

Magneto-optical effects arise as a consequence of frequency dispersion: with an external magnetic field or an internal exchange field, one may induce an unbalanced shift of the density of states involved in optical transitions caused by either right- or left-circularly polarized light. Magneto-optical effects are currently described within quantum theory as the interaction of circularly polarized photons with electron spins through spin–orbit interaction. The leading terms in magneto-optical effects are thus proportional to pure electric dipole (E1-E1) transitions. Magneto-optical activity then requires time-reversal symmetry breaking, but it does not require the sample to satisfy any specific symmetry condition. On the other hand, natural optical activity is a consequence of spatial dispersion,[3] i.e., it implies a dependence of the transition probabilities on a space variable $\boldsymbol{r}$ and on the light wavevector $\boldsymbol{k}$. As a linear first-order spatial dispersion effect with $\boldsymbol{k}$, natural optical activity is inherently associated with transition probabilities that mix multipoles of opposite parity, for example, electric dipole–magnetic dipole (E1-M1) or electric dipole–electric quadrupole (E1-E2). The principle of Curie's symmetry states that this can only happen in systems with broken inversion symmetry. Optical activity effects that are *even* with respect to time reversal symmetry, such as optical rotation or circular dichroism, are commonly referred to as natural optical activity or reciprocal optical activity. Time-reversal *odd* optical activity phenomena do also exist, even though the subject is not so-well documented: the corresponding phenomena are most often referred to as nonreciprocal optical activity effects.

These new effects are fundamentally different from magneto-optical effects and can only be observed in parity-odd magneto-electric media. The origin of optical activity is not identical in optical and in X-ray spectral regions. The E1-M1 interference terms dominate optical activity measured with visible light, however, become vanishingly small in X-ray spectroscopy because magnetic dipole transitions (M1) from deep core levels are forbidden, at least in nonrelativistic theories. Fortunately, in the X-ray range there is a significant contribution from the E1-E2 interference term,[4] which, in contrast, plays only a marginal role at optical wavelengths. It may be worth emphasizing here that the electric dipole–electric octupole interference term mixes multipole moments of the same parity and hence cannot give rise to any optical activity effect.

XNCD requires the breaking of space inversion (the inversion symmetry is not a symmetry of the system), which occurs in non-centrosymmetric materials. Unlike optical activity where multiple mechanisms contribute,[5] XNCD is mainly dominated by a single contribution. At K and L_1 edges, XNCD presents only if p and d orbitals overlap,[6] which can uniquely measure the mixing of even and odd orbitals. XNCD arise from an interference of electric dipole and quadrupole transitions. The idea that interference between electric dipole and electric quadrupole transitions gives rise to natural circular dichroism in the X-ray range was put forward in 1988.[4] The validity of this idea was confirmed by recent experiments [5,7] and calculations.[6] The basic formula of XNCD can be expressed in term of an odd second rank tensor, and the sum rules can derived to link XNCD to the mixing of odd and even orbitals on the absorbing site in the ground state. XNCD can also be simulated using a multiple-scattering theory. The paths responsible for the signal can be identified though scattering path analysis. XNCD is sensitive to the relative orientation of the cluster shells.[8]

XNLD

Contrary to magnetic linear dichroism (XMLD) where magnetostriction works, XNLD refers to the pure ligand field effect.[9] The linear dichroism is consequent to the non-spherosymmetric charge distribution of the excited atom due to a low symmetry ligand field, or to magnetization (magnetostriction). XNLD experiments are performed by using the linearly polarized radiation of which ***E*** vector is oriented with different angles with respect to anisotropic samples. XNLD has a very broad application in intrinsically low symmetry systems like 2D interfaces and monolayer at surfaces. In that case the anisotropy of the empty states parallel versus perpendicular to the surface (interface) is directly probed by impinging the linearly polarized light at normal incidence in one experiment and at glancing incidence in a second experiment or exploring also intermediate incidence angles. When ***E*** is aligned along an orbital it gives maximum absorption, when it is otherwise oriented in a nodal plane it gives zero intensity. There is a simple formulation and definition of *XNLD* as "search light" effect of ***E*** with respect to the empty orbitals in the *low-dimension* system, equivalently one can describe the effect as probing the angular part of the matrix elements. If the X-absorption intensity is measured as a function of angle between ***E*** and a symmetry axis, *e.g.* the surface/interface normal, then $\mu(\theta) = \mu^{\parallel}\cos^2\theta - \mu^{\perp}\sin^2\theta$. If the charge distribution has a node perpendicular to the symmetry axis, then $\mu^{\perp} = 0$, and the XNLD depends only on $\cos^2\theta$. This is the case for a uniaxial system.

14.2 Theory

14.2.1 Optical activity tensor formalism

Due to its spatial dispersion effects, X-ray detected optical activity can be most easily described using a complex gyration tensor, which takes the breakdown of the usual electric dipole approximation into consideration in radiation–matter interactions. For a transverse polarized wave of wavevector $\boldsymbol{k}$, the forward scattered amplitude is most conveniently described with a complex tensor $\boldsymbol{a}^*$, which can be expanded in series as

$$a_{ij}^* = \beta_{ij}^* + \gamma_{ijl}^* k_l + \left(\frac{1}{4} Q_{illj}^* + \frac{1}{4} \eta_{illj}^*\right) k_l^2 + \cdots. \tag{14.1}$$

The first term refers to the rank-2 electric dipole complex polarizability tensor ($\beta_{ij}^* \propto E1E1$). This fully symmetric term describes natural linear birefringence as well as natural linear dichroism effects. In an external magnetic field or in the presence of spontaneous magnetic order, this tensor has an antisymmetric part that is nonzero: the latter component is at the origin of the Faraday effect and of the magnetic circular dichroism. There are as well additional "magnetic" contributions to the symmetrical part, but the latter contributions are quadratic in the magnetization: the corresponding terms describe the Cotton–Mouton effect and magnetic linear dichroism.

The last term in expression 14.1 involves two complex rank-4 tensors: the pure electric quadrupole polarizability tensor $Q_{illj}^* \propto E2E2$ and the electric dipole–electric octupole interference term $\eta_{illj}^* \propto E2E3$. Many recent experimental results [10] support the view that the electric quadrupole (E2E2) contribution should not be neglected in X-ray absorption spectroscopy because they are responsible for weak natural X-ray linear dichroisms and for additional signatures in X-ray magnetic linear or circular dichroism spectra.[11] Typically, the electric quadrupole contribution can give rise to a detectable optical anisotropy in cubic crystals, which can be measured either with visible light [12,13] or with X-rays.[14] It may be worth emphasizing here that the electric dipole– electric octupole interference term mixes multipole moments of the same parity and hence cannot give rise to any optical activity effect.

The present chapter mainly focuses on the second term of expression 14.1, which is often identified as the complex, rank-3 gyration tensor ($\gamma^* \propto$ E1E2 or E1M1). In the absence of any external magnetic field (or spontaneous magnetic order), γ^* is fully antisymmetric and describes natural optical activity effects, that is, the reciprocal rotation of the plane of polarization, circular birefringence, and natural circular dichroism. One should emphasize here again that the elements of the gyration tensor γ^* are associated with transition probabilities that mix multipoles of opposite parity: those terms vanish except in non-centrosymmetric systems.

14.2.2 Origin of XNCD

The quantum mechanical approach to light absorption involves evaluating matrix elements of a radiation–molecule interaction operator between initial and final states, subject to the conservation of energy. The cross section for a photon absorption process in which there is an electronic transition from an initial state $\langle g|$ to a final state $|f\rangle$ is given by

$$\sigma = (4\pi^2\alpha_0\hbar/e^2\omega)|\langle f|H_I|g\rangle|^2\delta(E_f - E_g - \hbar\omega), \qquad (14.2)$$

for radiation of angular frequency ω. α_0 is the fine structure constant and the δ-function takes care of energy conservation. The interaction operator is essentially the scalar product of the electron momentum $\boldsymbol{p}$ and the vector potential $\boldsymbol{A}$ of the radiation field, $H_I \equiv \frac{e}{m}\boldsymbol{p}\cdot\boldsymbol{A}$, with $\boldsymbol{A} = \boldsymbol{\epsilon}e^{i\boldsymbol{k}\cdot\boldsymbol{r}}$ for photons with wave vector $\boldsymbol{k}$ and polarization vector $\boldsymbol{\epsilon}$ interacting with an electron at position $\boldsymbol{r}$. The non-relativistic Hamiltonian can be expressed as $H = H_0 + H_I$. H_0 is the non-interacting part, and the interaction part H_I between the field above and an electron of charge is

$$H_I = \frac{e}{m}\left[\boldsymbol{p}\cdot\boldsymbol{A} + \boldsymbol{S}\cdot\boldsymbol{B} + \frac{e}{2}\boldsymbol{A}^2\right], \qquad (14.3)$$

As discussed in the previous chapter, the formal theories of XANES and EXAFS are essentially the same, and are both interpreted by Fermi's golden rule, *i.e.*, $\mu(\omega) \sim \sum_F |\langle f|\hat{O}|g\rangle|^2\delta(\varepsilon_f - \varepsilon_g - \omega)$, where $|g\rangle$ and $|f\rangle$ are the ground and final many-body electronic states. When an effective single-particle description of the spectrum is reasonable, the absorption rate can be expressed

$$R_{fg} = \frac{2\pi}{\hbar}|\langle f|\hat{O}|g\rangle|^2\delta(\varepsilon_f - \varepsilon_g - \omega), \qquad (14.4)$$

where δ is the Kronecker symbol, the transition between two electronic levels normalized to one $(\langle\varphi_{g,f}|\varphi_{g,f}\rangle = 1)$, the transition operator $\hat{O} = (\epsilon\cdot p + iS\cdot(k\times\epsilon))e^{ik\cdot r}$. Using the Taylor expansion of the exponential up to the third order for the potential vector and to the first order for the magnetic field, and commutator relations, one obtains $\langle\varphi_f|\hat{O}|\varphi_g\rangle = i\frac{m}{\hbar}(\varepsilon_f - \varepsilon_g)\ \langle\varphi_f|\hat{o}|\varphi_g\rangle$ with the operator expanded as $\hat{o} = \hat{o}_{E1} + \hat{o}_{E2} + \hat{o}_{E3} + \hat{o}_{M1} + \cdots$, they correspond to the electric dipole (E1), electric quadruple (E2), magnetic dipole (M1) transition channels:

$$\hat{o}_{E1} = \epsilon \cdot r,$$

$$\hat{o}_{E2} = \frac{i}{2}(\epsilon \cdot r)(k \cdot r),$$

$$\hat{o}_{E3} = -\frac{1}{6}(\epsilon \cdot r)(k \cdot r)^2,$$

$$\hat{o}_{M1} = \frac{\hbar}{2\mathrm{m}}(\varepsilon_f - \varepsilon_i)(k \times \epsilon) \cdot (L + 2S), \tag{14.5}$$

Note that the third order also gives a term proportional to $\hat{o}_{E1}$ but with a factor $\alpha^2(\varepsilon_f - \varepsilon_i)/12$, in atomic units, which can be ignored. In practice, dipole transitions highly dominate in X-ray absorption edges. Other electric channels increase with energy. Typically, E2 must be taken into account from 4.5 keV (above the Sc *K*-edge). It can have also an important effect at the $L_{2,3}$ edges of heavier elements. E2 typically represents some percentages of E1, but interest in them comes from the fact that this transition channel follows different selection rules. E3 and M1 remain challenging to detect in the X-ray energy range. M1 is nevertheless of fundamental interest because the crossed channels involving this could be a way to detect fundamental properties of materials.

Electric dipole and quadrupole transitions have been fully recorded in the X-ray range. Therefore, the interference between dipole and quadrupole transition appears to be greater than that of pure quadrupole transition. The absorption cross-section including dipole and quadrupole transitions is proportional to the sum over final states of $|\langle f|\epsilon \cdot r|g\rangle + (i/2)\langle f|\epsilon \cdot rk \cdot r|g\rangle|^2$.[10] In a non-magnetic sample, the ground and excited wavefunctions $|g\rangle$ and $|f\rangle$ can be chosen real. Thus, if the X-rays are linearly polarized, ϵ is real, the dipole and quadrupole contributions are orthogonal: they do not interfere. In other words, the interference term is observable only with circularly polarized light and the measurements of XNCD rely on reliable circularly polarized sources.

In the optical range, natural circular dichroism is dominated by the interference of electric dipole and magnetic dipole transitions. In the X-ray range, no magnetic dipole transition is clearly observed. For fully circularly polarized X-rays directed along the *z*-axis, the XNCD cross-section is

$$\sigma \propto \Sigma_f \left[|\langle g|x|f\rangle\langle f|yz|g\rangle - \langle g|y|f\rangle\langle f|xz|g\rangle|\delta\left(E_f - E_g - \hbar\omega\right)\right]. \tag{14.6}$$

The cross section would be zero if the sample has one of the following symmetries: a center of inversion, a minor plane containing the *z*-axis, or a mirror plane perpendicular to the *z-axis.*

A density of states is a scalar quantity describing the number of states for a given symmetry. For example, the *K*-edge absorption spectrum of a powder sample (up to matrix elements) measures the density of the p-states of the absorbing atom. However, the dichroic spectroscopies reveal the properties, not only of the density of states, but also of the states themselves through the projection operator $P(E) = \Sigma_\phi |\phi\rangle\delta(E - E_\phi)\langle\phi|$. The density of p states is the scalar $n_p(E) = \langle x|P(E)|x\rangle + \langle y|P(E)|y\rangle + \langle z|P(E)|z\rangle$. The various dichroisms give information on the non-diagonal elements of *P(E)*, or cross density of states. For instance, at the *K*-edge, magnetic circular dichroism is proportional to a vector (the dipole moment of the p shell) with *z*-component $m_z(E) = \langle x|P(E)|y\rangle - \langle y|P(E)|x\rangle$, natural linear dichroism is proportional to a rank 2 tensor (the quadrupole moment of the p shell) with elements such as $\langle x|P(E)|x\rangle - \langle y|P(E)|y\rangle$. Natural circular dichroism at the *K*-edge is proportional to some non-diagonal matrix elements of *P*(*E*) between p and d states. More precisely, XNCD is proportional to a rank 2 tensor (the p-d cross density of states) with elements such as $M(E) = \langle x|P(E)|yz\rangle - \langle y|P(E)|xz\rangle$. Contrary to the quadrupole moment, this rank 2 tensor is odd under inversion.

The relation between the dichroic cross-sections and various ground state properties are established through sum rules: two multi-electronic and one mono-electronic sum rules. The first multi-electronic sum rule concerns the sum over the all the edges and reads

$$\int \frac{\sigma}{(\hbar\omega)^2} d(\hbar\omega) \propto \langle g|M|g\rangle, \tag{14.7}$$

where $|g\rangle$ is the multi-electronic ground state and M is the mixing operator

$$M = \sum_{i,j=1}^{N} z_i(x_i y_j - y_i x_j)$$

N is the number of electrons. Note that $\langle g|M|g\rangle = 0$ for a configuration containing only one electron or one hole in an open shell.

The second multi-electronic sum rule can be derived for each edge selectively.[6] For a K- or L_1-edge it is

$$\int \frac{\sigma}{(\hbar\omega)^2} d(\hbar\omega) \propto \langle g|M_{pd}|g\rangle, \tag{14.8}$$

where $M_{pd} = a_{xz}^{+}a_y - a_{yz}^{+}a_x$ in second quantized form. M_{pd} cannot be written as a linear combination of spherical harmonics, because the latter do not mix different orbitals. Notice that the operator M_{pd} does not look Hermitian. However, since the wavefunctions are real, $\langle g|M_{pd}|g\rangle$ is real.

Finally, it is possible to derive an alternative sum rule corresponding to a one-electron model. The one-electron formula for XNCD can be given in terms of the Green function G as

$$\sigma \propto \langle\phi_0|x' Im[G(r',r)]yz|\phi_0\rangle - \langle\phi_0|y' Im[G(r',r)]xz|\phi_0\rangle, \tag{14.9}$$

where $|\phi_0\rangle$ is the s core-state of the K- or L_1-edge absorption process. Using the operator identity

$$\int_{-\infty}^{+\infty} G(r',r;E)\, dE = -\frac{i}{\pi}\delta(r'-r),$$

the sum rule can be expressed as

$$\int_{-\infty}^{+\infty} \frac{\sigma(\omega)}{(\hbar\omega)^2} d\hbar\omega = 0, \tag{14.10}$$

which does not assume that the electric dipole and quadrupole reduced matrix elements are constant. However, the integral must be carried out over the whole energy range (of occupied and unoccupied states). Therefore, the sum rule says that the experimental sum over the unoccupied states is equal to minus the sum over occupied states, which is the quantity of interest.

It has already been observed for magnetic circular dichroism, that the dichroic spectrum is proportional to the orbital momentum at all energies in a one-electron approach.[15] The same behavior is observed for natural circular dichroism, and the XNCD spectrum is proportional to the p-d cross density of states at all energies. Therefore, the sum over energy of the XNCD spectrum is proportional to

$$-\int_{-\infty}^{E_F}[\langle x|P(E)|yz\rangle - \langle y|P(E)|xz\rangle]\, dE. \qquad (14.11)$$

A similar procedure can be applied to derive one-electron sum rules for other moments of the XNCD spectrum

$$\int_{-\infty}^{+\infty} \sigma(\omega)\,(\hbar\omega)^n d\hbar\omega, \qquad (14.12)$$

but these moments are not very interesting in the X-ray range, because the edge energy is much larger than the energy range over which XNCD is visible, and the various moments differ essentially by a power of $\hbar\omega_{\text{edge}}$.

E1-E2 interference

The link between the integrated intensity Σ_{XNCD} and a microscopic description is provided by the relation

$$\sigma_X^{\epsilon}(\omega) = (4\pi^2\alpha\hbar\omega)\left[\frac{i}{2}\sum_f \sum_{nn'}\langle g|\epsilon^* \cdot r_n|f\rangle\langle f|\epsilon \cdot r_{n'} k \cdot r_{n'}|g\rangle + c.c.\right]\delta\left(E_f - E_g - \hbar\omega\right), \qquad (14.13)$$

picking out the E1-E2 interference in the absorption cross section. The notation is as follows: $\hbar\omega$, k, and e represent energy, wave vector, and polarization of the photon; $|g\rangle$ and $|f\rangle$ denote ground and final states of the electron system, with energies E_g and E_f respectively; electrons are labeled by n and n'; $\alpha = e^2/\hbar c$.

A sum rule for XNCD, integrated over the two partners $j_{\pm} = l \pm \frac{1}{2}$, can now be written. It reads

$$\Sigma_{XNCD} = \int_{j_- + j_+} \frac{\sigma_X^{\epsilon}(\omega) - \sigma_X^{\epsilon^*}(\omega)}{(\hbar\omega)^2} d(\hbar\omega) = \frac{2\pi^2\alpha}{\hbar c}(3\cos^2\theta - 1)(2c + 1) \sum_{l,l'} R_{cl}^{(1)} R_{cl'}^{(2)} \alpha_{l'}(c,l) \left\langle g \middle| \sum_n i\left[\left(A_0^\dagger - A_0\right) l_0\right]_n^{l,l'} \middle| g \right\rangle, \quad (14.14)$$

where $l=c\pm1$, $l'=l\pm1$, $\cos\theta = \hat{k} \cdot \hat{z}$, with $\hat{z}$ the quantization axis, and

$$\alpha_{l+1}(c,l) = \frac{[4 + 3c(c+1) - l(3l+5)]\sqrt{(2l+1)(2l+3)}}{(c-l-3)(c+l+4)(c+l)^2(c+l+2)^2}$$

$$\alpha_{l-1}(c,l) = \frac{[6 + 3c(c+1) - l(3l+1)]\sqrt{(2l+1)(2l-1)}}{(c-l+3)(c+l-2)(c+l)^2(c+l+2)^2}$$

This Equation is consistent with Kuhn's natural dichroism sum rule.[16]

For simplicity, consider a single ion with a partially filled valence shell in configuration l^n; all other shells filled. The effect of spin-orbit interactions and/or crystal fields results in a deformation of the electronic cloud. Its multipolar expansion will contain spin and orbital moments characteristic of the symmetry of the deformation and described by one particle coupled tensors. Linear and circular X-ray dichroism from pure electric multipole transitions, provide a measure of the ground-state expectation value of these tensors.[17,18,19] As is well known, integrating over the two partners of a spin-orbit split inner shell singles out orbital moments, which can all be constructed by coupling the spherical components of l.

Inclusion of hybridization amounts to considering a valence shell given as a superposition of states with different l values. New electronic moments, stemming from l,l' pairs of states and probed by electric-multipole interferences, appear in this case. These intrinsic hybridization moments are expressible by way of coupled tensors, and for the spinless case they can be written out in terms of de Sitter generators. The E1-E2 interference is sensitive to space-odd tensors. The orbital pseudodeviator represents a space-odd intrinsic hybridization moment of rank 2. Space-even moments could be revealed by the E1-E3 interference, if ever observable. The hybridization moments can be expressed in terms of position and momentum operators, with $r \rightarrow n$ and $= n(\partial/\partial r) + 1/r\nabla_\Omega \rightarrow \nabla_\Omega$, since only angular variables are considered.

E1-M1 interference

An aspect of symmetry relevant to naturally active compounds is the absence/presence of mirror planes in the permitted point groups, with the ensuing difference between enantiomeric and non-enantiomeric systems. To properly elucidate this point, effective operators will be discussed for the E1-M1 interference, which governs the natural dichroism in the optical range. In the optical range (O), the absorption cross section is controlled by the E1-M1 interference and reads

$$\sigma_O^{\epsilon}(\omega) = (4\pi^2\alpha\hbar\omega)\left[\frac{i}{2}\sum_f \sum_{nn'}\langle g|\epsilon^* \cdot r_n|f\rangle\langle f|\epsilon \times k \cdot l_{n'}|g\rangle + c.c.\right]\delta(E_f - E_g - \hbar\omega). \quad (14.15)$$

Effective operators of rank zero and two are obtained by coupling the generators of O(3,2). Notice that two-particle operators are found in this case, as intra-shell transitions are considered. For the rank zero tensor, one has

$$\sum_{n\neq n'} \left[i(A_n - A_n^{\dagger}) \cdot l_{n'}\right]^{l,l'}$$

with $l' = l \pm 1$. This expression defines an orbital pseudoscalar, an irreducible tensor able to distinguish between enantiomeric and nonenantiomeric systems. Indeed, it does not branch to the totally symmetric representation in the allowed point groups which contain a mirror plane. The E1-M1 interference terms dominate optical activity measured with visible light but become vanishingly small in X-ray spectroscopy because magnetic dipole transitions (M1) from deep core levels are forbidden, at least in nonrelativistic theories.

Table 14.1 Normalized quadrupole moments for the one-particle 3d hole states

D_{4h}	orbital	m	w^{202}
b_1;b_2	d_{x2-y2}; d_{xy}	±2	1
e	d_{xz}, d_{yz}	±1	-0.5
a_1	d_{z2}		-1

14.2.3 X-ray natural linear dichroism

The integrated intensity as a function of the linear polarization direction reveals the shape of the charge distribution for the d electrons. The shape anisotropy can be expressed by the expectation value of the *l*-shell tensor operator $w^{202} = [3l_z^2 - l(l+1)]/[l(2l-1)]$ for the quadrupole moment.[20] According to the Wigner-Eckart theorem, the integrated intensity per hole from the 2p core level to the sublevel m of the 3d level ($l = 2$) is equal to $(4 - m^2)/6$ for z-polarized light ($\Delta m = 0$) and $(m^2 \pm 3m + 2)/12$ for x-, y-polarized light ($\Delta m = \pm 1$). This yields the sum rule for linear dichroism [21,22]

$$\rho_x + \rho_y - 2\rho_z = \tfrac{1}{2}\Sigma_m \langle \underline{n}_m \rangle m^2 - 1 \equiv \langle \underline{w}^{202} \rangle, \quad (14.16)$$

with the normalization

$$\rho_x + \rho_y + \rho_z = \Sigma_m \langle \underline{n}_m \rangle \equiv \langle \underline{n}_d \rangle, \quad (14.17)$$

where ρ_k is the integrated intensity over the 2p XAS with polarization vector along the k direction, n_m is the occupation number of the sublevel m, and n_d the total number of d holes. The underscore indicates that holes are counted instead of electrons. The one-electron values of $\underline{w}^{202}$ are given in Table 14.1.

14.3 Applications

14.3.1 XNCD applications

XNCD refers to the difference in X-ray absorption cross-sections measured with left- and right-circularly polarized photons independently of the presence or absence of any external magnetic field. In the absence of magnetism, linear dichroism arises from the charge anisotropy, as theoretically analyzed in detail by Brouder[23] for crystal fields lower than cubic symmetry, which was later generalized using group theory.[24] The linear dichroism in anisotropic crystals can be very large, as was, *e.g.*, observed at the Fe $L_{2,3}$ edge of gillespite $BaFeSi_4O_{10}$, where the Fe d^6 ion has as square planar coordination.[25] The origin of X-ray detected natural optical activity is the E1-E2 interference terms mixing multipoles of opposite parity, which exclusively exist in systems of broken inversion symmetry. The X-ray optical activity in langasite $La_3Ga_5SiO_{14}$ crystal (space group P321) can be described by gyration tensor formalism. In the $La_3Ga_5SiO_{14}$ crystal structure, La^{3+} ions are located at the A site, Ga^{3+} ions can take the B (Ga1), C (Ga3) and half of the D (Ga2) sites, whereas Si^{4+} ions occupy another half of the D site (see Figure 14.1a). The symmorphic space group P321 has no forbidden resonant reflections. The XNCD spectra were recorded for two experimental configurations with the X-ray wavevector of incident radiation is parallel to (a) the crystal *c*-axis ($\Delta\sigma_{XNCD}^{\parallel}$) and (b) the *a*-axis ($\Delta\sigma_{XNCD}^{\perp}$), respectively. The absence of the pre-edge features in both L_1-edge XANES and XNCD spectra suggests the delocalized d- or p–d hydbridized electronic states of Ga in langasite, since 3d-states of Ga are fully occupied, and quadrupolar transitions are to empty 4d-states. To obtain site-selective information, the XANES / XNCD spectra and the tensor components $\gamma_{xyz}^{''(s)}$ were calculated with the FDMNES program[26] for Ga in each position, and displayed in Figure 14.1c. The different spectral shapes reflect the dependence of tensor components on the local environment of the absorbing atom. The energy spectra corresponding to each of three nonequivalent Ga positions can be extracted by comparing the computation results with the experimental data. The difference between the spectra corresponding to each Ga position clearly reveals the difference in the density of the p–d mixed electronic state of Ga at positions 1a, 2d and 3f. The calculation also shows that the largest contribution of the Ga K edge to XNCD comes from the Ga at position 1a. Finally, the combination of ab initio calculations and high-quality experimental results proved to be a very valuable method for studying the local chirality of multi-component crystals (such as langasite).

The existence of a well-resolved pre-edge feature assigned to 1s →3d quadrupolar transitions is often found on the low energy side of the *K*-edge X-ray absorption spectra of transition metals. The presence of this signature is often retained as a favorable circumstance to detect a large XNCD signal. For example, the 1s →3d quadrupolar transitions appear in the nickel *K*-edge XANES and XNCD spectra of an enantiomeric single crystal α-$NiSO_4{\cdot}6H_2O$. The uniaxial crystal (each unit cell has four molecular formula units) belongs to the enantiomorphpous tetragonal space group, $P4_12_12$ or $P4_32_12$.[27] Due to the chiral arrangement of the achiral unit, the compound only shows natural optical activity in the crystalline state: This compound exhibits natural optical activity only in the crystalline state, due to the chiral arrangement of the nonchiral units: four (nearly) perfect $Ni(H_2O)_2C_6$

octahedrons are placed along the spiral axis parallel to the tetragonal axis of the crystal. That is the crystallographic *c*-axis, *i.e.*, the optical axis. Natural optical activity measurements on this crystal date back to the 1940s [28] and were reported from the infrared up to the UV range. There was even an attempt to detect some natural optical rotatory power in the X-ray range using Cu *Kα* radiation.[29] The latter experiment had, however, very little chance to yield any positive result, because it was performed far from the absorption edges of the atoms constituting the crystal.[30]

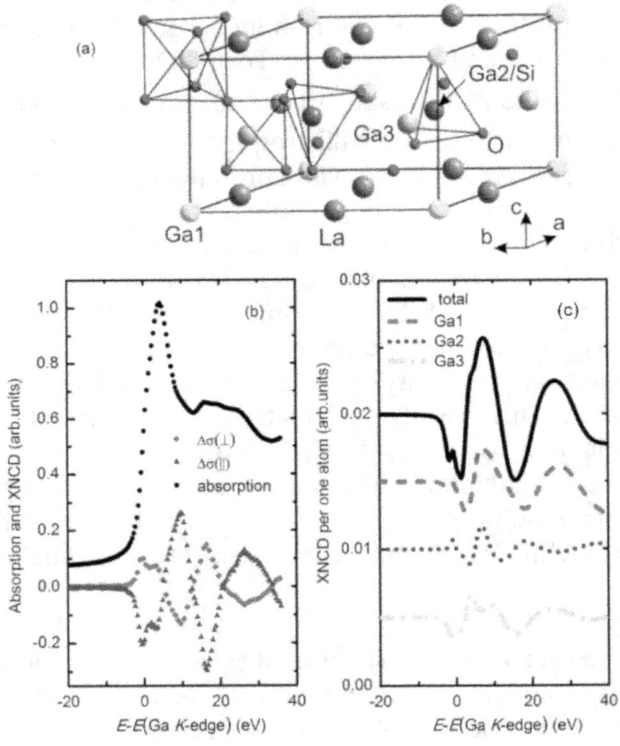

Figure 14.1 (a) Crystal structure of $La_3Ga_5SiO_{14}$. Ga1 represents Ga ions in the octahedral 1a position, Ga2 corresponds to gallium ions in the tetrahedral 2d position and Ga3 denotes Ga in the 3f tetrahedral position. (b) Absorption (dots) and XNCD [×50] spectra of langasite crystal near the Ga K-edge. (c) Calculated XNCD spectra ($\Delta\sigma^{\parallel}_{XNCD} = \gamma''_{xyz}$) of langasite crystal (total) and tensor components γ''_{xyz}spectra corresponding to Ga1 ($\gamma''^{(1a)}_{xyz}$), Ga2 ($\gamma''^{(2d)}_{xyz}$) and Ga3 ($\gamma''^{(3f)}_{xyz}$) positions near the Ga K-edge [Oreshko, et al. (2018) J. Synchrotron Rad. 25, 222–231, reproduced with permission from International Union of Crystallography].

The optical activity discussed in the previous sections is induced by pure transitions and has even parity. Optical transitions can be expanded in multipole terms, such as electric dipole (E1 = parity odd, time even), electric quadrupole (E2 = parity even, time even), magnetic dipole (M1 = parity even, time odd) and higher-order terms that can normally be neglected.[31] Optical activity induced by pure transitions must have even parity; only interference terms can give odd parity.[32] In the visible, the E1M1 interference allows the detection of natural circular dichroism or optical rotation in powdered samples or in solution and single crystals. However, in the case of X-ray absorption, which involves core to valence shell excitations, the M1 transitions are practically forbidden due to the restriction imposed by the monopole selection rule for the radial part. While in the visible region the E2 transitions are negligibly small, their magnitude increases with photon energy, so that for harder X-rays the E1E2 interference can become significant. In this way, XNCD at the I 2s and 2p edges of lithium iodate (a uniaxial laevorotatory crystal of α-$LiIO_3$, lacking inversion symmetry, which belongs to the enantiomorphous crystal class 6) were observed with the intense circularly polarized X-ray beams.[5] The measured dichroism was in agreement with theoretical model calculations.[6] A rather large XNCD signal (up to 6% of the edge jump) was observed at the iodine L_1 absorption edge. At the L_2 and L_3 absorption edges, the recorded XNCD spectra were found to have very similar spectral shapes and the same sign: this observation is at variance with XMCD spectra, which most often exhibit opposite signs at spin–orbit split edges.[5] This can easily be understood as the spin–orbit and exchange splitting are the driving forces in XMCD, while this is not the case in XNCD, as long as the signal is to be assigned to electric dipole–electric quadrupole E1-E2 interference terms. This interpretation was supported by a direct comparison of the experimental spectra with ab initio simulations carried out in the framework of the multiple scattering theory.[6,33] While XMCD requires parity even and time odd symmetry, XNCD requires parity odd and time even symmetry. Non-reciprocal linear dichroism, which is parity odd and time odd, which was observed in the low-temperature antiferromagnetic insulating phase of a Cr-doped V_2O_3 crystal in which one single antiferromagnetic domain was grown by magnetoelectric annealing.[34] This field has continued to flourish and more recent developments have been described in Ref. [35]. Sum rules for X-ray detected optical activity have been reviewed in Refs. [36].

14.3.2 Applications of XNLD

XNLD measures the difference in the X-ray absorption cross sections measured for two orthogonal (vertical and horizontal) polarization states of the incident X-ray beam. Application of XNLD sum rules to the dichroism provides information about the anisotropy of charge distribution (quadrupole / hexadecapole moments). The XNLD is due to an anisotropic charge distribution. Measuring the X-ray absorption of linearly polarized light for different angles of incidence, the electric field vector of the X-rays acts like a search light for the maximum and minimum unoccupied states. This can be used to study the orientation of molecules.[37] But this effect can be detected also for solids. For instance, at the $L_{3,2}$ absorption edges, the unoccupied d states are probed and the X-ray absorption intensity vanishes if the vector of the electric field lies along the d orbital nodal axis. One example of the XNLD effect measured at the Cu $L_{3,2}$ absorption edges of $La_{1.85}Sr_{0.15}CuO_4$ can be found in Ref. [38]. The X-ray magnetic linear dichroism occurs if the charge anisotropy is induced by an axial spin alignment. A first measurement yielding a significant XMLD was performed at the $M_{4,5}$ absorption edges of Tb in $Tb_3Fe_5O_{12}$.[39] The existence of the effect at the $M_{4,5}$ absorption edges of rare earths was predicted one year before, in 1985.[40]

XNLD at the Fe 2p absorption edge of gillespite can be used to understand the pressure induced transition. Gillespite ($BaFe^{2+}Si_4O_{10}$) is a rare, naturally-occurring sheet silicate with a structure consisting of silicate layers having Fe in square planar coordination linked by Ba. The analysis of the XAFS spectral shape revealed that the lowest energy level of gillespite has $^5A_1$ symmetry with the $^5B_2$ level at about 500 cm^{-1} higher energy. A strong polarization dependence in the double peak structures at Fe 2p absorption edges of gillespite is in a good agreement with atomic multiplet calculations for a d^6 $5A_1$ ground state in square planar symmetry. For pure $^5A_1$ and $^5B_2$ states $\langle \underline{w}^{202} \rangle$ is equal to 1 and - 1, respectively, so that a change in ground state symmetry can give a complete reversal of the quadrupole moment. The mixing of the one-electron states by electrostatic interaction and spin-orbit interaction reduces the quadrupole moment of the pure states. The quadrupole moment can be calculated by applying the sum rule to the integrated intensity of the linear dichroism. Thus, at T= 0 K where only the lowest level is occupied $\langle \underline{w}^{202} \rangle = 0.93$. On increasing temperature, the value of quadrupole moment is reduced due to the Boltzmann population over the $^5B_2$ levels. The calculated intensity ratio $\rho_x/\rho_z = 0.368$ at 300 K yields with $\underline{n}_d = 4$ a quadrupole moment $\langle \underline{w}^{202} \rangle = 0.83$. Uniaxial pressure along the z axis will increase the energy of the d_{z^2} state and lower the energy of the d_{xy} state, thereby changing the shape of the charge distribution. Considering that these energies are very close together one might expect that this already occurs at moderately high pressures. An alternative explanation for the color change in gillespite would be a pressure induced transition from high to low spin.[41] The branching ratio $I_{2P_{3/2}}/\left[I_{2P_{3/2}} + I_{2P_{1/2}}\right]$ of the isotropic spectrum can be utilized as a diagnostic tool for the spin state.[42] High-spin states have a larger branching ratio than low-spin states. A quintet spin Fe d^6 state has a branching ratio of *ca.* 0.75, whereas a triplet spin, which for a crystal field above 2 eV can become the lowest state, has a branching ratio in the region 0.6 to 0.65. The measured spectrum has a branching ratio around 0.77 which clearly indicates that the ground state has a quintet spin, with little or no mixing by low-spin states. The branching ratio for z-polarization is higher than for *xy*-polarization, indicating that spin-orbit correlation prefers to align the orbital and spin moments along the z-axis, which is expected for a $^5A_1$ ground state. The analysis of the branching ratio and the details of the spectral shape give evidence confirming that the pressure induced structural phase transition in gillespite has a crystal field origin rather than being due to a high to low spin transition.

XNLD can also be utilized to measure f orbital hybridization in rare earths through electric dipole-octupole interference. Hybridization of empty f and p states can be measured using the interference between E1-E3 transitions at the L_1 edge and then the ground state f-p mixing can be calculated by applying the XNLD sum rule. Contrary to the E1-E2 interference measured by XNLD, the E1-E3 interference also exists in the crystal of parity symmetry. Lanthanide and actinide based materials have a peculiar electronic structure that produces mixed valence, heavy fermion, or Kondo-like behaviors. It remains a challenge for both experiment and theory to understand the ground state properties such as the f orbital occupancy and the role of f/d orbitals in bonding.[43] For instance, $Gd_3Ga_5O_{12}$ (GGG) has a cubic crystal structure, Gd is expected to have highly localized 4f states. To analyze the ground state electronic structure, the hybridization of the Gd orbitals with the O ligand orbitals and the Gd 5d crystal field splitting can be quantified using first-principles calculations. The degree of hybridization between the Gd empty f and p states (*i.e.*, orbitals with same parity) can be directly measured from the XNLD signal. A very high quality single crystal of GGG cut along (110) was positioned at 45° with respect to the incident beam with linear horizontal polarization. It was rotated clockwise about the [110] axis by an angle α defined such that for α=90°, the [001] axis of the crystal is vertical (Figure 14.2a). The L_1 edge XNLD of Gd in $Gd_3Ga_5O_{12}$ were collected with high energy resolution fluorescence detection (5p→2s emission following resonant 2s excitation) for the study of the interference between electric dipole and octupole transitions.[44]

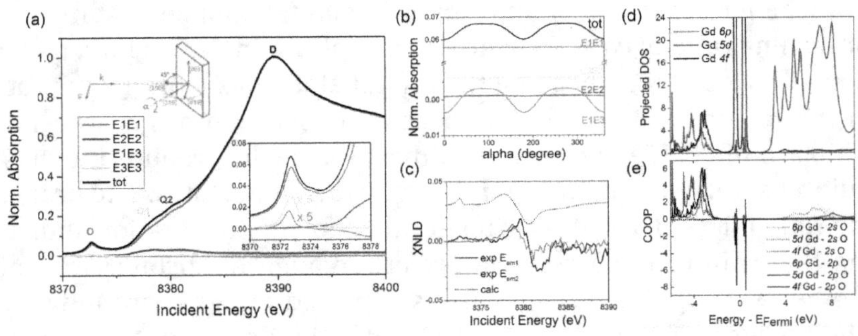

Figure 14.2 (a) XANES spectrum calculated for α=90° showing the different contributions. Inset: Zoom-in of the pre-edge. (b) Contribution of the different terms to the angular dependence of peak O. (c) Experimental XNLD measured for two different values of the emission energy (solid lines) and calculated XNLD (dots). The calculation has been shifted by 0.03 in the vertical axis for clarity. (d) Gd projected partial DOS and COOP. (e) (number of electrons/Ry) calculated in the ground state close to the Fermi level (eV) [Juhin, et al. (2019) Phys. Rev. Materials 3, 120801(R), reproduced with permission from American Physical Society].

The XANES spectra at Ga L_1-edge exhibit four spectral features labeled respectively **D** (main edge), **Q1** and **Q2** (rising edge), and **O** (pre-edge). Based on the relative energy position of the levels in atomic Gd, the peak **D** is assigned to the transitions from 2s to 6p empty states, the peaks **Q1** and **Q2** may involve transitions to the empty 5d states, while the peak **O** could arise from transitions involving empty 4f states. Note the near perfect match of pairs of spectra rotated by 180°, which suggests at this point a π periodicity for the total cross section. The XANES cross section can be written as

$$\sigma = (4\pi^2 \alpha_0 \hbar\omega)|\langle f|E1 + E2 + E3|g\rangle|^2 \delta(E_f - E_g - \hbar\omega). \quad (14.17)$$

Expansion of the above squared matrix element results in six terms, the three squared elements include σ(E1,E1) (dipole), σ(E2,E2) (quadrupole), and σ(E3,E3) (octupole). Since GGG crystallizes in a cubic structure, the dipole is isotropic [45] (*i.e.*, it has no angular dependence), |g⟩ and |f⟩ are parity invariant, implying that both σ_{cube}(E1,E2) and σ_{cube}(E2,E3) are zero, which leaves σ(E1,E3) as the only cross term (*i.e.*, the interference term). For a cubic crystal, the quadrupole sums one isotropic part $\sigma_{cube}^{iso}(E2, E2)$ and one anisotropic part $\sigma_{cube}^{dichro}(E2, E2)$.[46] The main contribution to intensity in the measured spectral range is the dipole, indicating that empty p states are involved in peaks **O**, **Q1**, **Q2**, and **D**. The quadrupole contributes only to the rising edge, *i.e.*, close to peaks **Q1** and **Q2**, consistent with the expected transitions to empty 5d states. In the region of peak **O**, the octupole and the interference term contribute little to the absorption cross section, empty 4f states hybridized with empty 6p states are reached. The theoretical angular dependence calculated at the energy of peak O with the contribution of the different terms suggests that the XNLD calculated for peak O is solely due to the interference term (Figure 14.2b).

The XNLD signal for XNLD=XAS(90°)−XAS(0°) (Figure 14.2c) is dominated by a derivative signal centered around 8380 eV that corresponds to the quadrupole XNLD. The energy difference between the maximum and the minimum (≈2.3 eV) gives an estimation of the cubic crystal field splitting between empty e_g-like 5d orbitals (probed at peak **Q1**) and empty t_{2g}-like 5d orbitals (probed at peak **Q2**) in the presence of the 2s core hole, which is close to the ground state value. The positive feature at 8372 eV originates solely from the interference term directly related to the 4f-6p mixing on Gd. Note that the shape and intensity of XNLD at peak **O** strongly depends on the emission energy used to detect HERFD-XAS spectra. The Gd 4f-6p admixture is quantified through the integral of the dipole-octupole XNLD using a sum rule that is easy to apply to experimental data. To quantify Gd 4f-6p mixing and Gd-O orbital hybridization in the ground state, by simplifying a general expression with the consideration of a cubic crystal and L_1-edge XANES, the sum rule for the integral of the XNLD of the interference term can be expressed as

$$\sum_{E1E3-XNLD}(\alpha) = \int \frac{\sigma_{cube}^{dichro}(E1,E3)(\alpha)}{(\hbar\omega)^3} d(\hbar\omega)$$
$$= \frac{\pi^2\alpha_0}{\hbar c}(-19 + 60\cos 2\alpha + 15\cos 4\alpha)D_1O_3\langle g|M_{130}^4|g\rangle$$

where D_1 and O_3 are the radial dipole and octupole integrals, respectively. The ground state hexadecapole moment $\langle g|M_{130}^4|g\rangle$ fully quantifies by itself the mixing of p and f orbitals. The integral over the small energy range where empty 4f states are probed (8370–8373 eV) makes its calculation tractable. In the general case of non-cubic crystals, $\Sigma_{E1E3\text{-}XNLD}$ is expressed as a combination of several ground state moments $\left\langle g\middle|M_{ll'\gamma}^{g}\middle|g\right\rangle$ where the operator $M_{ll'\gamma}^{g} = \sum_{mm'}(-1)^{l-m}\langle l-ml'-m'|g\gamma\rangle\, a_{lm}a_{l'm'}^{\dagger}$ measures the mixing of l and l' orbitals reached from an l_0 core orbital coupled at order g. In the case of the L_1-edge (l_0=0,l=1,l'=3), $M_{ll'\gamma}^{g}$ is analogous to a quadrupole moment for g=2, an octupole moment for g=3, and a hexadecapole moment for g=4: these moments are "cross terms" since they couple states with the same parity but with different values of angular momentum. In probe the 5d states below the continuum (8373–8388 eV), the quadrupole XNLD sum rule and the quadrupole XAS sum rule were applied, respectively,

$$\sum_{E2E2-XNLD}(\alpha) = \int \frac{\sigma_{cube}^{dichro}(E2,E2)(\alpha)}{(\hbar\omega)^3} d(\hbar\omega)$$
$$= \frac{3\pi\alpha_0}{2(\hbar c)^2}Q_2^2(-19 + 60\cos 2\alpha + 15\cos 4\alpha)\langle g|M_{220}^4|g\rangle$$

$$\sum_{E2E2-XAS} = \int \frac{\sigma_{cube}^{iso}(E2,E2)}{(\hbar\omega)^3} d(\hbar\omega) = \frac{\pi^2\alpha_0}{2(\hbar c)^2}Q_2^2\frac{10-n}{75}$$

where (10−n) is the number of 5d holes, Q_2 the radial quadrupole integral, and M_{220}^4 the (uncrossed) hexadecapole moment associated to 5d orbitals. The ratio of $\Sigma_{E2E2\text{-}XNLD}(\alpha)$ and $\Sigma_{E2E2\text{-}XAS}$ provides an absolute measurement of $\langle g|M_{220}^4|g\rangle$, which now allows the absolute determination of $\langle g|M_{130}^4|g\rangle$ by taking the ratio of [$\Sigma_{E1E3\text{-}XNLD}(90^\circ)-\Sigma_{E1E3\text{-}XNLD}(0^\circ)$] and [$\Sigma_{E2E2\text{-}XNLD}(90^\circ)-\Sigma_{E2E2\text{-}XNLD}(0^\circ)$], which is equal to $\frac{2D_1O_3\left\langle g\middle|M_{130}^4\middle|g\right\rangle}{3Q_2^2\left\langle g\middle|M_{220}^4\middle|g\right\rangle}$. Taking the values of the radial integrals extracted from FDMNES, application of the sum rules to the theoretical data of Figure 14.2c yields $\langle g|M^4_{130}|g\rangle = 1.1 \times 10^{-4}$(dimensionless).

Although the Gd 4f state has local characteristics, it is still mixed with the 6p state, which indicates that they may also participate in the bonding with the O ligand. Further evidence is given by the projected partial density of states (DOS) and crystal overlap orbital population (COOP) calculated close to the Fermi level for Gd in the ground state (Figures14.2 d,e). The two features near 0 eV in DOS correspond to the localized occupied and empty Gd 4f states. The negative Gd 4f-O 2p COOP shows considerable anti-adhesion properties, which is the main contribution, while Gd 4f-O 2s COOP exhibits about four times weaker bonding character. The Gd 5d state forms an energy band that extends over 7 eV, and its binding characteristics first combine with O 2s, and then combine with O 2p orbitals at higher energy.

14.4 Summary

The dependence of X-ray absorption on crystal and/or magnetic orientations with respect to the polarization of the photons, has increasingly attracted interests with the available polarized synchrotron radiation sources. The near-edge dichroism in crystals occurs when the X-ray energy approaches the value required to excite an inner-shell electron to an empty orbital of its atom's valence shell. This chapter introduced two types of dichroisms (XNCD and XNLD) caused by charge anisotropy. XNCD measures the difference in absorption between right and left circularly-polarized radiation. XNLD implies a difference in absorption between radiation with linear polarization parallel or perpendicular to a local symmetry axis. These natural optical effects in the X-ray regime are becoming increasingly popular. In contrast to the X-ray magneto-optical effects which require a time-reversal symmetry breaking due to a local exchange field, natural optical effects are present without such a symmetry-breaking field, *i.e.*, even with respect to time-reversal symmetry. However, some sort of internal symmetry breaking is required to bring about the natural optical effect.

Theoretical analysis requires the use of extended symmetry groups to resolve interference effects. The extended symmetry is determined by a de Sitter group, O(3,2) (which is a non-compact version of rotation group in five dimensions O(5). In the O(3,2) framework, a microscopic picture of the single electron effect accessible by X-ray dichroism can be expressed with the vector term in gyrotropy. By probing intrinsic hybridization moments, the interference dichroism provides as a powerful experimental tool to investigate electronic properties of centro- and non-centrosymmetric crystals. For centrosymmetric crystals, (1) time-odd electronic properties are detected by X-ray magnetic circular dichroism. For E_1 transitions, the familiar orbital sum rule (L_Z) is recovered. (2) Time-even electronic properties can be measured by X-ray linear dichroism. The quadrupolar orbital sum rule can be achieved for E_1 transitions. For non-centrosymmetric crystals, (3) time-even electronic properties are accessible by X-ray circular dichroism. The E_1-E_2 interference can be accessed through the orbital-pseudodeviator sum rule. In the presence of ferromagnetism, pure electric multipole transitions produce time-odd orbital tensors, usually measured by XMCD. These terms disappear when the magnetization direction is perpendicular to the photon wave vector. (4) time-odd electronic properties can be detected by X-ray linear dichroism. The E_1-E_2 interference will be reflected in the gyrotropic vector term and other electronic properties. In the case of a magnetic crystal, pure electric transitions also produce time-even effects. These terms can by distinguished by full angular-dependence analysis.

The XNLD and XMLD signals are even, under both time reversal and spatial inversion. In contrast, the XNCD effect has odd parity, as it is caused by a charge anisotropy lacking a center of inversion. Breaking inversion symmetry by magnetoelectric ordering yields the effect of X-ray non-reciprocal linear dichroism, which is a parity odd and time reversal odd property. Nonreciprocal optical activity (like X-ray nonreciprocal linear dichroism, NRLD) is concerned with time-reversal odd contributions. Both XNCD and NRLD stem mainly from the interference between E1-E2 transitions, which mix multipoles of opposite parity, thus requiring the breaking of space inversion. Both XNCD and NRLD phenomena stem from the interference between E1-E2 transitions that pump an inner-shell electron to the empty valence band. Thus, in order to detect their signals, the sample is required to have an ordered structure (crystal) with the breaking of space inversion. These symmetry requirements make XNCD and NRLD respectively sensitive to the electric and magnetoelectric properties of crystals. In case of pure electric multipole transitions, X-ray dichroism probes magnetic (orbital momentum and spin) and centrosymmetric charge order parameters. Space-inversion-odd electron operators can be utilized to describe the E1-E2 dichroism, which reveals the presence of parity nonconserving interactions.[47] Another type of dichroism that is odd under both spatial inversion and time reversal is the X-ray magnetochiral dichroism (XMχD) due to a chiral charge distribution and an axial spin alignment. The latter two types of dichroism are more complicated cases but have been experimentally proven.[34,48] For the readers' convenience, a summary of X-ray dichroic effects is provided in Table 14.2.

Table 14.2 Order parameters and X-ray dichroic effects (adapted from Carra, et al., (2003) Phys. Rev. B 67, 045111 with permission from American Physics Society)

Order parameter	Space inversion	Time reversal	X-ray dichroism	X-ray transition
Charge	even	even	Linear	Pure E1 or E2
Magnetic	even	odd	Magnetic circular	Pure E1 or E2
Electric	odd	even	Natural circular	Interference E1-E2
Magnetoelectric	odd	odd	Nonreciprocal	Interference E1-E2

All these dichroisms have their sum rules relating the integrated spectra to the ground-state moments. For example, by applying the XNCD sum rule to integrated spectra, one can obtain the relate expectation value of an orbital pseudodeviator, namely a rank-two tensor which is even under time reversal and odd under space inversion. The corresponding "sum rule" relates the integrated dichroism to a ground state moment:

XMCD: average value of $\langle M \rangle$ the local magnetic moment of the absorber
XMLD: average value of $\langle M^2 \rangle$
XNCD: mixture between states with different parity (orbital pseudodeviator)
XNLD: anisotropy of charge distribution (quadrupole / hexadecapole moments)

These sum rules greatly improve the application of these dichroisms to practical problems. Although some dichroic effects are weak, the latest developments in brightness of X-ray sources provide suitable conditions for measuring the weak dichroisms, and hopefully their importance will further increase in the next decades.

---

[1] A. Rogalev, J. Goulon, F.Wilhelm, & A. Bosak, (2010) in *Magnetism and Synchrotron Radiation*, edited by E. Beaurepaire, H. Bulou, F. Scheurer, & J.-P. Kappler, Springer Proceedings in Physics Vol. 133 (Springer, Berlin), pp. 169–190.

[2] R. Sessoli, M.-E. Boulon, A. Caneschi, M. Mannini, L. Poggini, F. Wilhelm, & A. Rogalev, (2015) Strong magneto-chiral dichroism in a paramagnetic molecular helix observed by hard X-rays. *Nat. Phys.* 11, 69.

[3] L.D. Landau, E.M. Lifshitz, (1960) *Electrodynamics of Continuous Media*, (Pergamon Press, New York)

[4] J. Goulon, (1990) in *Rayonnement Synchrotron Polarisé, Electrons Polarisés et Magnétisme*, ed. by E. Beaurepaire, B. Carriére, & J.P. Kappler, (Strasbourg: IPCMS), p. 333.

[5] J. Goulon, C. Goulon-Ginet, A. Rogalev, V. Gotte, C. Malgrange, C. Brouder, & C. R. Natoli, (1998) X-ray natural circular dichroism in a uniaxial gyrotropic single crystal of $LiIO_3$. *J. Chem. Phys.* 108, 6394.

[6] C. R. Natoli, C. Brouder, P. Sainctavit, J. Goulon, C. GoulonGinet, & A. Rogalev, (1998) Calculation of X-ray natural circular dichroism. *Eur. Phys. J. B* 4, 1.

[7] L. Alagna, T. Prosperi, S.Turchini, J. Goulon, A. Rogalev, C. Goulon-Ginet, C. Natoli, IL. Peacock, & B. Stewart, (1998) X-Ray Natural Circular Dichroism. *Phys. Rev. Lett.* 80, 4799-4802.

[8] N. Bouldi, N. J. Vollmers, C. G. Delpy-Laplanche, Y. Joly, A. Juhin, Ph. Sainctavit, Ch. Brouder, M. Calandra, L. Paulatto, F. Mauri, & U. Gerstmann, (2017) X-ray magnetic and natural circular dichroism from first principles: Calculation of *K*- and L_1-edge spectra, *Phys. Rev. B* 96, 085123.

[9] J. Stohr, & H.C. Siegmann, (2006) *Magnetism From Fundamentals to Nanoscale Dynamics* (Springer), ISSN 0171–1873.

[10] Ch. Brouder, (1990) Angular dependence of X-ray absorption spectra. *J. Phys. Condens. Matter* 2, 701.

[11] R.-M. Galera, Y. Joly, A. Rogalev, & N. Binggeli, (2008) The quadrupolar contribution to X-ray magnetic circular dichroism in ReZn compounds. *J. Phys. Condens. Matter* 20, 395217.

[12] H.A. Lorentz, (1922) Double refraction by regular crystals, *Proceedings Koninklijke Akademie van Wetenschappen* 24: 333-339.

[13] J. Pastrnak, & K. Vedam, (1971) Optical anisotropy of silicon single crystals. *Phys. Rev. B* 3, 2567.

[14] G. Dräger, R. Frahm, G. Materlick, & O. Brümmer, (1988) On the multipole character of the X-ray transitions in the pre-edge structure of Fe *K* absorption spectra. An experimental study. *Phys. Status. Solidi. B* 146, 287.

[15] R. Wu, D. Wang, & A. J. Freeman, (1993) First principles investigation of the validity and range of applicability of the X-ray magnetic circular dichroism sum rule. *Phys. Rev. Lett.* 71, 3581-4.

[16] D. Y. Smith, (1976) Superconvergence and sum rules for the optical constants: Natural and magneto-optical activity. *Phys. Rev. B* 13, 5303.

[17] B. T. Thole, P. Carra, F. Sette, & G. van der Laan, (1992) X-ray circular dichroism as a probe of orbital magnetization. *Phys. Rev. Lett.* 68, 1943;

[18] P. Carra, B. T. Thole, M. Altarelli, & X. Wang, (1993) X-ray circular dichroism and local magnetic fields. *Phys. Rev. Lett.* 70, 694;

[19] P. Carra, H. Konig, B. T. Thole, & M. Altarelli, (1993) Magnetic X-ray dichroism: General features of dipolar and quadrupolar spectra. *Physica B* 192, 182.

[20] G. van der Laan, & B.T. Thole, (1995) Core hole polarization in resonant photoemission. *J. Phys. Condens Matter* 7, 9947, Appendix A.

[21] B.T. Thole, & G. van der Laan, (1993) Sum rules for magnetic dichroism in rare earth 4f photoemission. *Phys. Rev. Lett.* 70, 2499.

[22] G. van der Laan, (1994) Sum rules and fundamental spectra of magnetic X-ray dichroism in crystal field symmetry. *J. Phys. Soc. Japan* 63, 2395.

[23] C. Brouder, (1990) Angular dependence of X-ray absorption spectra. *J. Phys.: Condens. Matter* 2, 701.

[24] G. van der Laan, (1994) Sum Rules and Fundamental Spectra of Magnetic X-Ray Dichroism in Crystal Field Symmetry. *J. Phys. Soc. Jpn* 63, 2393.

[25] G. van der Laan, P.F. Schofield, G. Cressey, & C.M.B. Henderson, (1996) Natural linear dichroism at the Fe 2p absorption edge of gillespite. *Chem. Phys. Lett.* 252, 272.

[26] O. Bunau, & Y. Joly, (2012) Time-dependent density functional theory applied to X-ray absorption spectroscopy. . Phys. Rev. B, 85, 155121. http://neel.cnrs.fr/spip.php?article3137

[27] Ch. Brouder, C.R. Natoli, Ph. Sainctavit, J. Goulon, C. Goulon-Ginet, & A. Rogalev, (1999) Theory of X-ray natural circular dichroism. *J. Synchrotron Radiat.* 6, 261.

[28] L.R. Ingersoll, P. Rudnick, F.G. Slack, & N. Underwood, (1940) Optical Activity, Circular Dichroism, and Absorption of Crystalline Nickel Sulphate. *Phys. Rev.* 57, 1145.

[29] M. Hart, & A.R.D. Rodrigues, (1981) Optical activity and the Faraday effect at X-ray frequencies. *Phil. Mag. B* 43, 321.

[30] A. Rogalev, J. Goulon, F.Wilhelm, & A. Bosak, (2010) X-Ray Detected Optical Activity, in *Magnetism and Synchrotron Radiation: New Trends*, ed. E. Beaurepaire, H. Bulou, F. Scheurer, J.-P. Kappler, Chapter 6, ISBN 978-3-642-04497-7.

31 G. van der Laan, (2001) Non-reciprocal X-ray linear dichroism. *J. Synch. Rad.* 8, 1059.

32 Gerrit van der Laan, (2012) 25 years of magnetic X-ray dichroism, in *Magnetism and Synchrotron Radiation: Towards the Fourth Generation Light Sources,* ed. E. Beaurepaire, H. Bulou, L. Joly, F. Scheurer, Chapter 9. ISBN 978-3-319-03031-9

33 Ch. Brouder, C.R. Natoli, Ph. Sainctavit, J. Goulon, C. Goulon-Ginet, & A. Rogalev, (1999) Theory of X-ray natural circular dichroism. *J. Synchrotron Radiat.* 6, 261.

34 J. Goulon, A. Rogalev, C. Goulon-Ginet, G. Benayoun, L. Paolasini, C. Brouder, C. Malgrange, & P.A. Metcalf, (2000) First Observation of Nonreciprocal X-Ray Gyrotropy. *Phys. Rev. Lett.* 85, 4385.

35 A. Rogalev, J. Goulon, F. Wilhelm, & A. Bosak, in *Magnetism and synchrotron radiation, new trends,* ed. by E. Beaurepaire, H. Bulou, F. Scheurer, & J.P. Kappler (Springer, Berlin, 2010), Springer Proceedings in Physics 133, p. 169

36 J. Goulon, A. Rogalev, F. Wilhelm, C. Goulon-Ginet, P. Carra, I. Marri, & C. Brouder, (2003) X-ray optical activity: Applications of sum rules. *J. Exp. Theor. Phys.* 97, 402.

37 J. Stöhr, K. Baberschke, R. Jaeger, R. Treichler, & S. Brennan, (1981) Orientation of chemisorbed molecules from surface-absorption fine-structure measurements: CO and NO on Ni (100). *Phys. Rev. Lett.* 47, 381.

38 C.T. Chen, L.H. Tjeng, J. Kwo, H.L. Kao, P. Rudolf, F. Sette, & R.M. Fleming, (1992) Out-of-plane orbital characters of intrinsic and doped holes in $La_{2-x}Sr_xCuO_4$. *Phys. Rev. Lett.* 68, 2543.

39 G. van der Laan, B.T. Thole, G.A. Sawatzky, J.B. Goedkoop, J.C. Fuggle, J.-M. Esteva, R. Karnatak, J.P. Remeika, & H.A. Dabkowska, (1986) Experimental proof of magnetic X-ray dichroism. *Phys. Rev. B* 34, 6529.

40 B.T. Thole, G. van der Laan, & G.A. Sawatzky, (1985) Strong Magnetic Dichroism Predicted in the X-Ray Absorption Spectra of Magnetic Rare-Earth Materials. *Phys. Rev. Lett.* 55, 2086.

41 G. van der Laan , P.F. Schofield, G. Cressey, & C.M.B. Henderson, (1996) Natural linear dichroism at the Fe 2p absorption edge of gillespite. *Chemical Physics Letters* 252 , 272-276.

42 B.T. Thole, & G. van der Laan, (1988) Branching ratio in X-ray absorption spectroscopy. *Phys. Rev. B 38,* 3158.

43 A. Kerridge, (2017) Quantification of f-element covalency through analysis of the electron density: insights from simulation. *Chem. Commun.* 53, 6685.

44 A. Juhin, S.P. Collins, Y. Joly, M. Diaz-Lopez, K. Kvashnina, P. Glatzel, Ch. Brouder, & F. de Groot, (2019) Measurement of orbital hybridization in rare earths through electric dipole-octupole interference in X-ray absorption spectroscopy. Phys. Rev. Materials 3, 120801(R)

[45] C. Brouder, A. Juhin, A. Bordage, & M.-A. Arrio, (2008) Site symmetry and crystal symmetry: a spherical tensor analysis. *J. Phys.: Condens. Matter.* 20, 455205.

[46] D. Cabaret, C. Brouder, M.-A. Arrio, P. Sainctavit, Y. Joly, A. Rogalev, & J. Goulon, Natural linear dichroism in pyrite (FeS_2): Experiments and calculations. *J. Synchrotron Radiat.* 8, 460 (2001).

[47] P. Carra, Andrés Jerez, and Ivan Marri, (2003) X-ray dichroism in noncentrosymmetric crystals, *Phys. Rev. B* 67, 045111.

[48] J. Goulon, A. Rogalev, F. Wilhelm, C. Goulon-Ginet, P. Carra, D. Cabaret, & C. Brouder, (2002) X-ray magnetochiral dichroism: A new spectroscopic probe of parity nonconserving magnetic solids. *Phys. Rev. Lett.* 88, 237401.

Index

A

B

D

E

F

G

H

K

S

T

U

W

X

www.ingramcontent.com/pod-product-compliance
Lightning Source LLC
LaVergne TN
LVHW052222200926
848481LV00001B/188

* 9 7 8 1 9 2 5 8 2 3 8 8 2 *